VARIABLE RENEWABLE ENERGY AND THE ELECTRICITY GRID

The integration of renewable energy resources into the electricity grid presents an important challenge. This book provides a review and analysis of the technical and policy options available for managing variable energy resources such as wind and solar power. As well as being of value to government and industry policy makers and planners, the volume also provides a single source for scientists and engineers of the technical knowledge gained during the 4-year RenewElec (renewable electricity) project at Carnegie Mellon University, the University of Vermont, Vermont Law School, and the Van Ness Feldman environmental law firm.

The first part of the book discusses the options for large-scale integration of variable electric power generation, including issues of predictability, variability, and efficiency. The second part presents the scientific findings of the project. In the final part, the authors undertake a critical review of major quantitative regional and national wind integration studies in the United States. Based on comparisons among these studies, they suggest areas in which improvements in methods are warranted in future studies, areas in which additional research is needed to facilitate future improvements in wind integration studies, and how the research can be put into practice.

Jay Apt is a Professor in Carnegie Mellon University's Tepper School of Business and in the Carnegie Mellon Department of Engineering and Public Policy, USA. He directs the Carnegie Mellon Electricity Industry Center and the RenewElec Project.

Paulina Jaramillo is an Assistant Professor in the Department of Engineering and Public Policy at Carnegie Mellon University, USA. She is also the Executive Director of the RenewElec Project.

ABOUT RESOURCES FOR THE FUTURE *AND* RFF PRESS

Resources for the Future (RFF) improves environmental and natural resource policy making worldwide through independent social science research of the highest caliber. Founded in 1952, RFF pioneered the application of economics as a tool for developing more effective policy about the use and conservation of natural resources. Its scholars continue to employ social science methods to analyze critical issues concerning pollution control, energy policy, land and water use, hazardous waste, climate change, biodiversity, and the environmental challenges of developing countries.

RFF Press supports the mission of RFF by publishing book-length works that present a broad range of approaches to the study of natural resources and the environment. Its authors and editors include RFF staff, researchers from the larger academic and policy communities, and journalists. Audiences for publications by RFF Press include all of the participants in the policymaking process—scholars, the media, advocacy groups, nongovernmental organizations (NGOs), professionals in business and government, and the public.

VARIABLE RENEWABLE ENERGY AND THE ELECTRICITY GRID

Jay Apt and Paulina Jaramillo

With Jonathan R. Dowds, Michael Dworkin, Emily Fertig, Mark Handschy, Paul Hines, Eric Hittinger, Warren Katzenstein, Elizabeth Kirby, Colleen Lueken, Roger Lueken, Brandon Mauch, Jared Moore, M. Granger Morgan, Robert R. Nordhaus, David Luke Oates, Scott Peterson, Steven Rose, Deborah D. Stine, Allison Weis, and David Yaffe

Routledge
Taylor & Francis Group

LONDON AND NEW YORK

First published 2014
by RFF Press

2 Park Square, Milton Park, Abingdon, Oxfordshire OX14 4RN
52 Vanderbilt Avenue, New York, NY 10017

Routledge is an imprint of the Taylor & Francis Group, an informa business

First issued in paperback 2018

British Library Cataloguing-in-Publication Data
A catalogue record for this book is available from the British Library

Library of Congress Cataloging-in-Publication Data
Apt, Jay, author.
Variable renewable energy and the electricity grid / Jay Apt and Paulina
Jaramillo; with Deborah D. Stine, Jonathan R. Dowds, Michael Dworkin,
Emily Fertig, Mark Handschy, Paul Hines, Eric Hittinger, Warren Katzenstein,
Elizabeth Kirby, Colleen Lueken, Roger Lueken, Brandon Mauch, Jared Moore,
M. Granger Morgan, Robert R. Nordhaus, David Luke Oates, Scott Peterson,
Steven Rose, Allison Weis, and David Yaffe.
    pages cm
  Includes bibliographical references and index.
  1. Wind energy conversion systems.   2. Photovoltaic power generation.
3. Interconnected electric utility systems.   I. Jaramillo, Paulina,
author.   II. Title.
TK1541.A68   2014
333.793′2—dc23

ISBN: 978-0-415-73301-4 (hbk)
ISBN: 978-0-367-17330-2 (pbk)

Typeset in Bembo
by Apex CoVantage, LLC

CONTENTS

FIGURES

TABLES

AUTHORS

Jay Apt is a Professor in Carnegie Mellon University's Tepper School of Business and in the Carnegie Mellon Department of Engineering and Public Policy. He directs the Carnegie Mellon Electricity Industry Center and the RenewElec Project. He has received the NASA Distinguished Service Medal and the Metcalf Lifetime Achievement Award for significant contributions to engineering. He served on the board of the Electric Power Research Institute (EPRI). He holds degrees from Harvard (AB in physics) and MIT (PhD in physics).

Jonathan R. Dowds is a Research Specialist at the University of Vermont Transportation Research Center. His primary research interests are on energy use and climate change mitigation with a focus on the intersection between the transportation and electricity sectors. He holds an MS degree from the University of Vermont in natural resources.

Michael Dworkin is a Professor of Law and Director of the Institute for Energy and the Environment at Vermont Law School. He clerked for the DC Court of Appeals, represented the U.S. Environmental Protection Agency in appellate litigation, and was General Counsel and then Chairman of the Vermont Public Service Board. He is a Director on the boards of the Vermont Electric Power Company and the Vermont Energy Investment Corporation. He holds degrees from Middlebury College (BA) and Harvard Law School (JD) and was awarded the Kilmarx Award for sustained contributions to clean energy, good government, and the environment.

Emily Fertig is a Postdoctoral Fellow at the Pennsylvania State University. Her research focuses on the economics of large-scale energy storage for the purpose of wind power integration, methods of characterizing the variability of wind power output, and federal research and development policy in anticipation of future climate legislation. She holds degrees from Williams College (BA in geosciences) and Carnegie Mellon (PhD in engineering and public policy). Prior to beginning her PhD program, she worked for a Moscow-based publication on the Russian oil and gas industry and interned with the nonprofit Union of Concerned Scientists.

Mark Handschy is a Partner at Enduring Energy, LLC, where he advises clients in the energy utility and industrial sectors on increasing renewable energy footprints and innovative clean energy technology development. In 2009 and 2010, he was a senior advisor to the Undersecretary for Energy at the U.S. Department of Energy. Dr. Handschy received a PhD in physics from the University of Colorado at Boulder.

Paul Hines is an Assistant Professor in the School of Engineering at the University of Vermont. He is also a member of the Carnegie Mellon Electricity Industry Center adjunct research faculty and a Commissioner for the Burlington Electric Department. He received his PhD in engineering and public policy from Carnegie Mellon and an MS degree in electrical engineering from the University of Washington.

Eric Hittinger is an Assistant Professor of Public Policy at the Rochester Institute of Technology (RIT), as well as an Extended Faculty in the Golisiano Institute for Sustainability. He completed his PhD in engineering and public policy at Carnegie Mellon University.

Paulina Jaramillo is an Assistant Professor in the Department of Engineering and Public Policy at Carnegie Mellon University. She is also the Executive Director of the RenewElec Project. Dr. Jaramillo holds degrees from Florida International University (BS in civil and environmental engineering) and Carnegie Mellon University (MS and PhD in civil and environmental engineering).

Warren Katzenstein is a Senior Principal for DNV KEMA Energy & Sustainability. Previously he was an associate with the Brattle Group. He holds degrees from Harvey Mudd College (BS in engineering) and Carnegie Mellon University (PhD in engineering and public policy).

Elizabeth Kirby is a current master's student in electrical engineering at the University of Vermont, where she also earned her BS. Her research considers the impact of high penetrations of wind on the stability of the electric grid using a dynamic system model.

Colleen Lueken recently completed her PhD in the Department of Engineering and Public Policy at Carnegie Mellon University and the Instituto Superior Tecnico-Lisbon. Her PhD research focused on technical and policy issues surrounding grid integration of renewable energy. She also holds degrees from MIT

(BS in aerospace engineering) and the University of Maryland (MS in engineering and public policy). Dr. Lueken is an Associate at DC Energy, an energy trading firm in Washington, DC.

Roger Lueken is a current PhD student in the Department of Engineering and Public Policy at Carnegie Mellon University. Prior to coming to CMU, he worked at LMI as an energy and environmental consultant. In that role, he worked with clients in the federal government to improve energy efficiency and reduce greenhouse gas emissions. He received his master's in engineering and public policy from the University of Maryland.

Brandon Mauch recently completed his PhD in the Department of Engineering and Public Policy at Carnegie Mellon University and the Instituto Superior Tecnico-Lisbon. His research focused on technical and economic aspects of wind power and wind forecast uncertainty. He is a policy analyst for the Iowa Public Utilities Board.

Jared Moore is a graduate from Rose-Hulman Institute of Technology with a degree in mechanical engineering. He is currently a PhD student in the Engineering and Public Policy department at Carnegie Mellon University. He worked for WorleyParsons Group as a consultant, acting as the owner's engineer for dozens of utility-scale solar power projects in development in California and around the world.

M. Granger Morgan is Lord Chair Professor and Head of the Department of Engineering and Public Policy and a professor in the Department of Electrical and Computer Engineering at Carnegie Mellon University. He is a member of the U.S. National Academy of Sciences and a Fellow of the American Association for the Advancement of Science (AAAS), the Institute of Electrical and Electronics Engineers (IEEE), and the Society for Risk Analysis (SRA). He holds degrees from Harvard (BS), Cornell (MS), and the University of California at San Diego (UCSD) (PhD, applied physics).

Robert R. Nordhaus is a member of the law firm of Van Ness Feldman, LLP, and a professional lecturer in energy law at the George Washington University Law School. He was a member of the Energy Policy and Planning Office in the Carter White House and was counsel to the House Interstate and Foreign Commerce Committee. After 3 years as the Federal Energy Regulatory Commission's first general counsel, he joined Van Ness Feldman in 1981. He was appointed general counsel of the Department of Energy in 1993 by President Clinton and served until 1997, when he rejoined Van Ness Feldman.

David Luke Oates is a PhD student in the Department of Engineering and Public Policy at Carnegie Mellon University. He graduated with first-class honors from Queen's University, Canada, with a BS in engineering physics. He took an interest in public policy issues through involvement with Engineers Without Borders Canada, where he learned about the effects of Canada's aid policy on rural development in Africa.

Scott Peterson obtained his PhD from the Department of Engineering and Public Policy at Carnegie Mellon University. He also holds degrees from Texas A&M University (BS in chemistry and MS in environmental engineering). Dr. Peterson's PhD research focused on issues related to plug-in hybrid electric vehicles and in particular their effects on the interactions with the power grid.

Stephen Rose recently completed his PhD in the Department of Engineering and Public Policy at Carnegie Mellon University, where he continues working as a postdoctoral fellow. Dr. Rose researches issues of integrating wind power into the electrical grid. Before starting his PhD research, he worked for 5 years as a controls engineer at GE Wind Energy.

Deborah D. Stine is a Professor of the Practice in the Engineering and Public Policy Department and the Scott Institute for Energy Innovation at Carnegie Mellon University. She was executive director of the President's Council of Advisors on Science and Technology at the White House from 2009 to 2012. Dr. Stine has spent most of her career working at the National Academies, where she was director of the Christine Mirzayan Science and Technology Policy Fellowship Program and director of the Office of Special Projects.

Allison Weis is a PhD student in the Engineering and Public Policy department at Carnegie Mellon University. Her current project is on the integration of controlled charging of electric vehicles with high levels of wind penetration, including the economic and emissions consequences. Before coming to Carnegie Mellon, Dr. Weis received her BS in computer and electrical engineering at Olin College. She also lived in Germany for 1 year, studying renewable energy and working for a solar company planning large, commercial plants.

David Yaffe is a member of the law firm Van Ness Feldman, LLP. His energy practice focuses on federal regulation of the electric utility industry by the Federal Energy Regulatory Commission (FERC), electricity supply transactions, and litigation before the FERC and state courts and arbitration involving energy issues. He has particular experience in addressing the unique needs of and challenges facing public agencies. Mr. Yaffe is also a professional lecturer in energy law at the George Washington University Law School.

PREFACE

Reduction of pollution from electric power generation while maintaining robust supply of electricity has been a primary goal of energy policy in the United States since President Richard Nixon signed the Clean Air Act of 1970. Emissions from power plants of sulfur dioxide and soot (particulate matter of 10 microns in size and smaller) in 2012 were 20% of what they had been in 1970 and power plant emissions of nitrogen oxides were 35% of their 1970 total (EPA 2013a). However, emissions of gasses that trap heat in Earth's atmosphere (greenhouse gases) from electric power generation rose by a bit more than 60% during the same period (EPA 2013b).

One of the main drivers of the encouragement of renewable electric power production in the United States has been pollution reduction. While "renewable" and "low pollution" are not the same thing, state renewables portfolio standards (RPSs) that require a set amount of electricity from sources defined as renewable are in force in 29 states. Hawaii requires as much as 40% renewable electricity by 2030, and several require 20% before 2020.

Low-pollution hydroelectric generators produced 30% of U.S. electricity in 1950, but most suitable sites have already been used, and hydro's market share has been eroded to roughly 7% of all electric power produced as demand has increased and other sources like coal generators filled the need. Because rainfall varies from year to year, so does the water available for hydroelectric generation. Other renewable sources also are variable. The wind blows strongly at times, but calms can set in and persist for days. Clouds can cover the sun for a moment or hours.

There has been robust debate over the practicality of using variable renewable electric generation at large scale. Wind power produced 4% of U.S. electricity in 2013. What would be the consequences of increasing the market share of such a variable resource to, say, 30%? Proponents of renewables argue that large amounts of variable power can be easily accommodated in the present power system.

Opponents argue that even levels as low as 10% of generation by variable power can cause serious disruptions to power system operation. This gap has not been bridged, in part because the level of advocacy required to enact renewables requirements has not been compatible with rigorous systems analysis.

For many years, researchers at Carnegie Mellon University's Electricity Industry Center have done engineering and economic analyses of both conventional and low-pollution electric power generation. In 2009, we began a project to take an objective look at the technical, regulatory, and policy requirements and implications of high levels of renewable electricity in the United States. A team with technical and policy experts at Carnegie Mellon University, the University of Vermont, Vermont Law School, and the Washington, DC, law firm of Van Ness Feldman was formed to survey previous research on the integration of variable renewable electricity and undertake a series of engineering and economic analyses that can provide the basis for recommending the regulatory and policy measures needed to accommodate a large penetration of renewable generation.

We began by writing a series of 15 white papers on regulatory, legislative, and technical aspects of renewable power generation, including a preliminary review of previous integration studies. These were provided to participants in an integration and policy workshop held at Carnegie Mellon University in late 2010. Then a family of models was developed to assess the interactions between renewable generators and the power grid that the workshop identified as most critical. A first series of policy materials was then prepared. These included comments made to the Federal Energy Regulatory Commission's Notice of Proposed Rule Making on Variable Energy Resources. A second integration and policy workshop was held in the fall of 2011 to ensure that operational experience from systems with moderate levels of variable generation was incorporated and to focus new research on issues that emerged as important. A second round of more quantitative materials describing this research was prepared and presented to the National Association of Regulatory Utility Commissioners, the Department of Energy, and the Federal Energy Regulatory Commission as well as to staff of the U.S. Congress.

This book is a resource for policy makers, legislators, and power system operators considering the integration of large-scale renewable generation in the United States. The first section is a generally nontechnical discussion of the technical and policy options for integration at large scale of variable electric power generation. The second section is composed of 10 chapters presenting some of the scientific findings of the project. The final section is a critical review of major quantitative regional and national wind integration studies in the United States. Based on comparisons among these studies, we suggest areas in which improvements in methods are warranted in future studies and areas in which additional research is needed to facilitate future improvements in wind integration studies.

A much-expanded role for variable and intermittent renewables clearly is possible. But it will occur at an affordable price and with acceptable levels of security and reliability only if we adopt a systems approach that considers and anticipates the many changes in power system design and operation that will be required to make high market penetration possible.

As the United States continues on its present path of increasing the market share of renewable electricity, this book is designed to help guide legislators, power system operators, environmental NGOs, and the general public to the smooth adoption of significant amounts of electric generation from variable and intermittent sources of renewable power.

Pittsburgh, Pennsylvania April, 2014

References

EPA. (2013a). *National Emissions Inventory air pollutant emissions trends data, 1970–2012 Average annual emissions, all criteria—June 2013.* U.S. Environmental Protection Agency. www.epa.gov/ttn/chief/trends/trends06/national_tier1_caps.xlsx

EPA. (2013b). *Monthly energy report September 2013, Table 12.6.* U.S. Environmental Protection Agency. www.eia.gov/totalenergy/data/monthly/archive/00351309.pdf

ACKNOWLEDGMENTS

While the authors are solely responsible for the content of this book, we received help, critical advice, and support from a number of people. We thank Inês Azevedo, William Buchannan, Pedro Carvalho, Danielle Changala, Aimee Curtright, James Dominick, Gerard Doorman, Paul S. Fischbeck, Baruch Fischhoff, Brandon M. Grainger, W. Michael Griffin, Iris Grossmann, Colin Hagan, Ane Marte Heggedal, Enes Hoşgör, Mark Keyser, Matt Kocoloski, Lester Lave, Jeremy Michalek, Jennifer Miller, Dora Nakafuji, Terry Oliver, Gregory Reed, Todd Ryan, Vanessa Schweizer, Kyle Siler-Evans, Mitchell Small, Patti Steranchak, Henry Stern, Jonathan Voegele, Sharon Wagner, Jay Whitacre, and Angie Yanez. We also thank the 34 participants in the RenewElec workshop held October 21 and 22, 2010, and the 45 who participated in the second workshop on October 20 and 21, 2011.

Primary financial support for this work was provided by grants from the Doris Duke Charitable Foundation and the Richard King Mellon Foundation to the Tepper School of Business at Carnegie Mellon University. Additional support was provided by the Department of Energy[1] under Award Number DE-FE0002736, by the Electric Power Research Institute, the Heinz Endowments, by the Center for Climate and Energy Decision Making through a cooperative agreement between the National Science Foundation and Carnegie Mellon University (SES-0949710), and by the Carnegie Mellon Electricity Industry Center.

---

1. Neither the United States government nor any agency thereof, nor any of their employees, makes any warranty, express or implied, or assumes any legal liability or responsibility for the accuracy, completeness, or usefulness of any information, apparatus, product, or process disclosed, or represents that its use would not infringe privately owned rights. Reference herein to any specific commercial product, process, or service by trade name, trademark, manufacturer, or otherwise does not necessarily constitute or imply its endorsement, recommendation, or favoring by the United States government or any agency thereof. The views and opinions of authors expressed herein do not necessarily state or reflect those of the United States government or any agency thereof.

ABBREVIATIONS

AC	alternating current; average cost
ACE	area control error
AEPS	alternative energy portfolio standard
APEEP	Air Pollution Emissions Experiments and Policy model
AGC	automatic generation control
BA	balancing area, balancing authority
BES	balancing energy services market
BEV	battery electric vehicle
BOEM	U.S. Bureau of Ocean Energy Management
BPA	Bonneville Power Administration
CA	California
CAES	compressed air energy storage
CAFE	corporate average fuel efficiency
CAIR	Clean Air Interstate Rule
CAISO	California Independent Systems Operator
CCDF	complimentary cumulative distribution function
CDF	cumulative distribution function
CEC	California Energy Commission
CEMS	continuous emissions monitoring system
CFS	Climate Forecast System reanalysis project
CHIPS	Coupled Hurricane Intensity Prediction System
CO	Colorado
CO_2	carbon dioxide
CONE	cost of new entry
COV	coefficient of variation
CP	criteria pollutants
CRA	Charles River Associates

CSP	concentrating solar power
CSAPR	Cross-State Air Pollution Rule
CT	Connecticut
CT-CAES	combustion turbine-compressed air energy storage combined system
DC	direct current
DC	District of Columbia
DIR	Dispatchable Intermittent Resource Program
DLN	dry low NO_x
DOE	U.S. Department of Energy
DSCR	debt service coverage ratio (ratio of revenues to required debt payments)
DSIRE	Database of State Incentives for Renewable Energy
EEX	European Energy Exchange
EIA	U.S. Energy Information Agency
ELCC	effective load carrying capacity
EPA	U.S. Environmental Protection Agency
EPACT2005	Energy Policy Act of 2005
EPRI	Electric Power Research Institute
ERCOT	Electric Reliability Council of Texas
EWITS	Eastern Wind Integration and Transmission Study
EWEA	European Wind Energy Association
FACTS	Flexible AC Transmission System
Fe/Cr	iron/chromium
FERC	Federal Energy Regulatory Commission
FFT	fast Fourier transform
FRCC	Florida Reliability Coordinating Council
GA	Georgia
GHG	greenhouse gases
GSS	grid scale storage
GW	gigawatt
GWh	gigawatt-hours
HVDC	high-voltage direct current transmission cable
Hz	Hertz (formerly cycles per second)
kW	kilowatt
kWh	kilowatt-hours
ICAP	installed capacity
IEC	International Electrotechnical Commission
ISCC	integrated solar combined cycle
ISO	independent system operator
ISO-NE	New England ISO
ITC	investment tax credit
IWGSCC	Interagency Working Group on the Social Cost of Carbon
km	kilometer

KS	Kansas
lb	pound
LBNL	Lawrence Berkeley National Laboratory
LCOE	levelized cost of electricity
LGIA	Large Generator Interconnection Agreement
Li-ion	lithium-ion
LIDAR	light radar technology, used for measuring wind speed
LOL	lower operating limit
LOLP	loss of load probability
MAE	mean average error
MAPS	Multi-Area Production Simulation Software
MARS	multi-area reliability simulation
MASS	Mesoscale Atmospheric Simulation System
MATS	Mercury and Air Toxics Standard
MC	marginal cost
MD	Maryland
ME	Maine
MERRA	Modern-Era Retrospective Analysis for Research and Applications
MI	Michigan
MIBEL	the Iberian Electricity Market
MISO	midcontinent independent system operator
MO	Missouri
mpg	miles per gallon
m/s	meters per second
MT	Montana
MW	megawatts
MWh	megawatt-hours
MY	model year
NaCl	halite (mineral form of sodium chloride; rock salt)
NaS	sodium sulfur
NC	North Carolina
NCAR	National Center for Atmospheric Research
NCEP	National Center for Environmental Prediction
NCEP	National Commission on Energy Policy
NEEDS	EPA National Electric Energy Data System
NERC	North American Electric Reliability Council
NETL	National Energy Technology Laboratory of the Department of Energy
NGCC	natural gas combined cycle generator
NGCT	natural gas combustion turbine generator
NHTS	National Household Travel Survey
NHTSA	National Highway Traffic Safety Administration
NJ	New Jersey
NM	New Mexico

NO_x	nitrogen oxides
NPA	Nebraska Power Association
NPV	net present value
NREL	National Renewable Energy Laboratory of the Department of Energy
NSO	Nevada Solar One
NTTG	Northern Transmission Tier Group
NWP	numerical weather prediction
NYISO	New York Independent System Operator
NYSERDA	New York State Energy Research & Development Authority
O&M	Operations and Maintenance
OATT	Open Access Transmission Tariff
OECD	Organization for Economic Co-operation and Development
PA	Pennsylvania
PCS	production cost simulation
PDF	probability density function
PG&E	Pacific Gas & Electric
PHEV	plug-in hybrid electric vehicle
PHS	pumped hydroelectric storage
PIRP	Participating Intermittent Resource Program
PJM	the PJM Interconnection, Inc. (an RTO)
$PM_{2.5}$	particulate matter less than 2.5 micrometers in diameter
PNBEPH	Programa Nacional de Barragens de Elevado Potencial Hidroeléctrico, the Portuguese organization for hydroelectric power development
PNNL	Pacific Northwest National Laboratory
PPA	power purchase agreement
PSD	power spectral density
PSDS	positive sequence dynamic simulation
PSLF	positive sequence load flow
PTC	production tax credit
PV	photovoltaic
QSS	quasi-steady-state
R&D	research and development
REC	Renewable Energy Certificate
RenewElec	Renewable Electricity Project at Carnegie Mellon University
RI	Rhode Island
RMSE	root mean square error
RPS	renewables portfolio standards
RTO	regional transmission organization
SC	South Carolina
SCB	social costs and benefits
SCC	social cost of carbon
SCE	Southern California Edison

SCR	selective catalytic reduction
SERC	Southeast Reliability Council
SG&E	San Diego Gas & Electric
SIRPP	Southeast Inter-regional Participation Process
SMES	superconducting magnetic energy storage
SO_2	sulfur dioxide
SPP	Southwest Power Pool
T&D	transmission and distribution
UCED	unit commitment and economic dispatch
UCTE	Union for the Co-ordination of Transmission of Electricity (Europe)
U.K.	United Kingdom
U.S.	United States of America
V2G	vehicle-to-grid energy transfer
VER	variable energy resources
WA	Washington (State)
WECC	Western Electricity Coordinating Council
WI	Wisconsin
WinDS	Wind Deployment System
WRF	weather research and forecasting
WWSIS	Western Wind and Solar Integration Study
Zn/air	zinc/air
Zn/Br	zinc/bromide

EXECUTIVE SUMMARY

This book is a resource for policy makers, legislators, power system operators, and researchers who are interested in the issues arising from integration of large-scale renewable generation in the United States. It is the result of research performed during the 4-year RenewElec (short for renewable electricity) project by a team of technical and policy experts at Carnegie Mellon University, the University of Vermont, Vermont Law School, and the Washington environmental law firm of Van Ness Feldman. The first section is a generally nontechnical discussion of the technical and policy options for integration at large scale of variable electric power generation. The second section is composed of 10 chapters presenting some of the scientific findings of the project. The final section is a critical review of major quantitative regional and national wind integration studies in the United States. Based on comparisons among these studies, we suggest areas in which improvements in methods are warranted in future studies, and areas in which additional research is needed to facilitate future improvements in wind integration studies.

I.1 Background

Renewable energy as a percentage of electricity generation in the United States fell from 30% in 1950 to a low of 8% in 2001, as the market share of hydroelectric power was eroded by fossil fuel generators built to keep up with rapidly increasing demand for electricity[1] (Figure ES.1). By 2013, renewables' market share had increased to 13%, primarily due to policies that encouraged an increase in wind power's proportion of generation (Figure ES.2 and Figure ES.3).

A significantly expanded role for renewable energy resources can be achieved. Renewables may be able to regain or exceed the share of electric power they

---

1. Even though production from U.S. hydroelectric plants tripled from 1950 to 1973, demand for electricity grew nearly six-fold in the same period.

represented in 1950. Unlike natural gas, coal, or nuclear plants, these power sources do not produce power on demand; they are subject to the variability of nature's bounty at time scales from seconds to decades. The management of that variability is the subject of this book.

Our purpose is to provide government and industry policy makers and operators, scientists, and engineers with an understanding of the technical and policy options available for managing variable energy resources such as wind and solar power to produce electricity at a much larger scale than at present.

This book provides a summary of major challenges and opportunities for integrating variable power generation sources and meeting the goals set forward in state renewable portfolio standards.

It is likely that the proportion of electricity generation supplied by wind and solar power can increase by about tenfold, from the present 4% to 20 or 30%. California provides evidence to support the goals of the renewable portfolio standards. In 2012, its three major investor-owned utilities, Pacific Gas and Electric (PG&E), Southern California Edison (SCE), and San Diego Gas & Electric (SDG&E), were each able to meet a required mandate of supplying their customers with 20% of their energy from renewable sources (CA PUC, 2012). California has committed to a 33% renewable market share (CA PUC, 2011).

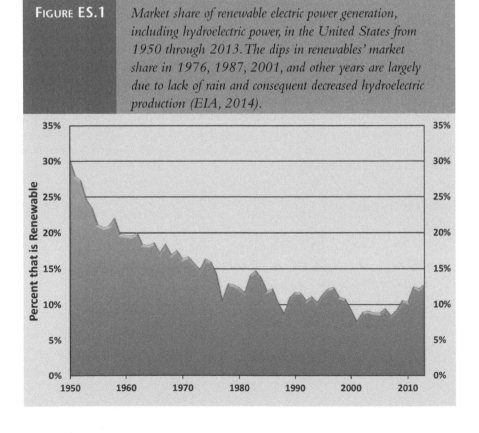

FIGURE ES.1   *Market share of renewable electric power generation, including hydroelectric power, in the United States from 1950 through 2013. The dips in renewables' market share in 1976, 1987, 2001, and other years are largely due to lack of rain and consequent decreased hydroelectric production (EIA, 2014).*

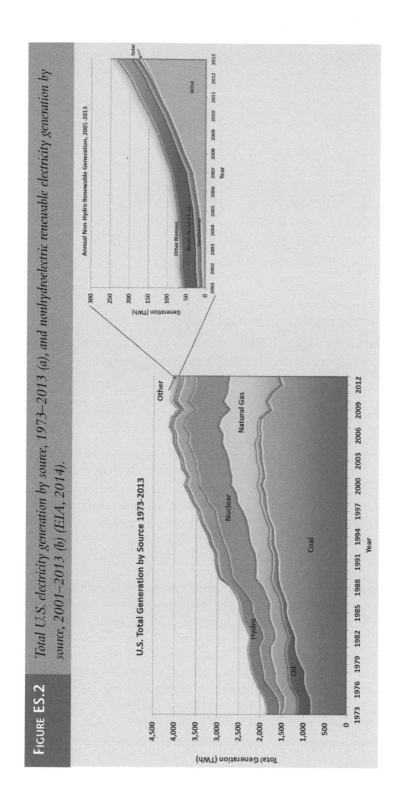

FIGURE ES.2 *Total U.S. electricity generation by source, 1973–2013 (a), and nonhydroelectric renewable electricity generation by source, 2001–2013 (b) (EIA, 2014).*

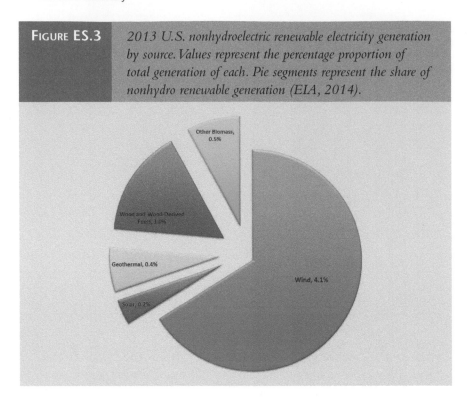

FIGURE ES.3 | *2013 U.S. nonhydroelectric renewable electricity generation by source. Values represent the percentage proportion of total generation of each. Pie segments represent the share of nonhydro renewable generation (EIA, 2014).*

Renewable electric power benefits society by reducing pollution from the electric power sector (if strict pollution standards have not already reduced both conventional pollution and greenhouse gas pollution). However, the variability of most renewable generation introduces system-level costs into the management of the electric grid; these arise from the need to manage the inherent variability.

The work presented here focuses mainly on wind and solar resources. Wind power generators are the lowest-cost and most widely available nonhydroelectric renewable power plants, so wind is expected to continue to dominate the growth in renewable energy in the United States.[2] In certain regions, solar electricity may present significant new contributions to the operations of the power grid. Geothermal power can be competitive in certain locations, and there is potential to use not only hydrogeothermal power (where nature supplies the hot water) but also enhanced geothermal power (where water is injected into hot underground rock and returned to the surface for use in generating electricity). We touch briefly on the siting issues that are emerging for enhanced geothermal power.

Successful deployment of the variable energy resources expected to be the major contributors to renewable power requires improved planning and operations, advanced technologies and infrastructure, and appropriate public policies.

---

2. Wind power production of electricity in the United States in 2013 was 18 times that of solar electric power production.

I.2 Challenges

The key challenges facing the increased use of wind and solar electric power are highlighted in this section.

I.2.1 Wind and solar power do not produce a consistent amount of power.

Figure ES.4 shows how renewable electricity production varies over the course of 2 separate days in California. Changes in power output also occur on shorter time scales. Figure ES.5 shows the variability of wind power output sampled every 5 minutes in the Bonneville Power Authority over the course of 8 days. Figure ES.6 shows

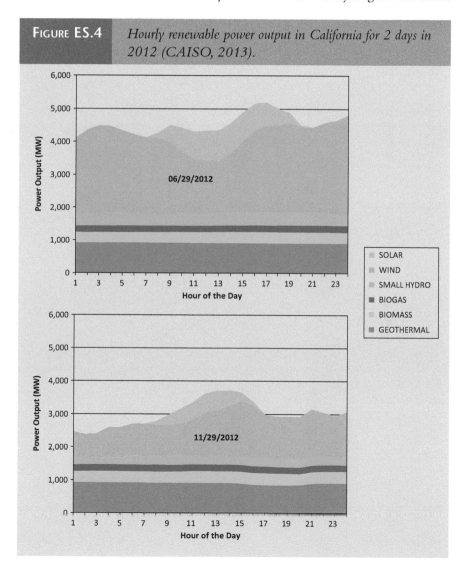

FIGURE ES.4 *Hourly renewable power output in California for 2 days in 2012 (CAISO, 2013).*

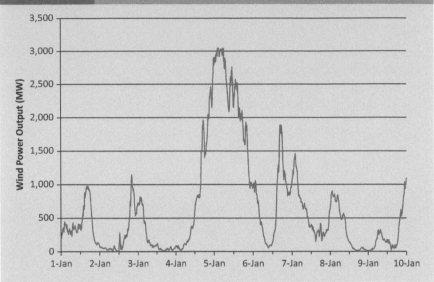

FIGURE ES.5 *Wind power output measured with 5-minute time resolution in the Bonneville Power Authority area, 1/1/2012 to 1/8/2012. Total wind generation nameplate capacity was 3,788 MW during this period (BPA, 2012).*

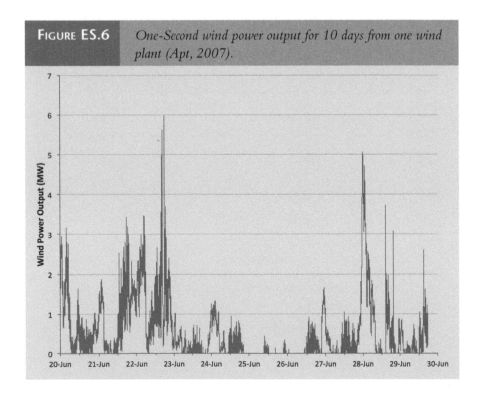

FIGURE ES.6 *One-Second wind power output for 10 days from one wind plant (Apt, 2007).*

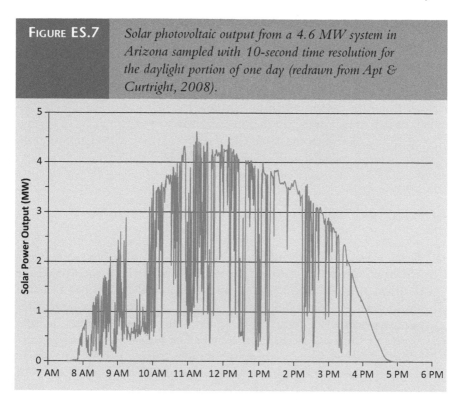

| FIGURE ES.7 | *Solar photovoltaic output from a 4.6 MW system in Arizona sampled with 10-second time resolution for the daylight portion of one day (redrawn from Apt & Curtright, 2008).* |

the 1-second variability in power output for 10 days in one wind plant in the Middle Atlantic region. Solar photovoltaic power has been shown to be more variable than wind power (this is discussed in Chapters 2 and 8); Figure ES.7 shows the output of a large utility-scale solar photovoltaic array in Arizona, sampled every 10 seconds.

I.2.2 The sun does not shine at night, and there are cloudy days; there are also days-long lulls in wind power.

In addition, just as there are rainy years and drought years for hydroelectricity, research indicates there will be windier years and calmer years for wind power. The most important consequence of year-to-year variability is that it determines the debt-to-equity ratio of wind power projects, which affects the profitability for their owners (see Chapter 11).

I.3 Opportunities to respond to those challenges

Grid operators have responded to increased variability in many ways. We concentrate here on a number of responses that are available to integrate variable renewable generators at large scale:

- Better prediction of variability and strategies to reduce it
- Changes in the operation of power plants, reserves, transmission systems, and storage

- Improved planning of renewable capacity expansion planning
- Implementation of new regulatory paradigms, rate structures, and standards.

Chapter 6 discusses in detail recommendations (based on the research described in Part II of this book) for policies that can be adopted by key decision makers to increase renewable energy's proportion of power supplied to the grid. These are summarized in a figure at the conclusion of this executive summary.

Here we describe briefly the results of our research in each of the four categories of opportunities to respond to the challenges posed by variable renewable electric power.

I.3.1 Better prediction of variability and strategies to reduce it

1. Forecasts of wind power in the United States systematically underpredict wind during periods of light wind and overpredict when there are strong winds (see Figure ES.8 and Chapter 9). This is important for those who manage the electricity grid, which incorporates power from a number of sources including wind power. It is the grid operators' responsibility to make sure power production instantaneously matches consumers' demand for electricity. In order to support large-scale deployment of wind and solar power, these operators can improve integration by correcting for forecasting errors. We have found that there is a straightforward mathematical framework to incorporate forecast uncertainty so that other generation sources can be scheduled in the day-ahead

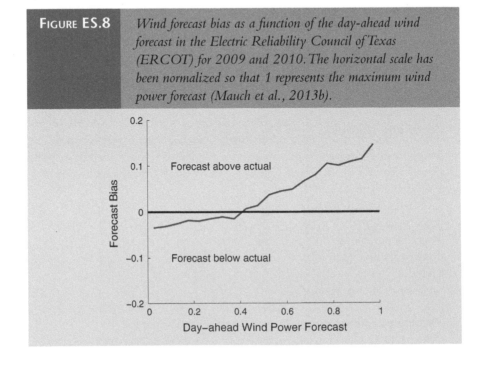

| FIGURE ES.8 | *Wind forecast bias as a function of the day-ahead wind forecast in the Electric Reliability Council of Texas (ERCOT) for 2009 and 2010. The horizontal scale has been normalized so that 1 represents the maximum wind power forecast (Mauch et al., 2013b).* |

time frame to match the variability of wind (and solar) power and fill in the deficit that results from variability. Distributions of the errors in wind forecasts are not bell-shaped (normal) curves; thus, predictions based on normal curves underpredict extreme forecast errors that are infrequent but have large consequences for grid reliability (see Chapter 9).

2. Net load is the demand for electricity minus the renewable production. Almost all the studies reviewed for this project used the standard deviation of net load step changes (the difference in the net load from one time period to the next) to estimate the need for regulation and load-following reserves. Doing so implicitly assumes that load and wind are uncorrelated and that the data fit Gaussian statistical models. Neither assumption is accurate. Use of appropriate statistics should allow for increasing statistical accuracy, and thus more insightful results, in future integration studies (see Chapter 17).

3. Electrically combining the output of several wind plants in a region can reduce variability in the aggregate power output, but the amount of reduction is dependent on the time scale involved. The small variability at intervals of an hour or shorter can be reduced by 95%, but the large variability at 12 hours and longer is reduced by only 50%. It is the strong fluctuations at long time scales that require the most compensating resources. There are quickly diminishing returns as more plants are interconnected (see Chapters 3 and 12; Katzenstein, Fertig, & Apt, 2010).

4. Aggregating wind power generated over large geographical areas is also beneficial for reducing variability and increasing economic efficiency, but the costs of interconnection are likely to be higher than building new natural gas combined-cycle plants within each of the areas (Fertig, Apt, Jaramillo, & Katzenstein, 2012). Thus, large new investments in transmission systems designed to interconnect large areas of the country are neither required nor desirable to decrease the variability of electric power generated from wind. Decreased transmission costs could change this conclusion (see Chapter 3).

I.3.2 Changing operations of power plants, reserves, transmission systems, and storage

1. The character of power variability from wind and solar power is such that the strongest power fluctuations occur slowly over many hours or days. Thus, slow-responding generators, such as coal and most combined-cycle gas plants that take a long time to change their power output (called slow-ramping plants), can compensate for most of the variability (see Chapters 2 and 8; Apt, 2007).

2. Fast-ramping power generators (those that are able to reduce or increase their power output over short periods of time, such as hydroelectric generators, natural gas turbines, newly designed flexible combined-cycle gas plants, and batteries) can play a role, because they are better suited for balancing higher-frequency variability. For example, a very small complement of batteries can reduce wind power variability to the electricity transmission grid and increase the economic integration of wind power (see Chapters 3 and 14; Hittinger et al., 2010).

At least one U.S. electricity market is considering market products for ramping services. That should be encouraged, and the ramping resources should not be limited to generators but should also include storage and verifiable demand response.

3. The use of fast-ramping gas plants can mitigate some of the high-frequency variability of wind. Continuous ramping of gas plants can increase the emissions from the power plants and thus reduce the emission benefits generally associated with wind (Katzenstein & Apt, 2009). New gas plant technology, like Siemens H-Class and GE's Flex 50 combined-cycle technology, can mitigate this effect. Coal plants can be cycled to manage the low-frequency variability of wind while incurring smaller emission penalties than those incurred by gas plants that compensate for wind's or solar's variability (see Chapter 13).

4. It is now possible to use accurate statistical methods to procure an economically efficient amount of generation a day ahead as a reserve for net load (load-wind) variability (Figure ES.9). As the amount of wind power in the system

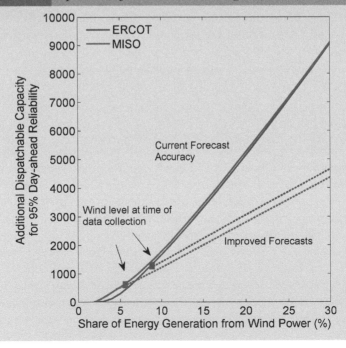

FIGURE ES.9   *Additional day-ahead reserve capacity for a range of wind penetration values in ERCOT and MISO. In this figure, the horizontal axis is the percentage of load served by wind power. Solid lines assume no change in wind forecast accuracy. Dashed lines show the effect of improved forecasts that may be possible if new wind plants are added far enough away from existing ones to allow low correlation of their output. Reprinted with permission from Mauch and colleagues (2013a).*

increases, operators are likely to have to also increase the amount of day-ahead reserve available. The proper level of such reserves depends on the levels of load and wind that are forecast (see Chapters 2 and 10).

5. Some grid codes incentivize operation of wind turbines somewhat below their rated capacity (e.g., Denmark, Ireland, Great Britain, and Germany; Rose & Apt, 2013) to allow them to increase power to somewhat compensate for wind variability. This practice is less economically efficient than compensating with a natural gas turbine is. In cases where it is required (perhaps when natural gas prices are very high), the requirement should not be uniformly spread across all wind turbines but instead placed on the lowest number required to achieve the desired compensation. It is reasonable for grid operators to require that wind plants have the capability to dynamically limit output for up-regulation capacity, since this is inexpensive. However, that capability should rarely be used because it is quite expensive to operate a wind plant that way (see Chapters 3 and 15).

6. Compressed-air energy storage (CAES) does not appear likely to be profitable in the bulk of the United States unless the market price differentials more than double or capital costs substantially decrease. Indeed, the one new CAES project in the United States is in Texas, where price spikes in the energy market may make storage economical. Better wind forecasting will not help CAES profitability. Subsidies to make a wind+CAES plant break even (have a net present value of zero) would need to be ~$100/tonne of avoided CO_2 (see Chapters 3 and 14, section B.3).

7. Portugal is expanding pumped hydro capacity to support wind by building 636 MW of new pumped hydroelectric storage (PHS), a 60% increase, for a system with a peak load of 9 to 10 GW. PHS in Portugal and in Norway (the latter to support German wind) is currently unprofitable based on energy arbitrage – that is, storing electricity when prices are low and selling it when prices are high. However, the likely increase in short-term price variability as Germany expands its wind power capacity could render Norwegian PHS profitable in the near future. The type of owner who operates grid-scale storage has a large effect on whether consumers, generation plant owners, or storage owners capture the benefits of storage. If operated in a manner that reduces price spreads, new storage may eliminate the need for some of the most inefficient and rarely used capacity (see Chapter 14, section 14.3.2).

8. Grid-scale storage can provide substantial benefits for renewables integration and can directly benefit consumers by reducing prices in the wholesale energy market and by avoiding the need to keep expensive and rarely used plants on retainer. However, large-scale deployment of storage for energy arbitrage is likely uneconomical for the storage owner under current market rules. Even with capital costs as low as $150/kWh, storage may not be profitable for the storage owner in the world's largest electricity market, PJM, where storage capacity of 4% of peak load already exists. Increasing efficiency or reducing Operations and Maintenance (O&M) costs is not sufficient to make arbitrage

profitable. Storage can provide other services to the grid not currently captured in the energy markets (see Chapters 3 and 14, Section 14.3).

9. If passenger vehicles containing large batteries achieve significant market penetration, a relatively small number of them are likely to be useful in maintaining the frequency of the power grid at 60 Hz, if the vehicle warranties are changed to permit use for grid stabilization. There are no market mechanisms that would incentivize a large number of vehicle owners to charge when solar panels produce power during the day and discharge after the sun has set, but an individual electric vehicle owner may find this useful if the vehicle is connected to a residential solar system. Use of vehicle batteries after they have degraded to a point at which they are no longer suitable for driving may be practical. It does not appear likely that varying the charging of vehicle batteries will be of much help in integrating variable renewables (see Chapters 3 and 14, Section 14.3.5).

I.3.3 Improved siting of renewable capacity expansion activities

1. When wind or solar energy displaces conventional generation, the reduction in emissions varies dramatically across the United States. If the goal of renewable power is pollution reduction (including displacing CO_2 from power plants), it is much better to locate the facilities in the Mid-Atlantic states than in the Southwest or West because generators that emit a good deal of pollution are located in heavily populated locations there. While the Southwest has the greatest solar resource, a solar panel in New Jersey displaces significantly more criteria air pollutants than a panel in Arizona, resulting in 14 times more health and environmental benefits. A wind turbine in West Virginia displaces twice as much carbon dioxide as the same turbine in California. Depending on location, the combined health, environmental, and climate benefits from wind or solar range from $10 to $100 per megawatt-hour, and the sites with the highest energy output do not yield the greatest social benefits in many cases. As a result, national production-based subsidies for wind and solar energy are poorly aligned with health, environmental, and climate benefits (see Chapter 4; Siler-Evans et al., 2013).

2. One of the incentives some electricity markets provide for generators is a retainer payment for the capacity that they provide (generally at times when demand is the highest). Capacity credit calculations in wind integration studies should be based on at least 5 years of wind data (see Chapter 17).

3. Although hurricanes can pose a risk to offshore wind turbines, making small changes such as having emergency power to yaw the turbine nacelle rapidly into the wind can improve survivability. The risks can also be reduced by strengthening turbine designs. In addition, it is possible to predict which offshore areas are the least risky for wind turbines prior to construction (see Chapters 4 and 16; Rose et al., 2012, 2013).

4. Traditional geothermal power systems are located in areas where there is sufficient naturally occurring heat, water, and rock permeability to extract energy (e.g., a geyser). An alternative type of geothermal energy, called enhanced geothermal systems, uses a process called hydraulic stimulation to generate energy from dry and impermeable rock. The challenge of this process is that small earthquakes may occur as the water is injected, causing public concern. This concern must be taken into account in the siting process for these facilities (see Chapter 4; Hoşgör et al., 2013).

I.3.4 Implementation of new regulatory paradigms, rate structures, and standards

1. One of the most significant barriers to the widespread adoption of renewable electricity is the extensive transmission infrastructure required to carry wind resources from their geographically isolated locations to major load centers. The recently promulgated U.S. Federal Energy Regulatory Commission (FERC) Order 1000 is expected to significantly alter transmission planning processes and cost allocation and make it somewhat easier to build transmission lines for renewable electricity, highlighting the influence that regulation can have on the large-scale deployment of variable energy resources like wind and solar (see Chapter 5).

2. FERC Order 764 (finalized in June 2012) reduces the interval in which a generator must forecast its output to the grid from 1 hour to 15 minutes. While all Independent System Operators (ISOs)/Regional Transmission Operators (RTOs) have already moved to 5-minute scheduling, most of the balancing areas in the West and the Southeast still use hourly scheduling. During the rule-making process, RenewElec urged FERC to require 5-minute scheduling in areas with significant VER integration needs instead of stopping at 15 minutes. Intrahour scheduling enhances the ability to incorporate variable energy resources, as it provides a much closer matching with real outputs; it also benefits nonvariable sources. More details are given in Chapter 5.

3. RenewElec research indicates that there is little economic motivation to consolidate balancing areas today or at the scale of renewables mandated by current RPS legislation. Many of the renewables integration benefits thought to be associated with balancing area consolidation can likely be achieved at lower cost through better balancing-area coordination (see Chapter 5).

4. Existing decommissioning requirements for wind plants are likely to be insufficient, and appropriate bonding requirements may need to be established to guarantee the proper decommissioning of wind turbines at the end of their life. While some jurisdictions have established bonding requirements for wind plants, these have generally been at the very low end of projected decommissioning costs (Changala et al., 2012). The states that have

established decommissioning regulations for wind projects require bonding amounts on a case-by-case basis (see Chapter 5). Evaluating decommissioning bonding requirements on a case-by-case basis will most likely produce more accurate bond amounts. Although both oil/gas and wind decommissioning regulations are insufficient to cover the actual costs of decommissioning, fossil fuel well requirements are generally lower than are those for wind. Equitable regulatory burdens are necessary for all energy activities so that each resource is subject to environmental accountability and proportional regulatory burdens.

5. Wind droughts and other long-term weather phenomena (see Chapter 11) must be considered in regulatory agency review of wind generation proposals.
6. Data on forecast and actual power production for variable energy resources (VER) are usually proprietary. Advances in the theory that underlies improved forecasts, efficient use of storage, and cost-effective siting requires that high-quality data be available to research teams. The results of RenewElec research suggest that FERC should require VER data to be made available for research purposes within 6 months of when it is submitted to transmission providers, in a sufficiently anonymous format so that the data could not be used for commercial format (see Chapter 5).

1.4 Summary

Reaching a 20 to 30% wind power renewable portfolio standard goal can be facilitated by changes in the management and regulation of the power system. Accurately assessing and preparing for the operational effects of renewable generation can ensure that renewables can play a much larger role in the U.S. power system. The actions outlined here and summarized in Figure ES.10 and discussed in detail in Chapter 6 can help reach the goal of increasing renewable energy's proportion of power supplied to the grid.

Short- and long-term strategies to increase renewable energy's proportion of power supplied to the grid.

Short-Term Strategies

Grid operators should incorporate wind forecast uncertainty in unit commitment and dispatch decisions. Forecast uncertainty is strongly dependent on whether the forecast is for low or high wind. Load errors are also less accurate at high load.

Forecasters and grid operators should correct for wind forecast biases before using the data for unit commitment and dispatch. Present methods tend to over predict wind power output when the wind blows strongly and under predict in light winds.

Analysts conducting large-scale integration studies should recognize that wind and load are correlated and that they do not fit Gaussian (bell-shaped) curves. Analysts should also base capacity credit calculations on at least 5 years of wind data and consider increased capacity credit for plants that have low correlation with other plants.

Large ISO/RTOs should recognize that dynamically limiting all turbines within a wind plant below maximum generation to provide regulation capacity is typically more costly than other options available to enhance reliability. It is not costly to give wind plants the ability to dynamically limit, but its use should be very infrequent.

Electricity markets should allow fossil-fueled plant operators to incorporate cycling costs in their bids to limit excessive cycling of units when they are used to compensate for variability.

Electricity markets should strongly consider implementing a market for ramping services provided by generators, storage, and demand response to facilitate large-scale integration of variable renewable generators.

The financial community should allow developers to combine projects to reduce year-to-year variability's effect on debt-to-equity ratios.

FERC should require 5-minute scheduling in all areas with significant variable energy resource penetration. FERC should also increase data reporting requirements for variable generators. FERC should clarify the definition of variable energy resources to apply to facilities that make use of energy storage or other variability mitigation.

Long-Term Strategies

ISO/RTOs and planning agencies should recognize that large-scale geographic aggregation is not necessary to mitigate the variability of wind. While variability slightly decreases when large areas are combined, and firm capacity slightly increases, using natural gas turbines and/or RTO-RTO coordination market products to counter variability is likely to be more cost effective.

ISO/RTOs should develop strategies to compensate energy storage operators for the benefits they provide to electricity consumers. Storage can significantly reduce high-frequency variability of wind and solar and better match these resources to load.

Local, state, and federal governments should establish appropriate decommissioning requirements for wind and solar power plants. **States** should set wind and solar plant decommissioning bonds to adequately cover end-of-life costs on a case-by-case basis.

Regulatory agencies and insurers should provide incentives for the development of renewable resources in areas with lowest risks of hazards like earthquakes and hurricanes.

Legislative bodies and regulatory agencies should provide incentives to site wind and solar power plants where benefits from displacing pollution are highest.

Standards organizations such as the IEC should consider requiring wind turbines in higher-risk hurricane areas to have the capability to yaw into the wind even without grid power.

References

Apt, J. (2007). The spectrum of power from wind turbines. *Journal of Power Sources, 169*(2), 369–374.

Apt, J., & Curtright, A. (2008). *The spectrum of power from utility-scale wind farms and solar photovoltaic arrays.* Carnegie Mellon Electricity Industry Center working paper, CEIC-08-04, http://wpweb2.tepper.cmu.edu/electricity/papers/ceic-08-04.asp

BPA. (2012). *Wind generation and total load in the BPA balancing authority.* http://transmission.bpa.gov/Business/Operations/Wind/

CAISO. (2013). *California ISO renewables watch.* www.caiso.com/market/Pages/ReportsBulletins/DailyRenewablesWatch.aspx

CA PUC. (2011). *Decision 11–12–020.* California Public Utilities Commission, http://docs.cpuc.ca.gov/WORD_PDF/FINAL_DECISION/154695.PDF

CA PUC. (2012). *Renewable energy portfolio standard, quarterly report, 1st and 2nd quarter 2012.* California Public Utilities Commission, www.cpuc.ca.gov/NR/rdonlyres/2060A18B-CB42-4B4B-A426-E3BDC01BDCA2/0/2012_Q1Q2_RPSReport.pdf

Changala, D., Dworkin, M., Apt, J., & Jaramillo, P. (2012). Comparative analysis of conventional oil and gas and wind project decommissioning regulations on federal, state, and county lands. *Electricity Journal, 25*(1), 29–45.

EIA. (2014). *Electric power monthly with data for December 2013.* U.S. Energy Information Agency, www.eia.gov/electricity/monthly/current_year/february2014.pdf

Fertig, E., Apt, J., Jaramillo, P., & Katzenstein, W. (2012). The effect of long-distance interconnection on wind power variability. *Environmental Research Letters, 7*(3), 034017.

Hittinger, E., Whitacre, J.F., & Apt, J. (2010). Compensating for wind variability using co-located natural gas generation and energy storage. *Energy Systems, 1*(4), 417–439.

Hoşgör, E., Apt, J., & Fischhoff, B. (2013). Incorporating seismic concerns in site selection for enhanced geothermal power generation. *Journal of Risk Research, 16*(8), 1021–1036.

Katzenstein, W., & Apt, J. (2009). Air emissions due to wind and solar power. *Environmental Science and Technology, 43*(2), 253–258.

Katzenstein, W., Fertig, E., & Apt, J. (2010). The variability of interconnected wind plants. *Energy Policy, 38*(8), 4400–4410.

Mauch, B., Apt, J., Carvalho, P.M.S., & Jaramillo, P. (2013a). What day-ahead reserves are needed in electric grids with high levels of wind power? *Environmental Research Letters, 8*(3). DOI: 10.1088/1748-9326/8/3/034013

Mauch, B., Apt, J., Carvalho, P.M.S., & Small, M.J. (2013b). An effective method for modeling wind power forecast uncertainty. *Energy Systems, 4*(4), 393–417.

Rose, S., & Apt, J. (2013). The cost of curtailing wind turbines for secondary frequency regulation capacity. *Energy Systems,* in press, DOI: 10.1007/s12667-013-0093-1

Rose, S., Jaramillo, P., Small, M.J., & Apt, J. (2013). Quantifying the hurricane catastrophe risk to offshore wind power. *Risk Analysis, 33*(12), 2126–2141. DOI:10.1111/risa.12085

Rose, S., Jaramillo, P., Small, M.J., Grossmann, I., & Apt, J. (2012). Quantifying the hurricane risk to offshore wind turbines. *Proceedings of the National Academy of Sciences, 109*(9), 3247–3252.

Siler-Evans, K., Azevedo, I., Morgan, M.G., & Apt, J. (2013). Regional variations in the health, environmental, and climate benefits of wind and solar generation. *Proceedings of the National Academy of Sciences, 110*(29), 11768–11773.

PART I
Technical and policy options

1

OVERVIEW

The purpose of this book is to provide government and industry policymakers, scientists, and engineers with an understanding of the technical and policy options available for managing variable energy resources such as wind and solar power to produce electricity. The text is informed by knowledge gained during Carnegie Mellon University's RenewElec (short for renewable electricity) project that began in 2010, based on years of preceding work.

Renewable energy resources tend to have variable output power on several time scales. Hydroelectric power's production, for example, varies by season. On longer time scales, it is lower in drought years and greater in wet years. Wind power varies on time scales of seconds to years. Solar power and geothermal power are also variable. The U.S. Federal Energy Regulatory Commission (FERC) calls such generators "variable energy resources" (VER).

Renewable energy as a source for electricity generation in the United States fell from 30% in 1950 to a low of 8% in 2001, as the market share of hydroelectric power was eroded by fossil fuel generators built to keep up with the rapidly increasing demand for electricity (Figure 1.1). Even though hydroelectric generation tripled between 1950 and 1973, demand for electricity increased nearly six-fold. By 2013, renewables' market share had increased to 13%, primarily due to policies that encouraged an increase in wind power's contribution.

A significantly expanded role for variable energy resources can be achieved. It is likely that renewables can regain or exceed the share of electric power they represented in 1950. In order to do this, the United States should adopt a systems approach that considers and anticipates the changes in power system design and operations that will be required while doing so at an affordable price and with acceptable levels of security and reliability.

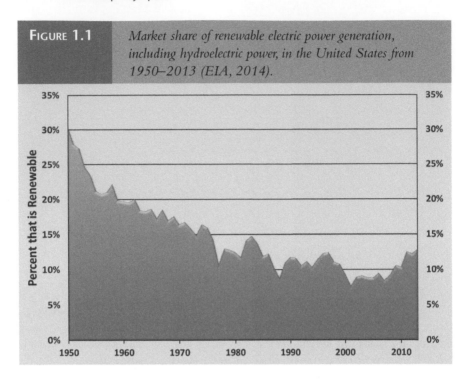

FIGURE 1.1 *Market share of renewable electric power generation, including hydroelectric power, in the United States from 1950–2013 (EIA, 2014).*

The RenewElec project[1] was created as an interdisciplinary project led by Carnegie Mellon University to facilitate dramatic increases in the use of electric generation from variable sources of renewable power in a way that:

- Is cost-effective
- Provides reliable electricity supply with a socially acceptable level of local or large-scale outages
- Allows a smooth transition using the architecture and operation of the present power system
- Allows and supports competitive markets with equitable rate structures
- Is environmentally benign
- Is socially equitable.

This book, based on empirical evidence and research by our PhD students and faculty, provides a summary of major challenges and opportunities for integrating variable power generation sources and meeting the goals set forward in the state renewable portfolio standards.

---

1. The RenewElec project (www.RenewElec.org) was supported by the Doris Duke Charitable Foundation, the Richard King Mellon Foundation, the Heinz Endowments, the U.S. Department of Energy, the Electric Power Research Institute, and the U.S. National Science Foundation.

In this chapter, we describe renewable electricity and provide information on renewable electricity's current and potential contribution to electricity generation. We also describe the national and state policies currently in place that influence renewable electricity's contribution to the electricity grid, as well as the major challenges and opportunities it faces in doing so. Last, we provide an overview of the remainder of this book.

1.1 What is renewable electricity?

Renewable electricity is generally defined as derived from any energy resource that is theoretically inexhaustible in periods of interest to humans and is thus constantly replenished on time scales of days to decades. Renewable electricity can be derived directly from the sun, such as thermal, photoelectric, and photochemical energy; indirectly from the sun, such as hydroelectric, wind, and photosynthetic energy stored in biomass; or from natural processes in the environment, such as geothermal and tidal energy. Renewable power sources generally have lower environmental damages than coal, natural gas, and oil power, particularly lower emissions of conventional air pollutants and greenhouse gases. However, these resources are not entirely free of environmental externalities. Large hydropower reservoirs, for example, are a source of methane emissions that contribute to climate change, and land-use changes caused by many power generators are a source of environmental concern.

The U.S. Environmental Protection Agency (EPA) defines a further subset of renewable power as green power, which consists of resources that do not directly emit greenhouse gas. These green power sources include wind, solar, geothermal, and biomass. Figure 1.2 shows a schematic of the different power source classifications described by the EPA.

FIGURE **1.2**   *Classification of power sources (redrawn from EPA, 2010).*

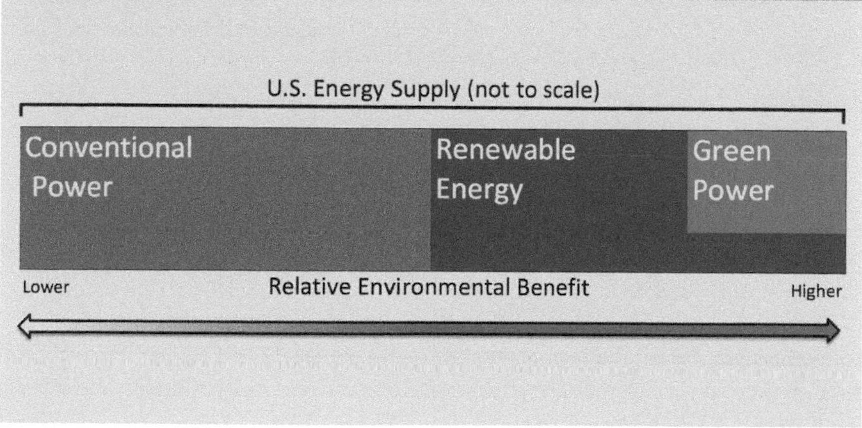

U.S. Energy Supply (not to scale)

| Conventional Power | Renewable Energy | Green Power |

Lower   Relative Environmental Benefit   Higher

1.2 What is renewable electricity's current contribution to electricity generation?

Renewable energy as a source for electricity generation is increasing at a rapid rate in the Unites States and elsewhere. In 2008, renewables, including hydroelectric power, constituted only about 9% of all U.S. electricity generation (see Figure 1.3). By 2013, the share had increased to 12.9%, primarily due to an increase in wind power's contribution (see Figure 1.4).

1.3 Why has renewable electricity grown in the United States?

Two policy changes at the state and national levels have led to the rapid increase of renewable energy: renewable portfolio standards (RPSs) at the state level and federal tax credits. The latter include the production tax credit (PTC) used primarily by wind developers and investment tax credit (ITC) used primarily by solar developers.

Although there is no U.S. national renewable energy standard, 29 states and the District of Columbia have RPSs requiring that some percentage of their electric power come from sources defined as renewable (DSIRE, 2011). The language in state RPSs can vary significantly, and while some states have a single (primary) standard, some states have several types. Colorado, for example, has a primary standard that applies to investor-owned utilities and a secondary standard that applies to electric cooperatives (DSIRE, 2011). Similarly, some states have "set-asides" that require that certain technologies be used to meet a given RPS level. Figure 1.5 shows the primary RPS requirements and solar set-asides that apply to large investor-owned utilities, as well as the target year for the final target.

In the United States, three major constituencies have advocated for growth in sources of renewable energy and energy efficiency measures, including an:

1. Environmentalist constituency worried about fossil fuels' contribution to climate change and pollution
2. Energy security constituency worried about the national security and the need to reduce dependence on foreign fossil fuels, limit demand, and lower cost (Pew Charitable Trusts, 2012; U.S. DOD, 2012)
3. Economic vitality constituency that is worried that high energy prices, market volatility, and disruptions to the energy supply will threaten the national economy and result in lost jobs (Long, 2008).

A detailed discussion of the economic rationale for the policies adopted by various governments is not the focus of this book; however, we now provide a brief discussion of the different mechanisms that are available to policy makers. Generation of electric power produces not only electricity but also pollutants that enter the environment, both during the manufacturing of the generator and during its operation. Economists call these costs "externalities." For conventional pollutants, these costs

FIGURE 1.3

(a) Total U.S. electricity generation by source, 1973–2013, and (b) nonhydroelectric renewable electricity generation by source, 2001–2013 (EIA, 2014).

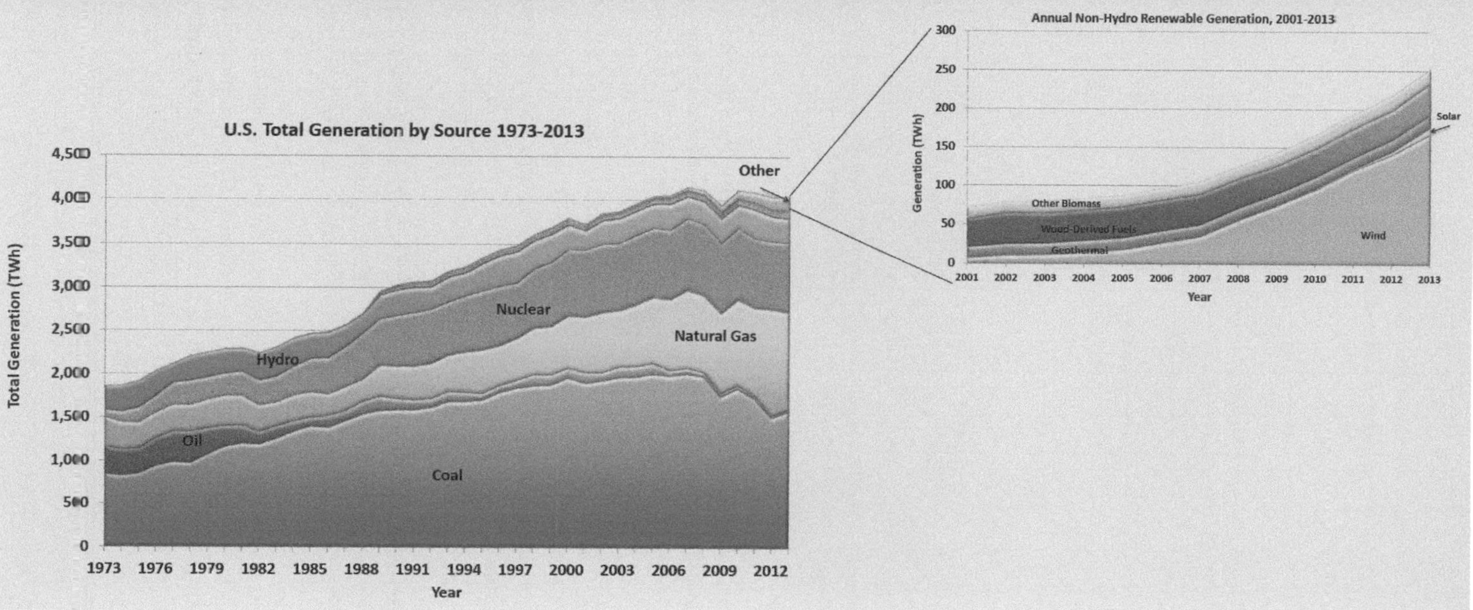

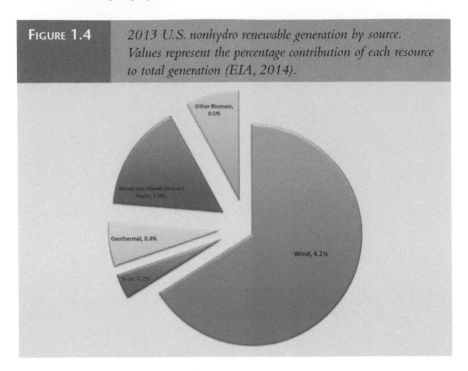

FIGURE 1.4	*2013 U.S. nonhydro renewable generation by source. Values represent the percentage contribution of each resource to total generation (EIA, 2014).*

can be estimated by observing the human health effects. Such estimates for greenhouse gas pollution are quite uncertain in magnitude and timing (see Chapter 7).

If the costs of this pollution are not included in the price of electricity, an economist would say that the artificially low prices cause customers to consume more power than the economically efficient amount. A number of jurisdictions forbid some of the pollution (command-and-control regulation). Economists realized that this sort of regulation can lead to retiring a generator before its useful life is reached, and so cap-and-trade regulations have been instituted (notably for nitric oxide and nitrogen dioxide [NO_x] emissions) to allow the pollution from older plants to be offset by newer and cleaner plants.

A few jurisdictions (for example, the United Kingdom for a time) levied a pollution fee on all electric power sold, encouraging less use and thus less pollution. Other jurisdictions have subsidized the introduction of low-polluting power (for example, with a PTC, ITC, or feed-in tariff) by an amount that is roughly equal to the externality costs. While this policy reduces pollution, its costs are borne by the taxpayers rather than by the users of electricity, and economists object that thus the price of power is being artificially reduced, leading to overconsumption.

An RPS, like command-and-control regulation, both lowers pollution and has costs that are borne by the electricity consumer, encouraging use of an economically efficient amount of electric power.

However, an RPS may not be efficient if the power sources included are restricted to those deemed "green" by its framers. "Renewable" and "low-carbon" or "low-pollution" are not synonyms, and the former generally does not permit low carbon

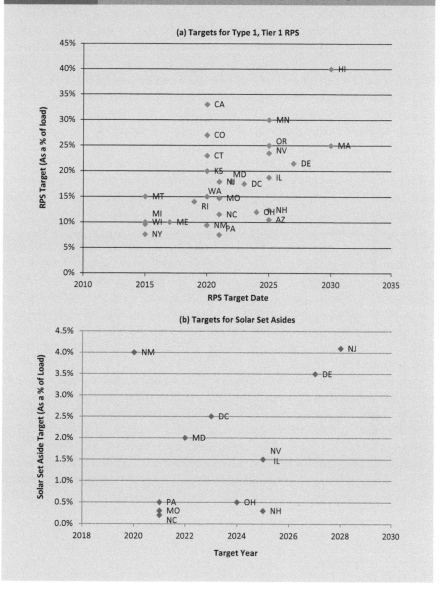

FIGURE **1.5**   *The final targets for primary, Tier 1 RPS targets in the final target year (a) and solar-set asides (b). This figure includes the contribution of all the renewable resources covered in each state's RPS (DSIRE, 2013a).*[2]

---

2. Pennsylvania has an alternative energy portfolio standard (AEPS), which mandates that a percentage of electricity in the state come from qualified resources. Unlike a renewable portfolio standard, the AEPS allows for the use of resources like waste coal and low-pollution-emissions coal technology.

generation from nuclear or large hydroelectric plants (Apt, Keith, & Morgan, 2007). In addition, it has been pointed out that renewables policies enhance net social welfare only when externalities are not properly addressed by regulation (Edenhofer, Knopf, & Luderer, 2013).

In the United States, a combination of command-and-control regulation and subsidies has led to the rapid increase of renewable energy; there are RPSs at the state level, as well as the federal PTC and ITC. An RPS is not an effective tool without penalties for noncompliance. Figure 1.6 shows penalties and alternative compliance payments that have been established for primary renewables standards that affect wind development and those for not meeting the solar set-asides.

The production tax credit, a per-kilowatt-hour (kWh) tax credit for electricity generated by wind, geothermal, and other qualified energy resources,[3] was first introduced in the Energy Policy Act of 1992, and it first expired in 1999. Since then it has been reinstituted and expired several times.[4] Most recently, it was extended for 1 year at the end of 2012. The PTC currently provides a tax credit of 2.3 cents/kWh of wind energy produced during the first 10 years of the wind plants covered, and it thus supports the economic viability of wind energy projects (DSIRE, 2013b).

Figure 1.7 shows that when the PTC is active, it has a demonstrable effect on yearly wind power installation. The renewable investment tax credit, a tax credit for investments in solar energy, fuel cells, small wind systems, geothermal systems, natural gas microturbines, and combined heat and power generators, has been in effect since 1978 with many modifications and extensions (Sherlock, 2012). The current ITCs are legislated to expire at the end of 2016 (DSIRE, 2013a). Because solar electric power generation systems are very capital intensive, it has generally been more economically advantageous for a developer to take advantage of the ITC than the PTC for such systems.

Seventeen of the states and the District of Columbia[5] shown in Figure 1.5 must meet their RPS within the decade (by 2023). If all of these states are to meet their standards, a total of 90 gigawatts (GW) of qualifying renewables must be available by 2023. The historical rate of construction of installed wind capacity for the United States as a whole has followed an exponential curve, as shown in Figure 1.7. An exponential curve also fits the aggregate growth of installed wind capacity in the 17 states that must meet their standard by 2023 (Figure 1.8). Following the same growth rate for qualifying renewable capacity as predicted by the historical wind

---

3. Landfill gas, wind, biomass, hydroelectric, geothermal electric, municipal solid waste, hydrokinetic power (i.e., flowing water), anaerobic digestion, small hydroelectric, tidal energy, wave energy, ocean thermal power.
4. For a history of the PTC, see www.eia.gov/forecasts/archive/aeo05/issues.html.
5. CA, CO, CT, DC, KS, MD, ME, MI, MO, MT, NC, NJ, NM, NY, PA, RI, WA, WI.

FIGURE 1.6

Wind-related (a) and solar-related (b) penalties for noncompliance with RPS requirements (DSIRE, 2013a). The horizontal lines show the load-weighted average prices for three major electricity-trading hubs for 2013 (EIA, 2013). We note that a real options approach would take into account uncertainty in future electricity prices.

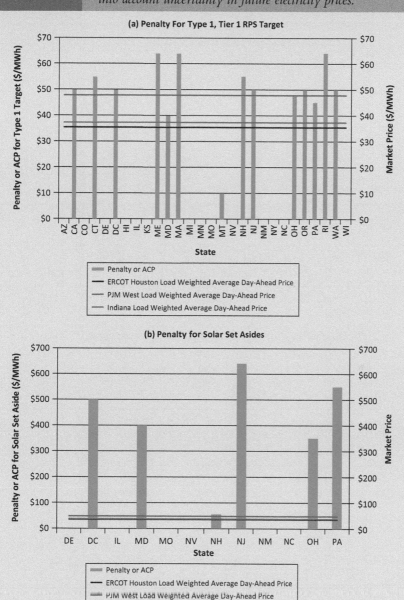

FIGURE 1.7

U.S. cumulative (left scale) and incremental (right scale) wind power installations (a) and solar power installations (b), 1999–2013. Black triangles and red squares indicate the status of the federal production tax credit (PTC) supporting wind projects (a) and the investment tax credits (ITC) supporting solar projects (b) in a given year (EIA, 2013 and FERC, 2014).

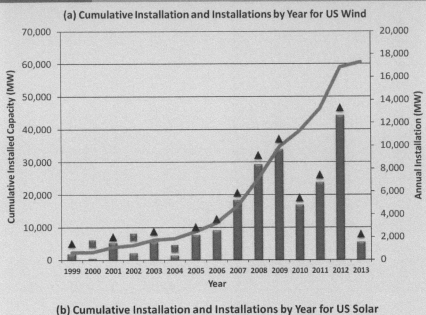

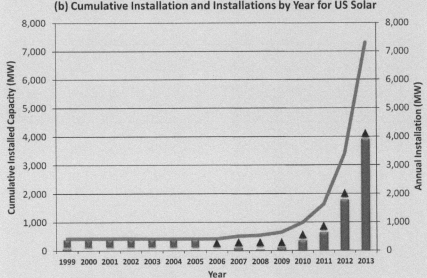

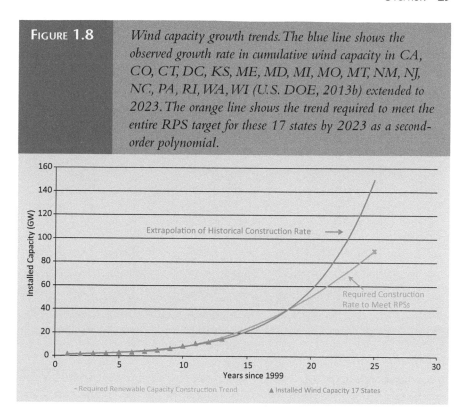

Wind capacity growth trends. The blue line shows the observed growth rate in cumulative wind capacity in CA, CO, CT, DC, KS, ME, MD, MI, MO, MT, NM, NJ, NC, PA, RI, WA, WI (U.S. DOE, 2013b) extended to 2023. The orange line shows the trend required to meet the entire RPS target for these 17 states by 2023 as a second-order polynomial.

installation data in these states would lead to an installed capacity of roughly 150 GW, more than sufficient to meet the aggregate targets. Figure 1.8 shows the trend that would be needed to reach the 90 GW target by 2023. This figure suggests that, if past construction rates are maintained, it appears feasible to build enough capacity to meet the combined requirements of the 17 states.

However, the growth rate in installed wind capacity has not been constant for individual states, and some states may require higher installation rates than others as they move to meet their targets. Expiration of tax incentives would likely significantly decrease the construction rate.

1.4 Wind and solar power

This book focuses on wind and solar resources. Wind is the lowest-cost and most widely available nonhydroelectric renewable resource, so it is expected to continue to dominate the growth in renewable energy. Solar's contribution to the power grid is likely to grow (in 2013, it was 18 times less than that of wind).

The majority of wood and wood-derived fueled generation is used in combined heat and power generation at pulp and paper processing facilities. These pulp and paper mills using the kraft sulfate pulping process produce black liquor that is used

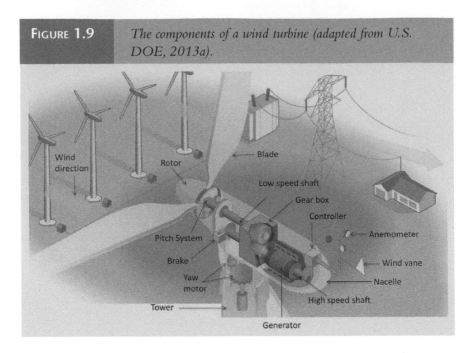

| FIGURE 1.9 | *The components of a wind turbine (adapted from U.S. DOE, 2013a).* |

in cogeneration facilities. Mill wood waste is also used both in cogeneration facilities and to cofire power plants where the majority of the fuel is from coal. There are currently nine U.S. power plants using biomass cofiring. The heat input from biomass cofiring is responsible for a combined total of approximately 70 megawatts (MW) of capacity (EIA, 2000). Municipal solid waste-to-energy plants and landfill methane are the next largest components of the "other biomass" category. Total capacity of all biomass generation facilities in the United States is approximately 11 GW, a bit less than 23% of all nonhydroelectric renewable generation capacity, and about 1% of the 1,025 GW U.S. generation capacity (EIA, 2012).

Geothermal power can be competitive in certain locations, and there is potential to overcome the limited availability of hydrogeothermal power (where nature supplies the hot water) by using enhanced geothermal power (where water is injected into hot underground rock and returned to the surface for use in generating electricity). Total capacity of existing U.S. geothermal plants (all of them hydrogeothermal) is 2.4 GW (EIA, 2012).

While biomass and geothermal power are subject to some variability, both are more constant than wind or solar power. Thus we concentrate our attention here on the issues involved with integration of the latter two sources.

Figure 1.9 shows a schematic of the components of a utility-scale wind turbine. The energy in the wind turns propeller-like blades that are attached to the main shaft, which spins a generator located inside the nacelle (turbine housing) to create electricity. Turbine blades rotate around the horizontal axis. The entire turbine rotates along a vertical axis to track changes in wind direction. Once generated, the wind power is transmitted through the electricity transmission and distribution system to consumers.

There are many different mechanisms to convert sunlight into energy. Figure 1.10(a) provides an illustration of a solar parabolic trough collector, which is the most common type of a concentrated solar power (CSP) system. In such a system, the receiver tube is positioned along the focal line of each parabola-shaped reflector. The tube is fixed to the mirror structure, and the heated fluid – usually a heat-transfer fluid – flows through and out of the field of solar mirrors to a heat exchanger, used to create steam. The largest individual trough system at the time of this writing generates 280 MW of electricity. Solar heliostat systems that focus sunlight to achieve very high temperatures and thus good efficiency are in operation, including one with a capacity of 390 MW. Individual systems can be colocated in power parks. Their capacity would be constrained only by the transmission capacity of nearby power lines and the availability of contiguous land.

Photovoltaic (PV) systems are the most common technology for generating solar power. PV cells convert sunlight into electricity when sunlight absorbed by the semiconductor materials in the PV cells creates a mobile electron-hole pair. PV cells are connected together to form modules, which in turn can be connected to form PV systems able to generate large amounts of electricity. Figure 1.10(b) illustrates how a solar photovoltaic system is connected to the grid.

1.5 Why are there challenges in integrating wind and solar power into the power grid?

The electric power system consists of three elements: generation, transmission, and distribution. Power can be generated in many ways – with nonrenewable sources such as oil, natural gas, coal, and nuclear power, and renewable energy sources such as hydropower, wind, solar, geothermal, and tidal power. Once generated, the power is transmitted through a high-voltage grid and then distributed to customers (their demand for electricity is often called the load); see Figure 1.11.

The grid is managed through an operations system to ensure that sufficient power is generated and that power is transmitted to the right places at the right times to meet the load. When customer demand varies, generation varies, or parts of the system become unavailable due to, for example, breakdowns or severe weather, the system must respond very quickly to keep from overloading circuits or causing a blackout.

A few aspects of the nature of the power system are of importance to an understanding of integrating variable renewable generators.

Currently, there are limited opportunities for storing electricity, so that the supply of power generation and demand must be matched instantaneously. The existing power system heavily relies on power plants that have controllable power output. Natural gas plants, for example, can be turned on and off as needed, or their output can be changed to balance changes in power supply. Wind and solar power

FIGURE **1.10** *Two types of solar electric power production, (a) concentrating solar power (CSP; U.S. DOE, 1996) and (b) solar photovoltaic (PV).*

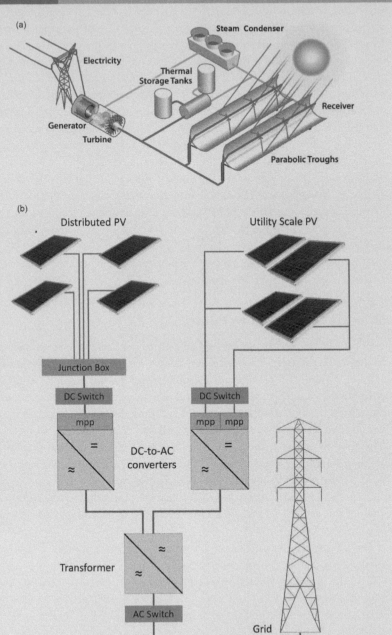

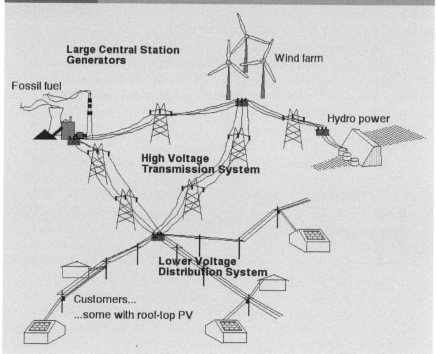

FIGURE **1.11** *Illustrative diagram of the power grid, with its central station generators, distributed generation, and customer loads. Illustration by M. Granger Morgan, Carnegie Mellon University.*

are not as controllable as conventional generation resources, and they thus present some challenges to the operation of the power grid.

The match between generation and demand is done on several time scales (EPRI, 1995). In real time (a fraction of a second to minutes), system frequency (60 Hz in the United States) and voltages are maintained within close tolerances by adjusting the power output of a few large generators to compensate for the variability of load or renewable generation. Also in that time frame, the system is managed so that a failure of the single most critical component (for example, a transmission line or large generation plant) will not disrupt the system. In emergencies, power can be cut to some customers to preserve the bulk of the system.

In the short term (a few hours to a week), generators are selected to produce power based on their operating costs and also considering startup and shutdown costs and how long it takes to get a generator running. For renewable generators, the forecast for wind, solar, or water production is used to schedule their operations. For scheduling both fossil and renewable generators, transmission bottlenecks are taken into consideration. These bottlenecks can cause generators to be scheduled somewhat out of the lowest-cost merit order. In this time frame, reserve generators that might be needed when the wind calms, clouds cover the sun, or a failure happens in a fossil generator are scheduled to be running at idle power.

On longer time scales (a year and longer), firms make decisions about whether to build plants and transmission lines, electricity markets change in response to stakeholder concerns, public utility commissions make decisions to approve or decline capital project expenditures and rates, and state and federal governments decide on environmental regulations and incentives.

As system operators in several industrial nations became familiar with the operation of power grids containing variable renewable generators (first hydroelectric power, then wind and solar power), they learned how to cope with variability. Hydroelectric power's variability arises from year-to-year rainfall differences and from season-to-season constraints on river flow and requirements to keep enough water flowing for uses other than power. Hydroelectricity supplied 6.6% of U.S. electric energy in 2013.

The key challenges facing the increased use of wind and solar power are:

* Wind and solar power do not produce a consistent amount of power. As an illustration, Figure 1.12 shows how renewable electricity production varied over the course of 2 separate days in California. Changes in power output also occur at shorter time scales. Figure 1.13 shows the variability of wind power output sampled every 5 minutes in the Pacific Northwest's Bonneville Power Authority between January 1 and 8, 2012. Figure 1.14 shows the 1-second variability in power output for 10 days in one wind plant in the Middle Atlantic. Solar photovoltaic power has been shown to be more variable at short time scales than wind power is; Figure 1.15 shows the output of a large utility-scale solar PV array in Arizona sampled every 10 seconds. Research described in Chapter 8 has found that the character of the variability (specifically, the ratio of fast variations to slow ones) does not change from location to location.
* The sun does not shine at night, and there are cloudy days; there are also days-long lulls in wind power. Just as there are rainy years and drought years for hydroelectricity, preliminary research indicates there are likely to be windier years and calmer years for wind power (see Chapter 11).

1.6 What might be the potential contribution of renewable energy to total electricity generation?

Wind and solar power are still at fairly low market penetrations in the United States (4.1% for wind and 0.2% for solar in 2013). In Germany, wind supplied 8.4% of electricity in 2013 and solar supplied 5.3%. It appears quite feasible to substantially increase the market share of wind and solar power in the United States.

The National Renewable Energy Laboratory (NREL) *Renewable Electricity Futures Study (RE Futures)* investigated "the extent to which renewable energy

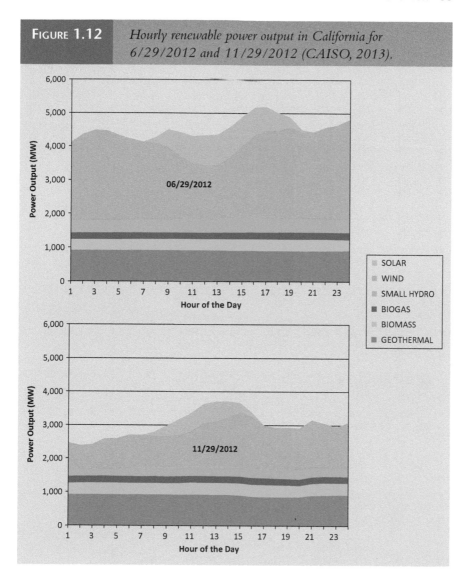

FIGURE 1.12 *Hourly renewable power output in California for 6/29/2012 and 11/29/2012 (CAISO, 2013).*

supply can meet the electricity demands of the continental United States over the next several decades" (NREL, 2012). Its key findings are that:

- Increased electric system flexibility, needed to enable electricity supply to balance demand with high levels of renewable generation, can come from a portfolio of supply-side and demand-side options, including flexible conventional generation, grid storage, new transmission, more responsive loads, and changes in power system operations.

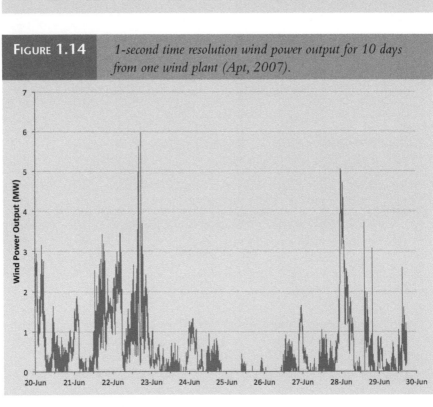

FIGURE 1.13 *5-minute time resolution wind power output in BPA, 1/1/2012 to 1/8/2012 (BPA, 2012).*

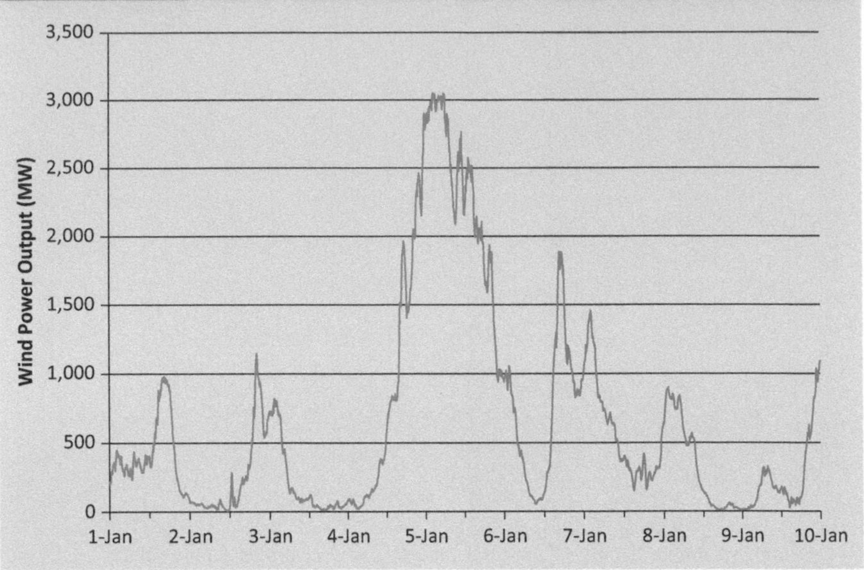

FIGURE 1.14 *1-second time resolution wind power output for 10 days from one wind plant (Apt, 2007).*

- The abundance and diversity of U.S. renewable energy resources can support multiple combinations of renewable technologies that result in deep reductions in electric sector greenhouse gas emissions and water use.
- The direct incremental cost associated with high renewable generation is comparable to published cost estimates of other clean energy scenarios.

Accurately assessing and preparing for the operational effects of renewable generation can ensure that renewables can play a much enlarged role.

1.7 What are the major challenges and opportunities for integrating variable power generation sources into the electricity grid?

Variability, described earlier, is the major challenge influencing the ability to incorporate variable renewable energy sources like wind and solar power into the electricity grid. Based on its research, the RenewElec project has found that the primary opportunities to respond to these challenges include:

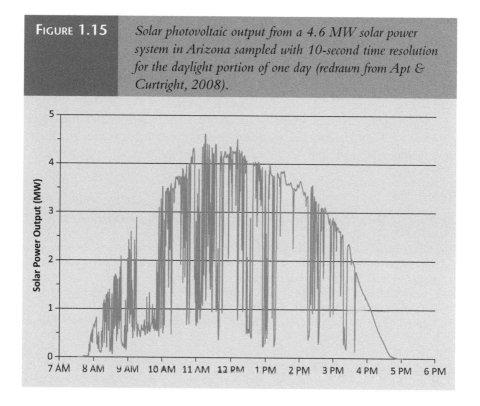

FIGURE 1.15 *Solar photovoltaic output from a 4.6 MW solar power system in Arizona sampled with 10-second time resolution for the daylight portion of one day (redrawn from Apt & Curtright, 2008).*

- Better prediction of variability
- Changes in the operation of power plants, reserves, transmission systems, and storage
- Improved planning of renewable capacity expansion
- Implementation of new regulatory paradigms, rate structures, and standards.

These opportunities will be discussed throughout the remainder of this book.

1.8 Overview of this book

This book has three parts. The first part gives background and discusses technical and policy options. In the first part, Chapter 2 describes the characteristics of wind and solar power short-term and long-term variability; why managing variability is important and the role of prediction in doing so; the accuracy of current wind power predictions; and how variability predictions can be used to support the operations of the power grid. Chapter 3 discusses the strategies that can reduce variability, including the use of aggregation, demand control, and storage, as well as the challenges in implementing these strategies. Chapter 4 focuses on how planning for renewable capacity expansion activities can be improved by taking into consideration hurricanes, pollution reduction, and facility siting. Chapter 5 discusses some of the existing regulations, rate structures, and standards and those that might be put into place. Chapter 6 brings all this information together to identify short-term and long-term strategies to manage variable energy resources in a way that will increase renewable electricity's contribution to the grid.

The second part of the book provides detailed scientific findings on issues covered at a more general level in the first part of the book.

The third part is a critical review of studies of the integration of large-scale wind power into the electric grid, with an emphasis on the techniques that are best suited to such studies.

References

Apt, J. (2007). The spectrum of power from wind turbines. *Journal of Power Sources, 169*(2), 369–374.

Apt, J., & Curtright, A. (2008). *The spectrum of power from utility-scale wind farms and solar photovoltaic arrays.* Carnegie Mellon Electricity Industry Center working paper, CEIC-08-04, http://wpweb2.tepper.cmu.edu/electricity/papers/ceic-08-04.asp

Apt, J., Keith, D.W., & Morgan, M.G. (2007). Promoting low-carbon electricity production. *Issues in Science and Technology, 24*(3), 37–44.

BPA. (2012). *Wind generation and total load in the BPA balancing authority.* http://transmission.bpa.gov/Business/Operations/Wind/

CAISO. (2013). *California ISO renewables watch.* www.caiso.com/market/Pages/ReportsBulletins/DailyRenewablesWatch.aspx

DSIRE. (2011). *Quantitative RPS data project.* www.dsireusa.org/rpsdata/RPSFieldDefinitionsApril2011.pdf

DSIRE. (2013a). *Federal incentives/policies for renewables and efficiency.* www.dsireusa.org/incentives/incentive.cfm?Incentive_Code=US02F

DSIRE. (2013b). *Renewable electricity production tax credit (PTC).* www.dsireusa.org/incentives/incentive.cfm?Incentive_Code=US13F

Edenhofer, O., Knopf, B., & Luderer, G. (2013). Reaping the benefits of renewables in a nonoptimal world. *Proceedings of the National Academies of Sciences, 110*(9), 11,666–11,667.

EIA. (2000). *Biomass for electricity generation.* U.S. Energy Information Agency. www.eia.gov/oiaf/analysispaper/biomass/pdf/tbl1.pdf

EIA. (2012). *Renewable energy annual table 1.12.* U.S. Energy Information Agency. www.eia.gov/renewable/annual/trends/xls/table1_12.xls

EIA. (2013). *Electric power annual 2012 chapter 4.* U.S. Energy Information Agency. www.eia.gov/electricity/annual/pdf/epa.pdf

EIA. (2014). *Electric power monthly with data for December 2013.* U.S. Energy Information Agency. www.eia.gov/electricity/monthly/current_year/february2014.pdf

EPA. (2010). *Green power market.* www.epa.gov/greenpower/gpmarket/

EPRI. (1995). *A primer on electric power flow for economists and utility planners.* EPRI report, TR-104604. www.epri.com/search/Pages/results.aspx?k=A%20Primer%20on%20Electric%20Power%20Flow%20for%20Economists%20and%20Utility%20Planners

FERC. (2014). *Federal energy regulatory commission energy infrastructure update for December 2013.* www.ferc.gov/legal/staff-reports/2013/dec-energy-infrastructure.pdf

Long, J.C.S. (2008). A blind man's guide to energy policy. *Issues in Science and Technology,* www.issues.org/24.2/long.html

NREL. (2012). *Renewable electricity futures study.* National Renewable Energy Laboratory. www.nrel.gov/analysis/re_futures/

Pew Charitable Trusts. (2012). *Energy innovation seen as needed to reduce dependence on foreign oil, save money.* www.pewtrusts.org/news_room_detail.aspx?id=85899407170

Sherlock, M. (2012). *Impact of tax policies on the commercial application of renewable energy technology.* http://science.house.gov/sites/republicans.science.house.gov/files/documents/hearings/HHRG-112-SY21-WState-MSherlock-20120419.pdf

U.S. DOD. (2012). Defense, Interior Departments join forces on renewable energy. *Armed Forces Press Service,* August 6 2012, www.defense.gov/news/newsarticle.aspx?id=117413

U.S. DOE. (1996). Linear concentrator solar power plant illustration. www.eeremultimedia.energy.gov/solar/sites/default/files/graphic_csp_parabolictroughs_1996_high.jpg

U.S. DOE. (2013a). *How does a wind turbine work?* http://energy.gov/eere/wind/how-does-wind-turbine-work

U.S. DOE. (2013b). *Yearly installed wind capacity maps.* www.windpoweringamerica.gov/wind_installed_capacity.asp#yearly

2

VARIABILITY AND ITS PREDICTION

One way to support the integration of variable renewable resources into the grid is to efficiently schedule conventional generators to compensate for the variability of wind and solar resources. Having good predictions of this variability can support such efforts. This chapter describes why predictions of variability are important and how variability can be better predicted.

2.1 What is variability?

As described in Chapter 1, renewable energy resources such as hydroelectric, wind, and solar power are variable – that is, they do not produce a constant amount of power. Generators using fuels such as coal or natural gas deliver a generally constant amount of power when they are in operation (of course, they occasionally mechanically fail, as do wind and water turbines). Because the wind does not always blow, sometimes for days, clouds block the sun (and it does not shine at all at night), and the water that provides hydroelectric power with its resource is seasonally variable, restricted for nonpower reasons, and limited during droughts, these renewable resources produce variable power. Since storage is presently limited, demand and supply of electricity have to be instantaneously matched. Thus, this variability can make it challenging to incorporate renewable resources into the grid.

The variability of solar photovoltaic (PV) power output depends on sunshine intensity and also on the season (peaking in summer), time of day (peaking midday), and short-term weather (clouds that cover the sun for minutes or days; IEA, 2005). Figure 2.1 shows the variability of solar PV, solar thermal, and wind power output at different time scales. Solar thermal sources are variable, but less so than solar PV, at fast time scales because the working fluid used for this type of source retains its heat due to thermal inertia (Lueken, Cohen, & Apt, 2012).

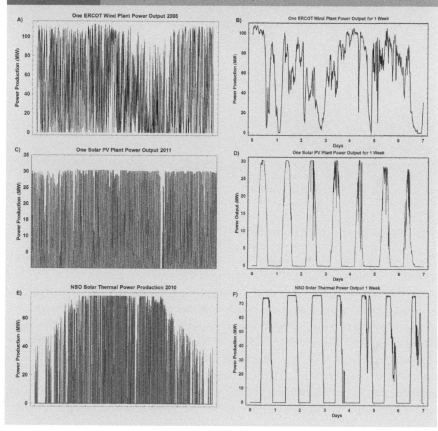

FIGURE 2.1 — *Solar PV, solar thermal, and wind power output data: (a) 2008 single ERCOT wind plant data; (b) 1 week of 2008 single ERCOT wind plant data; (c) 2005 Springerville PV data; (d) 1 week of 2005 Springerville PV data; (e) 2010 NSO solar thermal data (the data gaps near the beginning and end of the year represent times the plant was out of service); (f) 1 week of NSO solar thermal data (Lueken et al., 2012).*

The amount of power that can be produced by a wind turbine rises quickly with increasing wind speed over a substantial range of speeds (Figure 2.2). Thus, variations in wind speed cause substantial variations in power output. In addition, wind speed influences not only how much power the turbine generates but also whether it can operate at all. If wind speed is too low or too high, the turbine cannot operate either due to insufficient wind (wind speed too low to turn the blades) or too high a wind speed (potential to damage equipment).

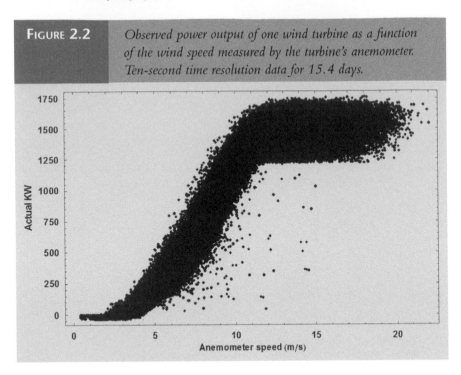

FIGURE 2.2 | *Observed power output of one wind turbine as a function of the wind speed measured by the turbine's anemometer. Ten-second time resolution data for 15.4 days.*

2.2 The importance of managing variability

The existing power grid was designed and has been operated to instantaneously match demand and supply of electricity using controllable resources like fossil fuel power plants that can change their power output on command. While the system already manages variability of demand and equipment that fails unexpectedly, the variability of large amounts of wind and solar is larger than the variability of demand. Managing this increased variability requires that grid operators use conventional resources or change operational procedures in order to balance wind and solar output. As indicated by the Northwest Power and Conservation Council (2007), for example, wind power system operators are required to "secure additional operating flexibility on several time scales to balance fluctuations and uncertainties in wind output." A number of methods are available to manage the existing variability in the grid, including:

- Load forecasting to predict future demand of electricity
- Scheduling and dispatching generating power plants
- Providing ancillary services, including load regulation (maintaining consistent voltage despite load changes), spinning reserve (running at a zero load and synchronized to the electric system), nonspinning reserve (generating capacity not currently running but capable of being connected to the bus and load within a specified time), replacement reserve, and voltage support (EIA, 2013; U.S. DOE, 2011).

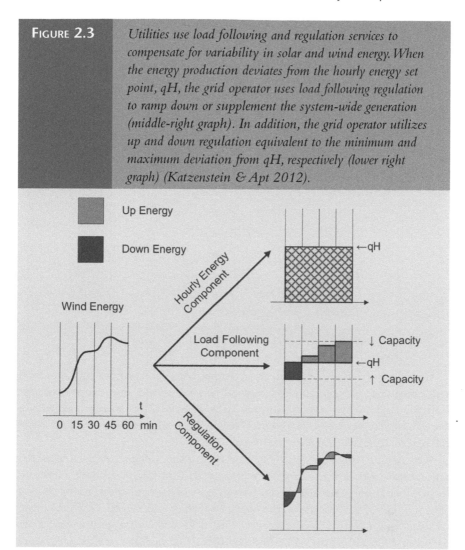

FIGURE 2.3 | *Utilities use load following and regulation services to compensate for variability in solar and wind energy. When the energy production deviates from the hourly energy set point, qH, the grid operator uses load following regulation to ramp down or supplement the system-wide generation (middle-right graph). In addition, the grid operator utilizes up and down regulation equivalent to the minimum and maximum deviation from qH, respectively (lower right graph) (Katzenstein & Apt 2012).*

These methods can be employed by a single vertically integrated electric utility or provided through a combination of market signals and nonmarket provisions of a grid operator. Figure 2.3 illustrates one representation of how some utilities respond when energy production from a variable source differs from what is expected. For moment-to-moment variations, regulation is used to maintain a consistent frequency. For larger variations, utilities have specific power plants that follow the load that can be brought on as needed on an hourly basis. Deploying these strategies to manage the variability of renewable resources likely increases the cost of providing electricity with wind and solar power.

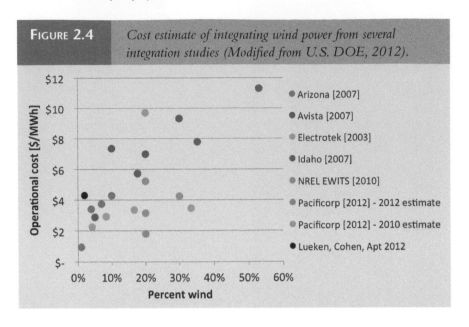

FIGURE 2.4   *Cost estimate of integrating wind power from several integration studies (Modified from U.S. DOE, 2012).*

Both peer-reviewed and commercial studies have estimated the cost associated with integrating variable sources like wind and solar power into the grid at large scale; Figure 2.4 shows several estimates (U.S. DOE, 2012). The vertical axis shows the estimated integration costs in $/MWh of wind, while the horizontal axis shows the percentage of electric energy derived from wind.

Recent estimates of the variability costs of wind and solar power quantified the ratio of variability at time scales that vary from slow (but large) day-long variations to fast (but small) second-by-second variations. Using a method similar to that shown in Figure 2.3 along with the observed market prices in ERCOT, the mean cost of 15-minute to 1-hour variations for 20 ERCOT wind plants was $9 per MWh in 2008 and $4 per MWh in 2009 (ERCOT ancillary service prices were considerably lower in the latter year; Katzenstein & Apt, 2012). A second peer-reviewed paper using the same method estimated the variability cost of solar PV generation at $8 to $11 per MWh, solar thermal generation at $5 per MWh, and ERCOT wind generation at $4 per MWh (Lueken et al., 2012; see Table 2.1). In all cases, the cost of variability consisted of both energy and regulation costs, with the regulation portion about half the cost of the energy portion. These costs are consistent with the ranges presented in Figure 2.4.

Although nonhydroelectric renewable energy sources generate only a bit less than 4% of U.S. electricity today, it appears quite feasible to increase the contribution of these resources to 20 or 30% in the near future. Better management of variability could thus reduce the cost of integrating renewable energy into the grid.

TABLE 2.1	Cost of variability of solar PV, solar thermal, and wind, using the average price of electricity in the CAISO nearest zone.			
	Solar thermal (NSO)	ERCOT wind	Solar PV (Springville, AZ, 5 MW)	Solar PV (20 MW + class)
Average cost of variability per MWh (2010)	$5.2	$4.3	$11.0	$7.9
Average hourly cost of variability per MW (2010)	$1.2	$1.4	$2.2	$2.0
Average cost of variability per MWh (2005)	$5.9	$5.0	$12.6	$9.9
Median cost of variability per MWh (2010)	$0.0	$2.2	$0.3	$0.2
Standard deviation cost of variability per MWh (2010)	$15.2	$9.0	$31.0	$18.5
Variability cost as a percentage of total cost of power (2010)	11.9%	10.2%	26.5%	18.9%
Capacity factor (or average capacity factor)	23%	34%	19%	25%

Source: Lueken et al., 2012.

2.3 Characteristics of wind and solar power variability

As discussed in Chapter 1, generation produced from wind and solar photovoltaic plants varies with time. Utility-scale renewable generators vary on both short- and long-term time scales, which can influence both power quality and reliability. Figure 2.1 shows the variation over 1 year and over 1 week for solar photovoltaic, solar thermal, and wind.

Understanding the variability characteristics of wind and solar power in a given situation is important, as it can enhance the ability of a power grid operator to manage energy supply activities in a particular region, thus reducing cost. The following sections discuss how variability can be measured, the importance of time scale, and how natural conditions can enhance the ability to manage variability.

2.3.1 Ways of measuring variability

One method commonly used to measure variability is to construct a histogram of the step changes in power output over time. This method gives the statistics of the size of jumps in power output at various time scales (for example, hourly). Chapter 8 shows examples of step changes. A complementary method, estimating the power spectral density (PSD), characterizes variability using power spectrum analysis (Apt, 2007; Katzenstein & Apt, 2012; Lueken et al., 2012). This method uses power output data from operational utility-scale plants to measure variability at different frequencies. Imagine the music from a symphony orchestra; the PSD would give the volume of the bass notes from the timpani compared to that of the treble notes from the piccolo. To conduct power spectral analysis, the frequency domain behavior of the time series of power output data from generation plants is used to estimate the PSD. From these data, the discrete Fourier transform of the time series is computed. This results in a quantitative measure of the ratio of fluctuations at high frequency to those at a low frequency.

Figure 2.5 shows the results of such an analysis of the power spectra of solar PV, wind, and solar thermal generation facilities. The first important result from such an analysis is that the variability of all three generation types is much stronger at low frequencies (say, those corresponding to time scales of several hours to several days) than it is at high frequencies (minutes to seconds). This feature of nature has enormous economic consequences. The spectral power of high-frequency variability is a thousand times smaller than that of low-frequency variability. If it were not, grid operators would need to employ many generators that can quickly change their output (for example, batteries or hydroelectric plants). Because the variability is sharply reduced at high frequencies, as first postulated in 1941 for turbulent fluids (Kolmogorov, 1941), need for these expensive generators that change their output power rapidly is also sharply reduced. In slightly more technical terms, there is a linear region in the spectrum of wind turbine output power over four orders of magnitude of frequency, between 30 seconds to 2.6 days, in which the power decreases as the frequency increases, as frequency to the negative 5/3 power (see Figure 2.6 for an example with 1-hour time resolution data; Apt, 2007).

The second result is that the variability characteristics of wind, solar PV, and solar thermal power generators are quite different from each other. By normalizing the spectra at a frequency corresponding to a range near 24 hours for each of the three power sources analyzed, the difference in the variability of each resource at high frequencies is shown clearly. These differences include:

- Solar photovoltaic electricity generation has approximately 100 times larger spectral power variations at approximately 15 minutes (10^{-3} Hz) than does solar thermal electricity generation. The thermal inertia of the working fluid in solar thermal collectors reduces the fluctuations caused by clouds flitting in front of the sun.
- Wind plant electricity generation is midway between solar PV and solar thermal at this time scale.
- Power spectra for all three sources for periods at frequencies corresponding to time periods greater than about 6 hours (4×10^{-5} Hz) are similar.

| FIGURE 2.5 | *Power spectra of solar PV, wind, and solar thermal generation facilities. The spectra have been normalized to one at a frequency corresponding to approximately 24 h. The Springerville, Arizona, PV plant has a capacity of 5 MW (Lueken et al., 2012).* |

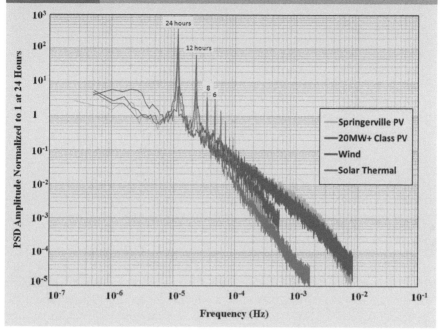

2.4 The role of prediction in managing variability and supporting grid operations

Grid operators have for many years used load forecasts to reduce the costs that arise from uncertainty in the demand for electricity. Similarly, wind and solar power forecasts that predict the amount of power at a time in the future support the integration of these resources into the grid. Currently, wind power forecasts are more widely used than solar forecasts are, so they are the focus of this section.

Wind forecasts are created using statistical models, physical models, or a hybrid of both. Some methods forecast wind power directly, while others forecast wind speeds and convert them to wind power. Statistical models rely on historical wind data from one or more wind plants to predict future values. Such statistical models generally employ time series (for example, Markov chains) or neural network analysis methods (Giebel, Brownsword, & Kariniotakis, 2003). Physical models, including numerical weather prediction (NWP) models, use meteorological data and fluid flow models of

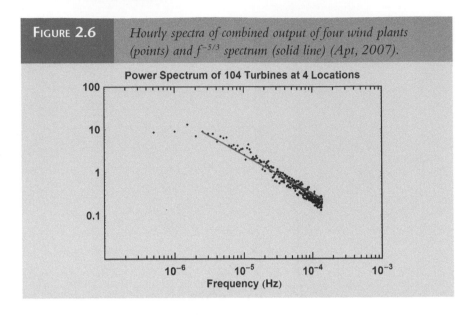

FIGURE 2.6 *Hourly spectra of combined output of four wind plants (points) and $f^{-5/3}$ spectrum (solid line) (Apt, 2007).*

the atmosphere to predict wind speeds (Landberg et al., 2003). Wind speed forecasts are then transformed to power forecasts using a transfer function to model wind plant output. In general, NWP models outperform statistical models for wind power predictions greater than 1 to 2 hours into the future, depending on local conditions. The main drawback of NWP models is their complexity and expense due to large computational needs. In a hybrid approach, NWP models are run every 3 hours, with statistical models used to predict variability at shorter time scales.

Grid managers use reserve generation to balance supply and demand. As the penetration of variable resources increases, the amount of reserve generation needed for large-scale wind and solar penetration will become highly dependent on the power forecasts. Unexpected increases in variable generation occasionally lead to curtailment of renewable power when these rapid generation changes cannot be balanced quickly enough. A more problematic occurrence is the unexpected loss of wind or solar power. An example of this occurred in Texas in February of 2008. A sudden change in weather occurred earlier than predicted by models then in operational use and dropped statewide wind generation by more than 1 GW in a short time (O'Grady, 2008). Wind power production in all of BPA dropped from 1.5 GW to 0 for a bit more than 11 days in January 2009. Although these are extreme examples, extremes drive the need for other resources and illustrate the value of understanding and using forecasts to prepare for these events. However, forecasts are never exact, and the uncertainty of predicted values must be understood in order to effectively use available forecasts.

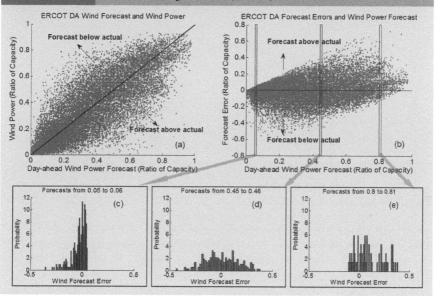

FIGURE 2.7 *ERCOT estimated uncurtailed wind power (a) and wind forecast errors (b) plotted against forecast wind power. Bottom plots show probability distributions of wind forecast errors corresponding to the three bins highlighted in (b). Forecast value ranges in the bottom plots are (c) 0.05 to 0.06, (d) 0.45 to 0.46, and (e) 0.8 to 0.81. All values are shown as ratios of installed wind capacity (Mauch, Apt, Carvalho, & Jaramillo, (2013b).*

2.5 Accuracy of current wind power predictions

Electric power system operators use wind power forecasts that range from a few minutes to several days to help them make decisions that are used for economic dispatch or real-time trading (up to 1 hour), and longer for unit commitment decisions and day-ahead market bids for wind plant operators. As part of their decision-making process, grid operators must also take into account the uncertainty in the wind forecast.

The uncertainty of wind power forecasts arising from the highly variable nature of wind speed over different time scales is amplified by the response of the wind turbine to changes in wind speed (Figure 2.2). Because the transformation of wind speed to wind power is nonlinear, the error will differ depending on whether the wind is at low or high speeds. Figure 2.7 and Figure 2.8 show that at low wind speeds, forecasts tend to predict less wind power than is actually generated, while at high wind speeds, the forecasts tend to predict more power than is actually generated.

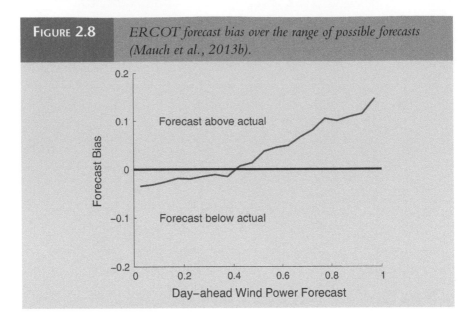

FIGURE 2.8   *ERCOT forecast bias over the range of possible forecasts (Mauch et al., 2013b).*

2.6 Improvement of grid operations from the incorporation of uncertainty

One common practice for estimating the uncertainty associated with wind power forecasts fits forecast errors to a specific distribution. Hodge and colleagues (2012), for example, fit a hyperbolic distribution to the forecast errors in ERCOT and CAISO. This method, however, assumes that forecast errors are independent of forecasted wind power, which is not what is observed (see the previous section). Forecast error distributions should be conditioned on the expected level of wind power. Using historical data for wind power forecasts and actual wind power output, Chapter 9 describes a method to model wind forecast errors conditioned on the forecast value by applying a logit (or logistic) transformation to the wind (Mauch et al., 2013b). Calculations of confidence intervals with this method use a model fit to the entire dataset while providing the ability to condition wind uncertainty on the wind forecast value for a given time period. Figure 2.9 shows the results of an analysis that uses this method to estimate the dispatchable capacity needed to account for day-ahead wind underforecast errors (forecasts that predict more wind than is produced in real time). This analysis is an example of incorporating forecast uncertainty to improve the operations of the grid by scheduling the appropriate amounts of dispatchable capacity to maintain reliability of the system.

2.7 Summary

This chapter has discussed the importance of estimating variability and methods to improve the forecasting of variability and the uncertainty associated with that forecast. In brief,

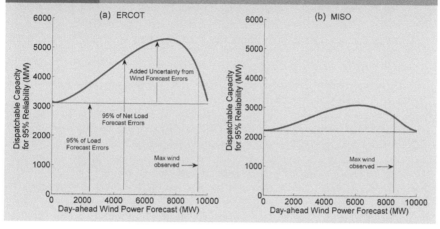

FIGURE 2.9   *Dispatchable generation capacity required to cover 95% of day-ahead net load underforecast errors in ERCOT and MISO when load forecasts are in the range 90% to 120% of mean load forecast. Each graph shows the 95th percentile of load forecast errors as a horizontal line. Net load forecast uncertainty converges to load forecast uncertainty at zero wind forecast and also at the maximum wind forecast (Mauch, et al., 2013a).*

- Renewable energy resources such as hydroelectric, wind, and solar power are variable – that is, they do not produce a consistent amount of power as is the case with conventional resources such as coal or natural gas. In addition, these variable resources cannot be "dispatched" like generators powered by coal, natural gas, or oil – that is, operators cannot just add more fuel to generate power as needed.

- Managing variability is important because it increases the cost of energy due to the need to "balance" that variability with more stable power sources like coal and natural gas.

- Utility-scale renewable energy systems fluctuate on both short- and long-term time scales, which can influence both power quality and reliability.

- The character of power fluctuations from wind and solar power is such that the largest power fluctuations occur slowly over many hours or days (i.e., low frequency). These are many times stronger than the fluctuations at short time scales. Thus, slow-responding generators, such as coal and older combined-cycle gas plants that take a long time to change their power output ("slow ramping"), can compensate for most of the variability.

- Forecasts of wind power in the United States systematically underpredict wind during periods of light winds and overpredict when there are strong winds. This is important for those who manage the electricity grid, which incorporates power from a number of sources, including wind power. Grid operators' responsibility is to make sure power production instantaneously matches consumers' demand for electricity. In order to support large-scale wind and solar into the electricity system, these operators can improve integration by correcting for forecasting errors. In addition, we have found that there is a simple mathematical framework to incorporate forecast uncertainty so that other generation sources can be scheduled to match the variability of wind (and solar) power and fill in the deficit that results from variability.

References

Apt, J. (2007). The spectrum of power from wind turbines. *Journal of Power Sources*, *169*(2), 369–374.

EIA. (2013). *Glossary*. www.eia.gov/tools/glossary/index.cfm

Giebel, G., Brownsword, R., & Kariniotakis, G. (2003). *State of the art on short-term wind power prediction*. ANEMOS deliverable report, D1.1.

Hodge, B., Florita, A., Orwig, K., Lew, D., & Milligan, M. (2012). A comparison of wind power and load forecasting distributions. *2012 World Renewable Energy Forum*. NREL/CP-5500-54384. www.nrel.gov/docs/fy12osti/54384.pdf

IEA. (2005). *Variability of wind power and other renewables: management options and strategies*. International Energy Agency. www.uwig.org/iea_report_on_variability.pdf

Katzenstein, W., & Apt, J. (2012). The cost of wind power variability. *Energy Policy*, *51*, 233–243.

Kolmogorov, A.N. (1941). The local structure of turbulence in incompressible viscous fluid for very large Reynolds numbers. *Dokl. Akad. Nauk. SSSR 30*, 301–305. Reprinted in *Proceedings of the Royal Society of London: Math. Phys. Sci.* (1991), *434*(1890), 9–13.

Landberg, L., Giebel, G., Nielsen, H.A., Nielsen, T.S., & Madsen, H. (2003). Short-term prediction – an overview. *Wind Energy*, *6*(3), 273–280.

Lueken, C., Cohen, G., & Apt, J. (2012). Costs of solar and wind power variability for reducing CO2 emissions. *Environmental Science and Technology*, *46*(17), 9761–9767.

Mauch, B., Apt, J., Carvalho, P.M.S., & Jaramillo, P. (2013a). What day-ahead reserves are needed in electric grids with high levels of wind power? *Environmental Research Letters*, *8*(3). DOI: 10.1088/1748-9326/8/3/034013

Mauch, B., Apt, J., Carvalho, P.M.S., & Small, M.J. (2013b). An effective method for modeling wind power forecast uncertainty. *Energy Systems*, *4*(4), 393–417.

Northwest Power and Conservation Council. (2007). *The northwest wind integration action plan*. www.uwig.org/nwwindintegrationactionplanfinal.pdf

O'Grady, E. (2008). Loss of wind causes Texas power grid emergency. *Reuters*. February 27, 2008. www.reuters.com/article/idUSN2749522920080228

U.S. DOE. (2011). *The role of electricity markets and market design in integrating solar generation*. www1.eere.energy.gov/solar/pdfs/50058.pdf

U.S. DOE. (2012). *2011 wind technologies market report*. www1.eere.energy.gov/wind/pdfs/2011_wind_technologies_market_report.pdf

3

STRATEGIES TO REDUCE OR MANAGE WIND AND SOLAR VARIABILITY

Chapter 2 discussed the nature of wind and solar variability and the importance of managing and predicting it. It also identified a number of strategies to improve predictions and therefore the reliability of power operations. Once the power output is predicted, with some variability in the error of the prediction, strategies are available to reduce and manage that variability. This chapter focuses on such strategies.

3.1 The importance of strategies to reduce or manage variability

Power grids generally include a mix of energy sources (see Figure 3.1). Traditionally, the grid included primarily sources with a relatively steady provision of power such as nuclear, coal, and natural gas. Industry professionals say that these energy sources can be dispatched. In dispatching, grid operators take actions to:

- Schedule generation by specific plants and other sources of supply to provide reliable power from the lowest-cost available generators and instantaneously meet varying demand
- Control operations and maintenance of high-voltage lines, substations, and equipment, including administration of safety procedures
- Operate the grid interconnection among generators and to customers
- Schedule energy transactions with other interconnected electric utilities (MGE, 2013).

As variable renewable energy sources such as solar and wind energy are increasingly added to the mix, those managing the grid must take additional actions to respond to this variability. These sources are not fully dispatchable. Predicting how much power they are likely to be able to contribute in a given time frame, as discussed in Chapter 2 and Chapter 9, and making plans in response to those predictions is important to ensure the reliable provision of electric power to customers.

At the present modest level of variable renewable generation in the United States, conventional dispatchable resources and renewable curtailments are the main tools used to counter variability. This chapter discusses four other strategies that

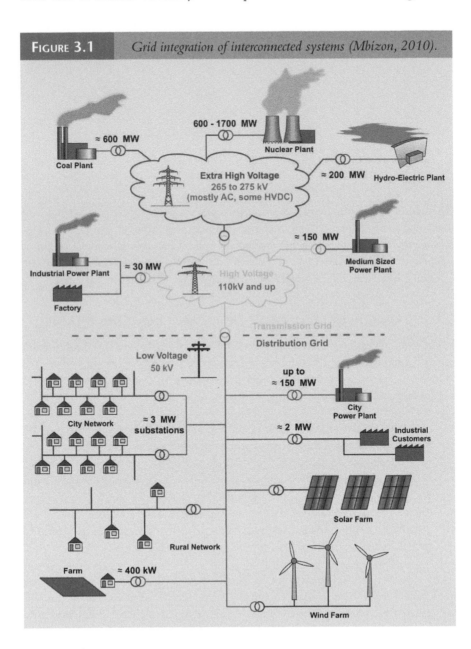

FIGURE 3.1   Grid integration of interconnected systems (Mbizon, 2010).

have the largest potential to aid independent system operators, regional transmission organizations, and others to reduce or manage wind and solar variability:

- Aggregation of wind plants within a region or large geographical areas
- Storage
- Load control (called demand-side management); and
- Grid codes and market programs.

3.2 Aggregation of wind plants to help manage variability

One strategy for managing variability is aggregating wind plants within a region or large geographical area. Aggregation is achieved by interconnecting wind plants with transmission lines.

Using the frequency analysis method described in Chapter 2, researchers examined 15-minute energy output data from 20 wind plants in the Electric Reliability Council of Texas (ERCOT) region (Katzenstein et al., 2010).

There are two important results from this analysis. First, the time scale is important when discussing the benefits of aggregation. Figure 3.2 and Figure 3.3 show that aggregation significantly reduces variability at time scales of an hour or shorter; there is very little reduction at long time scales. The small variability at time intervals of an hour or shorter can be reduced by 95%, but the large variability at 12

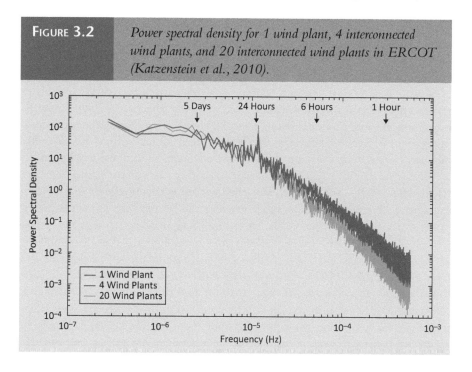

| FIGURE **3.2** | *Power spectral density for 1 wind plant, 4 interconnected wind plants, and 20 interconnected wind plants in ERCOT (Katzenstein et al., 2010).* |

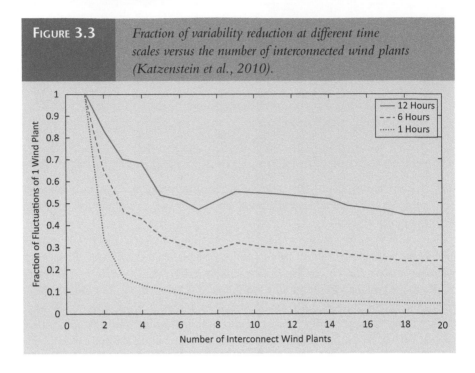

FIGURE 3.3 *Fraction of variability reduction at different time scales versus the number of interconnected wind plants (Katzenstein et al., 2010).*

hours and longer is reduced by only 50%. It is the strong fluctuations at long time scales that require the most compensating resources. These fluctuations at long intervals mean that slow-ramping plants such as natural gas combined-cycle and coal can be scheduled to compensate for the variability.

Second, as shown in Figure 3.2, although aggregating power from wind plants within a region reduces variability at a given time scale, there are quickly diminishing returns as more plants are interconnected. Figure 3.3 shows that interconnecting just four or five wind plants in ERCOT reduces the majority of the variability in power output, with only very small gains from adding additional wind plants.

Most wind plants in Texas are currently located in the western part of the state, where their wind power output may be highly correlated, reducing the benefits of aggregation (Figure 3.4).[1] To evaluate whether the Texas results are applicable to more widely dispersed wind plants, we examined aggregating wind power generated over larger geographical areas. Using time- and frequency-domain techniques (described earlier and in Chapter 8), this study used simultaneous wind power data from four regions: ERCOT, the Bonneville Power Authority (BPA), the Midcontinent Independent Systems Operator (MISO), and the California ISO (CAISO; see Figure 3.4) to see if increasing interregional transmission capacity is an effective means of smoothing wind power output (Fertig et al., 2012). As shown in Figure 3.5, the interconnected regions had 17% of installed wind capacity available

---

1. A few wind plants are located near the south Texas coast, with plans for additional installations there.

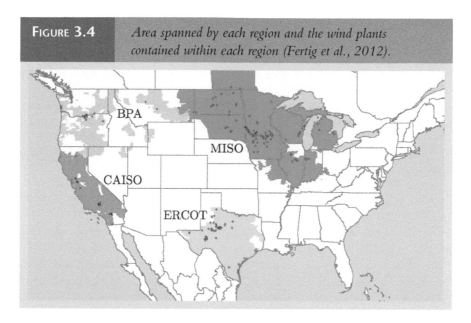

FIGURE 3.4 *Area spanned by each region and the wind plants contained within each region (Fertig et al., 2012).*

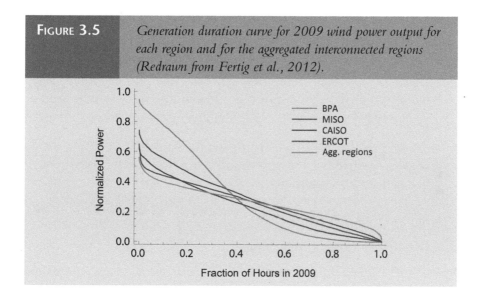

FIGURE 3.5 *Generation duration curve for 2009 wind power output for each region and for the aggregated interconnected regions (Redrawn from Fertig et al., 2012).*

79% of the time and 12% of capacity available 92% of the time. MISO had the firmest[2] wind power output and BPA the least firm.

While the peer-reviewed research found that there were some reductions in variability and an increase in the firm power that would be available from wind, the

---

2. "Firm power" is capacity available 79 to 92% of the time (Katzenstein et al., 2010).

paper compared the cost effectiveness of building new transmission to interconnect these regions to the cost effectiveness of building natural gas power plants to balance wind (Fertig et al., 2012). Their research found that local natural gas turbines would be more cost effective in balancing wind and providing firm power than building long transmission lines to interconnect widely spread regions (this conclusion could change if the costs of building new transmission capacity decreases substantially).

3.3 Using storage to help manage variability

Another way to manage renewable energy variability on the grid is energy storage. While grid-scale storage is presently limited to pumped hydroelectric facilities (with the exception of a few small installations of other technologies), there are prospects for affordable battery storage in the near future (Whitacre et al., 2012). Pumped hydroelectric storage capacity represents only around 2% of the United States' electric generation capacity (EIA, 2013) and requires special topography that is probably near maximum use in the United States. This section will discuss the potential benefits of storage should it become available at larger scale, as well as economics of battery storage, pumped hydroelectric storage, compressed air energy storage (CAES), and plug-in hybrid electric vehicle (PHEV) grid storage. More details are in Chapter 14.

At large scale, electricity storage could provide valuable services such as:

- facilitating power system balancing in systems that have renewable energy sources that produce power that varies and (in the case of solar at night or wind during days-long lulls) sometimes produces no power for extended periods
- improving grid optimization for bulk power production
- deferring investments in transmission and distribution (T&D) infrastructure to meet peak loads for a time; and
- providing ancillary services directly to grid/market operators (EAC, 2008).

3.3.1 Electricity storage technologies

Storage can be characterized by the period over which electric energy is stored and the amount of energy that can be stored. Some storage (for example, a flywheel or a capacitor) is very short term, while other types can store for hours or days.

Siting of such storage could be distributed throughout the grid, as shown in Figure 3.6.

As illustrated in Figure 3.7, there are five basic types of electrical energy storage systems: mechanical, electrochemical, chemical, electrical, and thermal. Figure 3.8 shows the stage of development for these options. Some, like pumped hydroelectric

FIGURE 3.6 *Potential grid storage types and locations (EIA, 2012).*

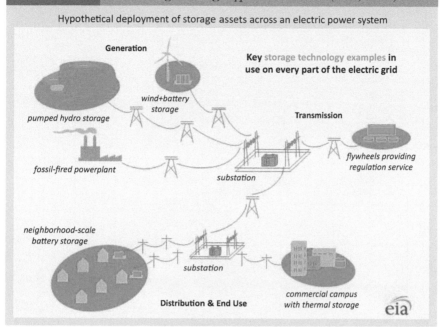

FIGURE 3.7 *Electrical energy storage systems by technology category (IEC, 2011).*

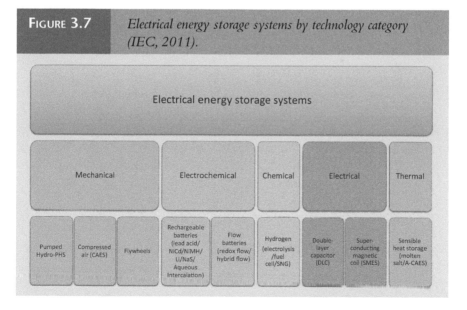

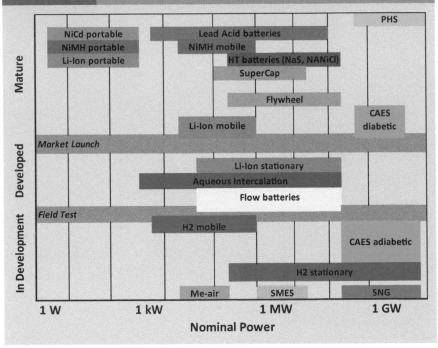

FIGURE **3.8**   *Electrical energy storage systems by development stage. Abbreviations: CAES: compressed air energy storage; H2: hydrogen; HT: high temperature batteries; Li-ion: lithium ion battery; Me-air: metal air battery; NaNiCl: sodium-nickel-chloride battery; NaS: sodium-sulfur battery; NiCd: nickel cadmium battery; NiMH: nickel metal hydride battery; PHS: pumped hydro storage; SMES: superconducting magnetic energy storage; SNG: synthetic natural gas; SuperCap: supercapacitor (IEC, 2011).*

storage (PHS), are mature technologies, while others, like flow batteries or superconducting magnetic energy storage (SMES), are in the development or field test stages. Table 3.1 shows the quantity of each type of electrical energy storage system used in the United States today.

3.3.2 What services can energy storage provide to the grid?

Wind and solar energy vary on both short (seconds) and long (days) time scales. Energy storage can enhance grid integration of these variable sources.

Figure 3.9 provides information on applications of storage technologies, while Table 3.2 shows more detailed information on the individual technologies.

TABLE 3.1	*Estimated installed capacity of energy storage in the U.S. grid, 2011.*
Storage Technology	**U.S. Installed Capacity (MW)**
Pumped hydroelectric	22,000
Compressed air	115
Lithium-ion batteries	54
Flywheels	28
Nickel cadmium batteries	26
Sodium sulfur batteries	18
Other (flow batteries, lead acid)	10
Thermal peak shifting (ice storage)	1,000
Total	23,251

Source: EAC, 2011.

3.3.3 What are the costs of storage?

Figure 3.7 and Table 3.2 show that electric energy storage encompasses a wide set of technologies with a wide set of applications. It is likely that it will take a mix of these options (along with dispatchable generators) to manage the variability of wind and solar power.

The costs of storage currently pose a barrier to large-scale deployment. Figure 3.10 shows the capital costs of different energy storage systems. Storage is frequently more expensive than using a dispatchable power plant, like a natural gas combined-cycle plant, to provide services to the grid.

It is likely that financially successful storage will derive revenue from providing several grid services, not just one. The next two sections provide several case studies that illustrate the opportunities and challenges energy storage provides when integrating renewable energy into small and large grids.

3.3.4 What are some of the opportunities storage provides to small-scale grids?

Grids can be isolated for geographic reasons when surrounded by mountains or water, as is the case for Hawaii. The engineering options and related economics for these small-scale grids can be quite different from large-scale grids such as PJM, a regional transmission organization that includes 13 states and the District of Columbia and is the world's largest electricity market.

For small-scale grids, one option for managing the variability of wind or solar power is to colocate natural gas or diesel generation with energy storage. These plants are able to start up quickly, so they work well in responding to hourly variability but have a difficult time counteracting the smaller variability that occurs on a second/

FIGURE 3.9 Applications for energy storage (Adapted from EPRI, 2010).

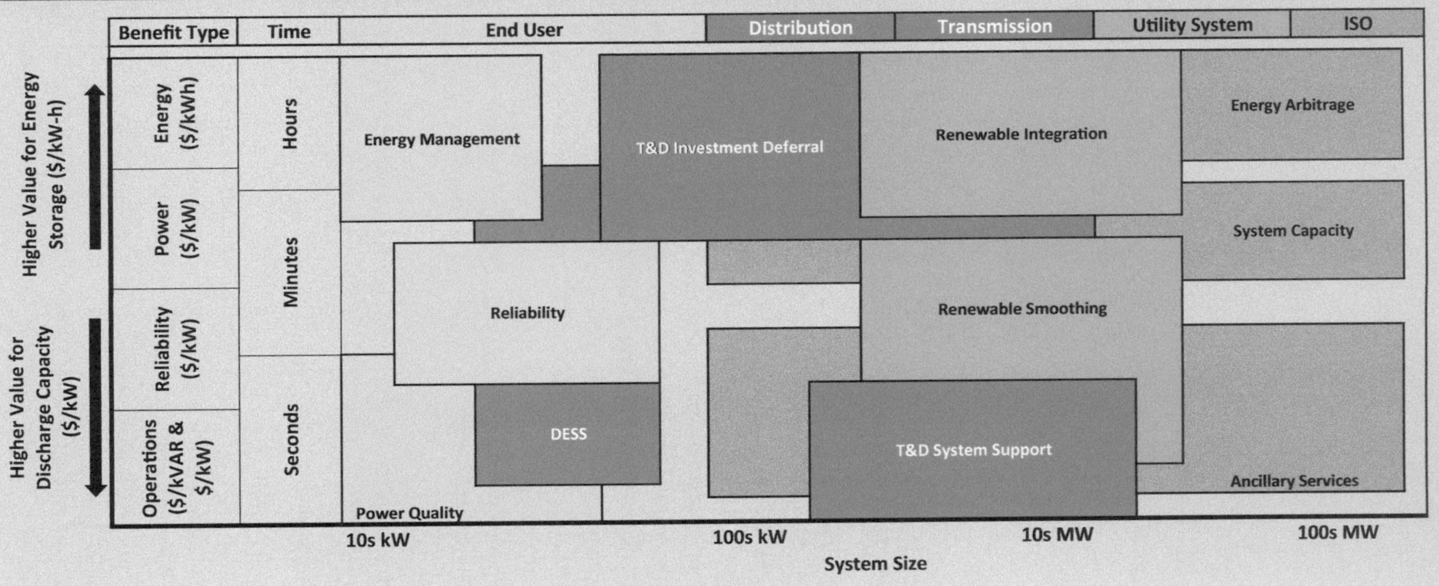

TABLE 3.2   *Characteristics of individual storage technologies for specific applications.*

Technology Option	Maturity	Capacity (MWh)	Power (MW)	Duration (hrs)	Total Cost ($/kW)	Cost ($/kWh)
Bulk Energy Storage to Support System and Renewables Integration						
Pumped hydro storage (PHS)	Mature	1,680–5,300	280–530	6–10	2,500–4,300	420–430
		5,400–14,000	900–1400	6–10	1,500–2,700	250–270
CT-CAES (underground)	Demo	1,440–3,600	180	8	960	120
				20	1,150	60
CAES (underground)	Commercial	1,080	135	8	1,000	125
		2,700		20	1,250	60
NaS battery	Commercial	300	50	6	3,100–3,300	520–550
Advanced lead-acid battery	Commercial	200	50	4	1,700–1,900	425–475
		250	20–50	5	4,600–4,900	920–980
	Demo	400	100	4	2,700	675
Vanadium redox flow battery	Demo	250	50	5	3,100–3,700	620–740
Zn/Br redox battery	Demo	250	50	5	1,450–1,750	290–350
Fe/Cr redox battery	R&D	250	50	5	1,800–1,900	360–380
Zn/air redox battery	R&D	250	50	5	1,440–1,700	290–340
Aqueous intercalation	Commercial (2014)	1–250	0.2–50	4 and longer	Depends on duration	300–400 (2015)

(Continued)

TABLE 3.2 Continued.

Technology Option	Maturity	Capacity (MWh)	Power (MW)	Duration (hrs)	Total Cost ($/kW)	Cost ($/kWh)
Energy Storage for ISO Fast Frequency Regulation and Renewables Integration						
Flywheel	Demo	5	20	0.25	1,950–2,200	7,800–8,800
Li-ion battery	Demo	0.25–25	1–100	0.25–1	1,085–1,550	4,340–6,200
Advanced lead-acid battery	Demo	0.25–50	1–100	0.25–1	950–1,590	2,770–3,800
Energy Storage for Utility T&D Grid Support Applications						
CAES (aboveground)	Demo	250	50	5	1,950–2,150	390–430
Advanced lead-acid battery	Demo	3.2–48	1–12	3.2–4	2,000–4,600	625–1150
NaS battery	Commercial	7.2	1	7.2	3,200–4,000	445–555
Vanadium redox flow battery	Demo	4–40	1–10	4	3,000–3,310	750–830
Zn/Br redox battery	Demo	5–50	1–10	5	1,670–2,015	340–1,350
Fe/Cr redox battery	R&D	4	1	4	1,200–1,600	300–400
Zn/air redox battery	R&D	5.4	1	5.4	1,750–1,900	325–350
Aqueous Intercalation	Commercial (2014)	1–250	0.2–50	4 and longer	Depends on duration	300–400 (2015)
Li-ion battery	Demo	4–24	1–10	2–4	1,800–4,100	900–1,700

Source: Adapted from EPRI, 2010.

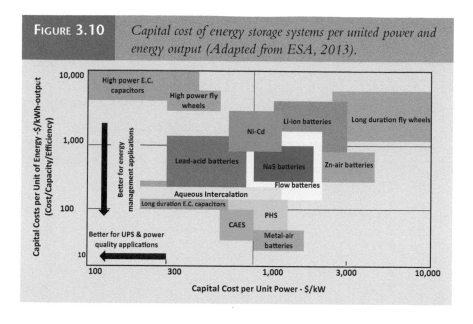

FIGURE 3.10 *Capital cost of energy storage systems per united power and energy output (Adapted from ESA, 2013).*

minute scale. Colocating a fast-ramping energy storage system, such as batteries, flywheels, or supercapacitors, that can respond on very short time scales to renewable power variability can support the operations of small-scale grids (Hittinger, Whitacre, & Apt, 2010). The research described here and more fully in Chapter 14 also has implications for all electrical systems seeking to integrate wind energy and informs potential incentive policies.

We modeled the power output of a system with 100 MW of natural gas capacity, 66 MW of wind capacity, and a sodium-sulfur (NaS) battery. The system has a target power output of 100 MW. Figure 3.11 shows the average cost of electricity from this hybrid system. In this model, batteries are not required until the delivered energy from wind exceeds 12% of the system total. The battery contribution to the electricity price is negligible relative to that of wind and no greater than that of the natural gas turbine. While the average costs are quite high at the higher wind levels, these costs may still be acceptable in systems with high wind requirements and high electricity prices – for example, in Hawaii.

Flywheels and supercapacitors could also support the integration of variable resources in small grids. Figure 3.12 shows the average cost of power in the wind/ gas/storage system for three different energy storage technologies. At lower wind penetrations, the cost contribution of energy storage is very small and the chosen technology has little effect on the average cost of power. At higher wind penetrations, when storage cost becomes important, NaS batteries dominate the other options, as would advanced batteries.

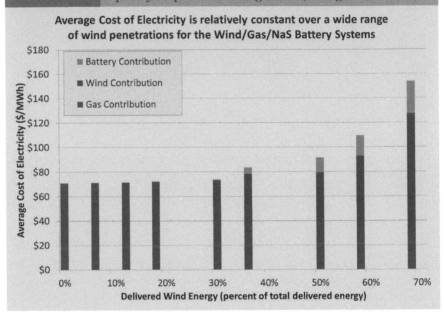

FIGURE 3.11 *Average cost of power in the wind/gas/NaS battery system as a function of delivered wind energy and divided according to system component. Each bar represents a system with 100 MW of gas generation and corresponds to the lowest cost of power for a particular wind/gas ratio (Hittinger et al., 2010).*

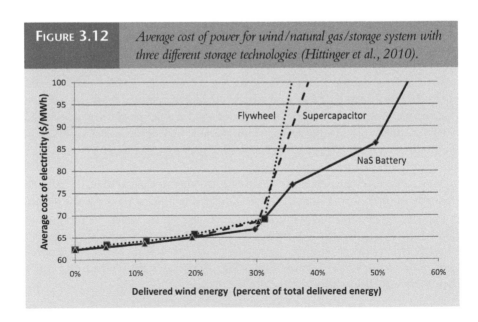

FIGURE 3.12 *Average cost of power for wind/natural gas/storage system with three different storage technologies (Hittinger et al., 2010).*

3.3.5 What are some of the challenges with large-scale deployment of energy storage?

Grid-scale storage can provide substantial benefits for renewables integration and can directly benefit consumers by avoiding the need to keep expensive and rarely used plants on retainer. However, large-scale deployment of storage for energy is likely uneconomical if applied in an energy arbitrage context.[3] Even with capital costs as low as $150/kWh, storage may not be profitable in PJM, where storage capacity of 4% of peak load already exists. Increasing battery efficiency or reducing operations and maintenance (O&M) costs is not sufficient to make arbitrage profitable.

We provide in Chapter 14 several case studies of the implementation of three large grid storage technologies (pumped hydroelectric storage, compressed air energy storage, and plug-in hybrid vehicles), which examined the profitability of large-scale storage projects under existing market conditions and found that, in general, none of these technologies are profitable for the storage owner. As previously described, storage can provide other services to the grid (Lueken & Apt, 2013). Developing mechanisms that compensate storage operators for the wide range of services they can provide to the grid will be critical in ensuring the economic viability of storage.

3.3.5.1 Pumped hydroelectric storage

Pumped hydroelectric storage (PHS) uses lower-cost, off-peak electricity (for example, wind power generated at night) to pump water from a lower-level to a higher-level elevation. When electricity is in high demand, the water is released and used in turbines to generate electricity (see Figure 3.13). The first PHS system in the United States was installed in 1929 with a capacity of 31 MW. Figure 3.14 is a photograph of the Seneca Pumped Storage Generating Station in northwest Pennsylvania, with a reservoir at the higher elevation. It can operate either as a 435 MW hydroelectric power plant or as a pumped storage facility (EIA, 2012). Pumped storage is a net consumer of energy, because more energy is used to pump the water uphill than is regained as it flows back down through the turbines. The round-trip efficiency of all U.S. pumped hydroelectric storage in 2011 was 79% (EIA, 2013).

While PHS is considered an established technology, the chief challenge to using it to support wind and solar resources is the scarcity of appropriate topography. Other challenges to PHS include the long construction times and the high costs of $1 to $2 billion for a 1,000 MW facility that can store about 8 hours of power. There currently is about 22 GW of PHS capacity in the United States (approximately 2% of total generation capacity) and 127 GW worldwide (Mandel, 2010;

---

3. In "energy arbitrage," power is purchased when the price is low and sold when the price is high. For example, during the summer, the peak demand and highest price is during the day when air conditioning is heavily used and businesses are in operation. Prices drop in the evening when energy is used less.

| FIGURE 3.13 | *Illustration of pumped hydroelectric storage system (NREL, 2012).* |

| FIGURE 3.14 | *The Seneca pumped hydroelectric reservoir in Pennsylvania (U.S. Army Corps of Engineers, 1993).* |

NREL, 2012). Although a small fraction of U.S. electricity generation, PHS is the dominant form of electric energy storage and is presently more economical than most other options for energy storage.

PHS has received significant attention in Portugal and Norway, where large amounts of wind power have been built in recent years. Portugal is expanding pumped hydro capacity to support wind by building 636 MW of new PHS, a 60% increase, in a system with a peak load of 9 to 10 GW. Wind provided 18% of Portugal's 2011 electricity. An analysis of the value of using PHS for energy arbitrage in the Iberic electricity market (see Chapter 14) shows that independent PHS operators would not achieve a positive net present value (Lueken, 2012).

Like Portugal, Germany would like to support its aggressive plans to expand wind power use, and storage operators might use energy arbitrage to profit from short-term price fluctuations created by the increase in wind power. Norway has considered adding to its PHS and building a transmission line to sell storage to Germany. However, investments in addition to Norwegian PHS would be profitable only if the price differential between on-peak and off-peak German energy prices significantly increases. This result further suggests that arbitrage is not sufficient for incentivizing investments in energy storage (see Chapter 14).

3.3.5.2 Compressed air energy storage

Compressed air energy storage (CAES) is another method to store electricity. As shown in Figure 3.15, excess energy from wind power is used to compress air, which is then stored underground in, for example, hollowed-out salt deposits (for grid-scale storage) or a carbon fiber storage tank (for smaller storage). When energy is needed, the compressed air is then released and heated, mixed with fuel, and used in a gas turbine to generate electricity. Because this processes uses natural gas energy as fuel, the degree to which this storage process is economical depends on the price of gas.

| FIGURE 3.15 | *Illustration of compressed air energy storage system (NREL, 2012).* |

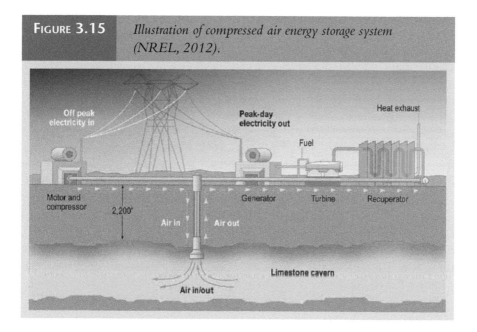

A CAES plant combined with a wind or solar plant could act as a baseload generator in place of fossil fuel and nuclear plants or could be dispatched to meet peak demand. The operating flexibility of CAES also enables the system to provide ancillary services such as frequency regulation, spinning reserve, capacity, voltage support, and black-start capability (Gyuk, 2004).

There are two existing CAES plants, located in Huntorf, Germany, and McIntosh, Alabama. Plans for two CAES plants in Texas were announced in 2012. Apex Energy is building the 317 MW Bethel Energy Center, and Chamisa Energy is building a 270 MW facility in the Texas panhandle. Both will use salt caverns for air storage and plan to earn revenue through intraday price arbitrage.

RenewElec research considered whether CAES is likely to be economical when used for energy arbitrage alone, with increased wind penetration and existing market structures. The results suggest (Chapter 14) that CAES does not appear likely to be profitable when used for energy arbitrage alone in ERCOT (and likely the United States in general) unless the market price differentials are more than double or capital costs substantially decrease. The social benefits might outweigh the private costs if air quality benefits were included.

3.3.5.3 Plug-in hybrid electric vehicle storage

Plug-in hybrid electric vehicles have also been proposed as an energy storage source. Vehicle-to-grid energy transfer (V2G) is the term used to describe the transfer of energy back and forth between plug-in vehicles and the grid. Since vehicles are parked for the majority of the day and night, they could potentially become energy storage devices. Conceptually, an owner of a vehicle could make a profit by storing energy when the price is low and selling while the price is high.

The potential economic implications of using these batteries for grid storage were examined for vehicles buying and selling power in three cities: Philadelphia; Rochester, NY; and Boston (Peterson, Whitacre, & Apt, 2010). The research found that if battery degradation is applied to batteries with a replacement cost of $5,000, the vehicle owner's annual profit from energy arbitrage can be as low as $10 and is never greater than $120. This amount of profit is likely to be insufficient to encourage vehicle owners to use their battery packs for electricity storage. If factors such as replacing peaking generators with the batteries is taken into consideration, the profit level might be $30 to $400, but only if the government offered an incentive for owners to use their vehicles for storage (see Chapter 14 for more details).

This lack of profit for large-scale energy storage does not mean that a few owners cannot make money from another energy service. V2G can help keep the grid frequency stable when wind, solar, or load fluctuations occur. It has been known for some time that a relatively small number of vehicles can profit by providing frequency regulation services to the grid (Letendre & Kempton, 2002). Tests of using vehicle batteries for frequency regulation at small scale have been successful (Kempton et al., 2008).

3.4 Load control to help manage variability

Load control refers to temporarily reducing consumer demand by giving the utility the ability to shut down equipment at short notice so power is available to other portions of the grid. For example, if residents do not need air conditioning at their home on a hot day because they are at the office, the power company could remotely increase the temperature set point of the home air conditioner (in consideration for which a payment is made to the customer). Similarly, controlling the rate of charge of vehicles connected to the grid is a form of load control. Load control, sometimes referred to as demand response, could be used to manage variability in power generation. The following sections briefly summarize two examples of direct load control: air conditioning and smart charging plug-in hybrid vehicles.

3.4.1 Air conditioning

Some electric utilities use direct load control of residential air conditioners to reduce peak demand within their service territory, either to provide services to the grid in times of contingencies or to reduce the amount of power they purchase from the open market. This same control might be used to help manage renewable power. Using historical load and temperature data to better forecast demand of residential air conditioners for three utilities, RenewElec researchers developed improvements on present methods of predicting air conditioner demand that can then be used to provide a resource for incorporation of wind and solar power into the grid through direct load control of these devices (Horowitz, Mauch, & Sowell, 2013). This can increase both cost effectiveness and grid reliability.

3.4.2 Smart plug-in hybrid electric vehicle charging

The concept of "smart charging" is that a plug-in vehicle's demand for electricity would be controlled and managed relative to the amount of electricity generated by variable wind sources. Doing so may help states meet their renewable energy portfolio standards (RPS; see Chapter 1). When there are high levels of wind generation, which in the United States typically occur at night when demand for other end uses is low, the vehicles could potentially be charged at a lower cost while balancing the variability of wind.

An analysis of the cost effectiveness of this scenario for a hypothetical system similar to the one that serves the New York region found that controlled charging significantly reduces the cost of the vehicle charging, but the savings with high wind penetration scenarios are not much larger than with low wind penetrations (see Chapter 14). Controlled charging does not provide much additional value in mitigating the variability of renewable energy in these cases (Weis, Jaramillo, & Michalek, 2014). As shown in Figure 3.16, the use of controlled charging reduces system cost and the effects of electric vehicles on the grid by 50 to 70% or $70 to $100 million per year if electric vehicle penetration is 10%.

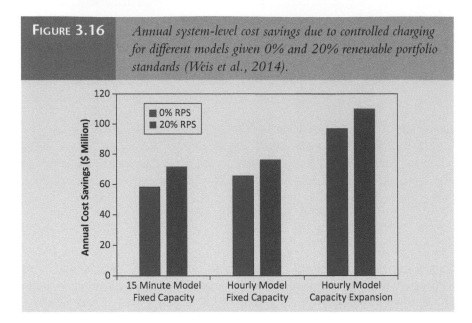

FIGURE 3.16 *Annual system-level cost savings due to controlled charging for different models given 0% and 20% renewable portfolio standards (Weis et al., 2014).*

Controlled vehicle charging was found to not help with the cost of integrating wind systems; the benefit for a 20% RPS is only slightly higher than if there were no RPS policy. This analysis assumed perfect information, no transmission constraints, and both 1-hour and 15-minute time resolution (it is possible that modulating vehicle charging may help smooth the small fluctuations at faster time scales).

3.5 Grid codes and market programs to help manage variability

Grid codes are developed by grid operators to set the technical requirements necessary for generators and other assets to enter the grid. The grid code varies for each grid operator.

Interstate grid operators are required by the U.S. Federal Energy Regulatory Commission (FERC) to treat all operators similarly. This can pose a challenge for variable energy resources. Specific requirements that differ for wind energy are forecasting and data (discussed earlier) and low voltage ride-through, a requirement by FERC that specifies how long a wind turbine needs to remain online during any event in which the voltage on a single leg or multiple legs of the interconnecting transmission line drops (Ryan, 2013).

One option, used in countries such as Denmark, Ireland, Great Britain, and Germany as part of their grid codes, is reducing the aerodynamic efficiency of wind turbines (by dynamically changing the angle at which the blades meet the wind) to decrease a wind plant's power output so that a reserve is created that can be available on demand. An analysis of using this strategy to manage the variability of wind

(Rose & Apt, 2013) found that this practice is less cost effective than compensating with a natural gas turbine (see Figure 3.17 and Chapter 15). In cases where it is required (perhaps when natural gas prices are very high), the requirement should not be uniformly spread across all wind turbines but instead placed on the lowest number required to achieve the desired compensation.

Market operators can also implement programs targeted to variable generators in order to better integrate the variability. Traditionally, wind power operators participating in the market have been must-run resources so the grid accepts all power generated as long as it is technically possible to do so. Wind resources are then price takers and do not bid into the market.

FIGURE 3.17   *Cost effectiveness of wind plant dynamic limiting for up-regulation as a function of opportunity cost of dynamic limiting; market data are from Texas. The vertical axis shows the percentage of times that operating wind turbines at less than their maximum efficiency so that their power can be increased when the wind blows less strongly is less than the prevailing market price for up-regulation. As the opportunity cost of dynamic limiting increases, up-regulation from dynamic limiting is less expensive than the market price for dynamic limiting in fewer 1-hour dispatch intervals (Rose & Apt, 2013).*

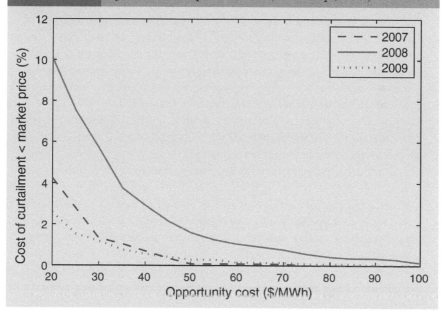

Two independent system operators in the United States have recently established programs allowing variable energy resources to be active market participants. MISO, for example, has created a Dispatchable Intermittent Resource Program (DIR), while CAISO operates an optional Participating Intermittent Resource Program (PIRP). The DIR program provides wind plants that have the technology to control their power output with the ability to make predictions of the expected power they will generate and then use those predictions to market their energy based on that forecast. There is no penalty as long as the dispatch interval is within an 8% tolerance band. The PIRP program provides the option for wind sources that install operation and meteorological data equipment to use hourly energy forecasts to provide information as to how much energy they will be able to market. Penalties occur on a monthly basis based on the net imbalances between scheduled and actual metered data, and there is a forecast fee and an export fee for energy sold outside the CAISO area.

Programs such as these could allow operators to use market-based mechanisms to more cost effectively balance wind and solar power. As more experience is gained through these programs in the coming years, this hypothesis can be tested.

3.6 Summary

In this chapter, we have discussed a number of strategies to reduce the effect of renewable energy variability using geographic aggregation, storage, load control, and grid codes. Among the key findings:

- Electrically combining the output of several wind plants in a region can reduce variability in the aggregate power output, but the amount of reduction is dependent on the time scale involved. The small variability at intervals of an hour or shorter can be reduced by 95%, but the large variability at 12 hours and longer is reduced by only 50%. It is the strong fluctuations at long time scales that require the most compensating resources. These fluctuations at long intervals mean that slow-ramping plants such as natural gas combined-cycle and coal can be scheduled to compensate for the variability.
- Large-scale interconnection of regions results in quickly diminishing returns in the reduction of variability beyond interconnecting about five wind plants and is more expensive than other strategies.
- Large-scale deployment of storage for energy arbitrage is likely unprofitable to the storage owner. Storage operators are likely to participate in several markets (for example, ancillary services as well as energy) to make a profit. There is a need for evaluating the full range of benefits of deploying storage into the grid and developing market structures or financial mechanisms that compensate storage providers for these benefits.
- Dynamically limiting wind power to maintain a reserve in each wind plant, as required by some grid codes, is less economically efficient than compensating with a natural gas turbine.

- There have been recent efforts by RTO/ISOs to develop mechanisms that allow variable resources to actively participate in energy markets with the goal of more cost effectively managing the variability of these resources. More experience with these programs is needed to understand their effects on system operations.

References

EAC. (2008). *Bottling electricity: Storage as a strategic tool for managing variability and capacity concerns in the modern grid.* Electricity Advisory Committee. http://energy.gov/sites/prod/files/oeprod/DocumentsandMedia/final-energy-storage_12-16-08.pdf

EAC. (2011). *Energy storage activities in the United States electricity grid.* Electricity Advisory Committee. www.doe.gov/sites/prod/files/oeprod/DocumentsandMedia/FINAL_DOE_Report-Storage_Activities_5-1-11.pdf

EIA. (2012). *Electricity storage: Location, location, location … and cost.* U.S. Energy Information Administration. www.eia.gov/todayinenergy/detail.cfm?id=6910

EIA. (2013). *Pumped storage provides grid reliability even with net generation loss.* U.S. Energy Information Administration. www.eia.gov/todayinenergy/detail.cfm?id=11991

EPRI. (2010). *Electric energy storage technology options: A white paper primer on applications, costs, and benefits.* EPRI report, 1020676. Palo Alto, CA: Electric Power Research Institute.

ESA. (2013). *Technology comparisons.* Electricity Storage Association. www.electricitystorage.org/technology/storage_technologies/technology_comparison

Fertig, E., Apt, J., Jaramillo, P., & Katzenstein, W. (2012). The effect of long-distance interconnection on wind power variability. *Environmental Research Letters*, 7(3), 034017.

Gyuk, I. (2004). *EPRI-DOE handbook supplement of energy storage for grid connected wind generation applications.* EPRI report, 1008703. Palo Alto, CA: Electric Power Research Institute.

Hittinger, E., Whitacre, J.F., & Apt, J. (2010). Compensating for wind variability using co-located natural gas generation and energy storage. *Energy Systems*, 1(4), 417–439.

Horowitz, S., Mauch, B., & Sowell, F. (2013). *Forecasting for direct load control in energy markets.* http://wpweb2.tepper.cmu.edu/electricity/papers/ceic-13-02.asp

IEC. (2011). *Electrical energy storage: A white paper.* International Electrochemical Commission. www.iec.ch/whitepaper/pdf/iecWP-energystorage-LR-en.pdf

Katzenstein, W., Fertig, E., & Apt, J. (2010). The variability of interconnected wind plants. *Energy Policy*, 38(8), 4400–4410.

Kempton, W., Udo, V., Huber, K., Komara, K., Letendre, S., Baker, S., Brunner, D., & Pearre, N. (2008). *A test of vehicle-to-grid (V2G) for energy storage and frequency regulation in the PJM system.* www.udel.edu/V2G/resources/test-v2g-in-pjm-jan09.pdf

Letendre, S.E., & Kempton, W. (2002). The V2G concept: a new model for power? *Public Utilities Fortnightly*, 140(4), 16–26.

Lueken, C. (2012). *Integrating variable renewables into the electric grid: An evaluation of challenges and potential solutions.* PhD thesis, Carnegie Mellon University. http://wpweb2.tepper.cmu.edu/electricity/theses/Colleen_Lueken_PhD_Thesis_2012.pdf

Lueken, R., & Apt, J. (2013). *The effects of bulk electricity storage on the PJM market.* Carnegie Mellon Electricity Industry Center. Working paper. http://wpweb2.tepper.cmu.edu/electricity/publications.htm

Mandel, J. (2010). DOE promotes pumped hydro as option for renewable power storage. *New York Times*, October 15, 2010. www.nytimes.com/gwire/2010/10/15/15greenwire-doe-promotes-pumped-hydro-as-option-for-renewa-51805.html?pagewanted=all

MBizon. (2010). *Electricity grid schematic.* http://commons.wikimedia.org/wiki/File:Electricity_ Grid_Schematic_English.svg

MGE. (2013). *Electric glossary.* Madison Gas and Electric. www.mge.com/about/electric/ glossary.htm

NREL. (2012). *Renewable electricity futures study.* National Renewable Energy Laboratory. www.nrel.gov/analysis/re_futures/

Peterson, S.B., Whitacre, J.F., & Apt, J. (2010). The economics of using PHEV battery packs for grid storage. *Journal of Power Sources, 195*(8), 2377–2384.

Rose, S., & Apt, J. (2013). The cost of curtailing wind turbines for secondary frequency regulation capacity. *Energy Systems,* in press, DOI: 10.1007/s12667-013-0093-1

Ryan, T. (2013). *ISO rules for intermittent generation.* https://wpweb2.tepper.cmu.edu/rlang/ RenewElec/ISO%20Intermittent%20Rules.pdf

U.S. Army Corps of Engineers. (1993). *USACE Kinzua Dam Downriver.* http://upload.wiki media.org/wikipedia/commons/8/88/USACE_Kinzua_Dam_downriver.jpg

Weis, A., Jaramillo, P., & Michalek, J. (2014). Estimating the potential of controlled plug-in hybrid electric vehicle charging to reduce operational and capacity expansion costs for electric power systems with high wind penetration. *Applied Energy, 115,* 190–204.

Whitacre, J.F., Wiley, T., Shanbhag, S., Wenzhou, Y., Mohamed, A., Chun, S.E., Weber, E., Blackwood, D., Lynch-Bell, E., Gulakowski. J., Smith, C., & Humphreys, D. (2012). An aqueous electrolyte, sodium ion functional, large format energy storage device for stationary applications. *Journal of Power Sources, 213,* 255–264.

4

IMPROVED PLANNING FOR RENEWABLE ENERGY CAPACITY EXPANSION

In previous chapters, we have discussed the need to address the variability of wind and modifications that can be made to system design and operations to respond to that variability. Another factor in policy and operation decisions, discussed in this chapter, is the siting of renewable energy facilities.

Many considerations go into the expansion of renewable energy facilities from an engineering perspective. Some considerations are clear: for example, the wind potential, as measured by wind speeds (see Figure 4.1), and solar potential (Figure 4.2). For solar energy, factors such as the local landscape and shadows, the amount of water vapor, clouds, dust, pollutants, and forest fires influence how much solar radiation reaches a solar power plant.

4.1 Strategies to improve planning of renewable energy plant expansions

The RenewElec project has examined three areas in which the strategies used to plan renewable energy plant expansions might be improved: offshore wind turbines and hurricanes, pollution reduction, and facility siting.

4.1.1 Offshore wind turbines and hurricanes

Some of the highest wind speeds in the United States are offshore (Figure 4.1), but currently the United States has no offshore wind plants. The U.S. Department of Energy has estimated that more than 50 GW of offshore wind energy will be needed if the United States is to achieve a 20% renewable energy level (U.S. DOE, 2008). Typhoons have caused wind turbines to buckle in Japan and China; hurricanes are likely to occasionally pose a similar risk. Because there are no offshore wind plants in the United States at the time of this writing, the RenewElec project used probabilistic

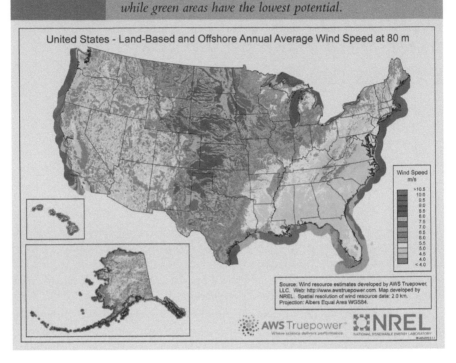

FIGURE 4.1   *United States wind speed map (U.S. DOE, 2013a). Purple shades show areas with the highest wind potential, while green areas have the lowest potential.*

risk assessment to estimate the number of turbines that would be destroyed by hurricanes in an offshore wind plant at four representative locations in the Atlantic and Gulf Coasts of the United States: Galveston County, Texas; Dare County, North Carolina; Atlantic County, New Jersey; and Dukes County, Massachusetts (Rose et al., 2012). Figure 4.3 provides the results of this analysis – the number of turbines that will buckle for different classes of hurricanes in a 50 5-MW wind turbine offshore plant over a 20-year period. A Category 2 hurricane similar to Hurricane Ike (Galveston area) has a 95% probability of buckling one or fewer towers and a Category 3 will buckle up to 12% of the towers (Figure 4.3). When the risk of multiple hurricanes is considered, the research found that Galveston County, TX, is the riskiest location, followed by Dare County, NC, with Atlantic County, NJ, and Dukes County, MA, being significantly less risky (Figure 4.4).

While these analyses showed the risk to individual wind plants, a larger concern for system operators is the number of turbines that might be simultaneously unavailable as a result of hurricanes. To evaluate this risk, a model was also developed to estimate the catastrophe risk to offshore wind power (Rose et al., 2013). This analysis showed that only a small fraction of offshore wind power in a region would be offline simultaneously because of tower buckling by hurricanes. However, the cumulative damage over several years can be significantly larger (see Chapter 16).

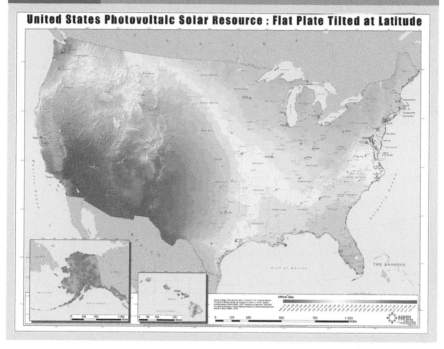

FIGURE 4.2 *United States photovoltaic solar resource map (U.S. DOE, 2013b). Red areas have the highest solar resources, while purple areas have the lowest.*

Figure 4.5 shows that the cumulative damage increases with the length of the period and that Texas is likely to see the most cumulative damage of all the hurricane-prone areas. For Texas and the entire Atlantic coast except Florida ("All U.S."), we predict a 10% probability that more than 0.7% of offshore wind power will be destroyed in any 2-year period, more than 2.6% in any 5-year period, and more than 5.2% in any 10-year period if the turbines cannot yaw. For Texas alone, there is a 10% probability that more than 0.8% of offshore wind power will be destroyed in any 2-year period, more than 4.3% in any 5-year period, and more than 9.2% in any 10-year period if the turbines cannot yaw. For the Southeast (GA, SC, NC), there is a 10% probability that more than 0.04% of offshore wind power will be destroyed in any 2-year period, more than 0.4% in any 5-year period, and more than 1.4% in any 10-year period if the turbines cannot yaw. Having available onsite power to allow turbines to yaw to point directly into the wind reduces these risks by at least a factor of 10. Figure 4.5 does not show results for the Mid-Atlantic and New England because the simulated cumulative damages are too small to estimate with the 3,285 simulated landfalling hurricanes we used in our simulations.

What actions might be taken to reduce the risk to offshore wind turbines from hurricanes? The analysis found that making small changes, such as having emergency power with a battery at a cost of approximately $30 to $40,000 to yaw the

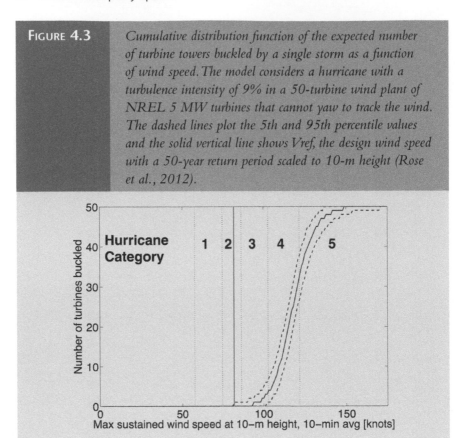

FIGURE 4.3   *Cumulative distribution function of the expected number of turbine towers buckled by a single storm as a function of wind speed. The model considers a hurricane with a turbulence intensity of 9% in a 50-turbine wind plant of NREL 5 MW turbines that cannot yaw to track the wind. The dashed lines plot the 5th and 95th percentile values and the solid vertical line shows Vref, the design wind speed with a 50-year return period scaled to 10-m height (Rose et al., 2012).*

turbine nacelle rapidly into the wind, could significantly improve survivability. In addition, as illustrated in Figure 4.4, it is possible to predict prior to construction the offshore areas with the least risk to wind turbines. The destruction of wind turbines not only causes an economic loss but also makes the grid less reliable.

4.1.2 Pollution reduction

Reduction in the criteria pollutants and greenhouse gas emissions, as discussed earlier, is one of the reasons policy makers take actions to support renewable energy activities.

4.1.2.1 Getting the most pollution reduction out of wind and solar power

When wind or solar energy displaces conventional generation like coal- and gas-powered electricity, the reduction in air pollution emissions and thus the social benefits varies dramatically across the United States. The Air Pollution Emissions

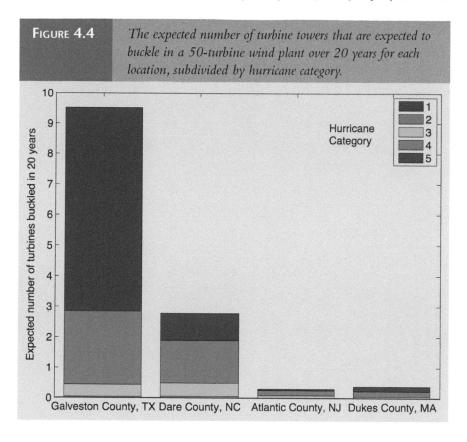

FIGURE 4.4   *The expected number of turbine towers that are expected to buckle in a 50-turbine wind plant over 20 years for each location, subdivided by hurricane category.*

Experiments and Policy (APEEP) model estimates marginal human health damages associated with the emissions of criteria air pollutants in each U.S. county (Muller & Mendelsohn, 2007). Using data from more than 1,400 fossil-based power plants, the human health damages estimated by the APEEP model, and a social cost of carbon of $20/ton CO_2, Siler-Evans and colleagues (2012) developed a regression model to estimate the damages from marginal electricity production by U.S. region. Using these values, the researchers then estimated the reduction in human health damages that occur when wind and solar power displace conventional generators in these regions.

While, on an energy basis, the southwest United States is more suitable for solar power (Figure 4.2), the reductions in damages associated with criteria air emissions are greater in the Midwest, and CO_2 reductions are greater in areas of the Great Plains. Similarly, wind power provides the largest reductions in damages associated with criteria air pollutants in the Midwest, even though this region does not have the best wind resources. Figure 4.6 summarizes these results, which highlight the tradeoffs associated with siting wind and solar energy projects to maximize energy production or to maximize social benefits.

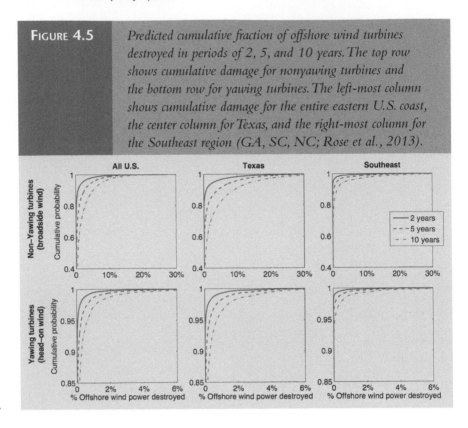

FIGURE 4.5   *Predicted cumulative fraction of offshore wind turbines destroyed in periods of 2, 5, and 10 years. The top row shows cumulative damage for nonyawing turbines and the bottom row for yawing turbines. The left-most column shows cumulative damage for the entire eastern U.S. coast, the center column for Texas, and the right-most column for the Southeast region (GA, SC, NC; Rose et al., 2013).*

Siler-Evans and colleagues (2013) write that

> Thirty percent of existing wind capacity is installed in Texas and California, where the combined health, environmental, and climate benefits from wind are among the lowest in the country. Less than 5% of existing wind capacity is in Indiana, Ohio, and West Virginia, where wind energy offers the greatest social benefits from displaced pollution.
>
> *(p. 11770)*

4.1.2.2 Use of coal and gas plants to mitigate variability

As discussed in Chapter 3, a number of strategies can be used to mitigate the variability of wind and solar energy. Planning for the use of coal and natural gas plants is important, as the pollution reduction effects change based on how plants are operated and the technology that is used. These vary regionally, as illustrated in the discussion in the previous section.

The use of fast-ramping gas plants can mitigate some of the high-frequency variability of wind and solar electric power. According to a RenewElec analysis

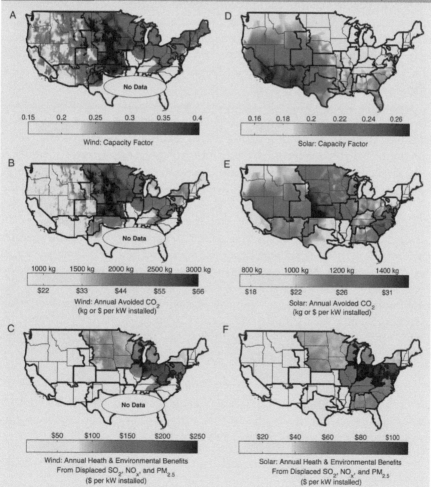

FIGURE 4.6   *Pollution reduction by wind turbines and solar panels. Performance of wind turbines (A–C) and solar panels (D–F) relative to three objectives: capacity factor, a measure of energy output (A and D); annual avoided CO_2 emissions and the corresponding social benefits, assuming a social cost of $20 per ton of CO_2 (B and E); and annual health and environmental benefits from displaced SO_2, NO_x, and $PM_{2.5}$ emissions (C and F). Because of data limitations, the eastern part of Texas and the Southeast are excluded from this assessment of wind energy. Monetary values are in 2010 dollars (Siler-Evans et al., 2013).*

(Katzenstein & Apt, 2009), buffering wind's variability with gas plants can increase the emissions from the natural gas plants and thus slightly reduce the emission benefits generally associated with wind (see Chapter 13). When 20 natural gas plants, with the required number of plants operating as spinning reserve, were used to balance 20% renewables, we found that 87 to 93% of the expected reduction in CO_2 emissions was achieved. NO_x emissions depend greatly on the type of control technology used; the emissions of some significantly increase when they are ramped up and down to follow variable generation.

If system operators recognize the potential for ancillary emissions from gas generators used to fill in variable renewable power, they can take steps to produce a greater displacement of emissions. Operation of NO_x controls with existing (but rarely used) firing modes that reduce emissions when ramping may be practical. On a time scale compatible with RPS implementation, design and market introduction of generators that are more appropriate from an emissions viewpoint to pair with variable renewable power plants may be feasible. In fact, recent developments in gas turbine technology suggest that this is the path forward: Siemens Frame H and GE's new FlexEfficiency portfolio of power plants have been designed and built to more efficiently balance variable power.

Coal plants can also be cycled[1] to manage the low-frequency variability of wind. Even though there are efficiency and emission penalties associated with cycling large thermal power plants, these are not very high: the observed reductions in emissions from coal power plants, accounting for cycling, are, on average, 97 to 98% of what the emission reductions would be if cycling were not accounted for (Oates & Jaramillo, 2013; Valentino et al., 2012; see the discussion in Chapter 13). Part of the difference between coal and gas is that coal plants can be idled at a lower power level than can natural gas plants. Further, the emission benefits associated with a 20% penetration of wind power can be larger than 20% in systems where coal dominates the power fleet. In addition to displacing coal with wind in these systems, some coal is also displaced with natural gas, thus resulting in higher reductions than would be expected using a linear assumption (see Chapter 13). This is further evidence that areas that rely on coal-based generation may benefit more from wind and solar than areas that rely on natural gas plants.

4.1.2.3 Hybrid solar-fossil plants

Plant decisions are often site-specific. One question addressed in the RenewElec project was whether a hybrid solar-fossil integrated solar combined cycle (ISCC) plant will mitigate CO_2 at a lower cost than solar-only power sources such as photovoltaics or concentrated solar power (Moore & Apt, 2013).

---

1. Cycling is when a generation unit is turned off or turned on in response to the need to respond to variability in the grid. Ramping is changing the output of a plant once it is running.

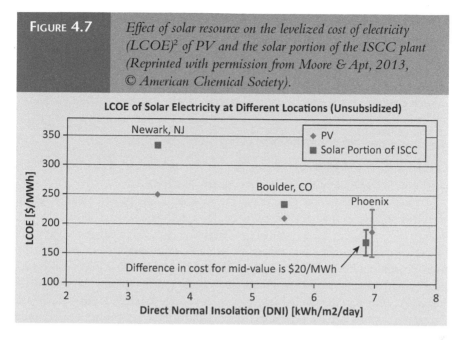

FIGURE 4.7 *Effect of solar resource on the levelized cost of electricity (LCOE)[2] of PV and the solar portion of the ISCC plant (Reprinted with permission from Moore & Apt, 2013, © American Chemical Society).*

ISCC are natural gas combined-cycle (NGCC) power plants hybridized with solar thermal energy to boost the output of the heat-recovery steam generator. The principal advantage of hybridization for solar power is the ability to directly offset fossil fuel energy without having to pay for a power block or transmission lines dedicated to solar energy. The solar field is made up of parabolic troughs, which have mirrors that reflect solar rays onto an evacuated glass tube carrying heat transfer fluid designed to absorb solar energy.

Researchers in the RenewElec Project found that in a few favorable locations, it is less expensive to meet the solar set-aside portion of an RPS with an ISCC plant than with solar photovoltaic (PV) or concentrated solar power (CSP) alone. Using a detailed thermodynamic model of the power plant to examine how much electricity would be produced and when, given the variability of the solar portion of the power plant, shows that the cost and amount of solar electricity generated varies by location (see Figure 4.7). The solar portion of ISCC produces solar electricity at lower cost than stand-alone solar power plants in Phoenix, Arizona. This is due to the strong solar resource and market conditions that provide an option for the solar portion of the power plant capacity factor to exceed 21%. For the other regions, it is less expensive to meet the solar portion of a solar RPS with a stand-alone PV plant; however, this does not include the cost of variability and additional transmission lines.

---

2. LCOE is equal to the lifetime costs (including construction, financing, fuel, maintenance, taxes, insurance, and incentives) divided by the system's lifetime expected power output (kWh).

The cost of mitigating CO_2 would be \$130/ton if a new coal-fired power plant were offset by an ISCC plant and \$250/ton if the solar portion of an ISCC offset a natural gas combined cycle plant. This is considerably less than that for photovoltaics or concentrated solar power, but far more than any of the carbon prices discussed by the 133th Congress (C2ES, 2013). Planning for the optimal location of such facilities is important if the goal is to mitigate CO_2.

4.1.3 A lesson from geothermal siting

Geothermal electric power generation can provide power with low variability and also satisfy the criteria of most renewable portfolio standards.

Traditional hydrothermal geothermal power systems are focused on areas where there is sufficient naturally occurring heat, water, and rock permeability to extract energy (e.g., a geyser). A new type of geothermal energy, called enhanced geothermal systems (EGS), uses a process called hydraulic stimulation to generate energy from dry and impermeable rock. The challenge of this process is that small earthquakes may occur as the rock is stimulated, causing public concern. The best geothermal resources in the United States are located in the western United States (see Figure 4.8), which also has substantial background seismic activity. Induced seismic activity in populated areas can thus create concern among stakeholders and may

FIGURE 4.8   *United States geothermal resource map (U.S. DOE, 2009). Red areas have the best geothermal resources.*

inhibit the successful siting of these projects. Indeed, this has already happened in California, Germany, and Switzerland, where projects have been shut down due to government or public concern about either the occurrence or potential of earthquakes (Glanz, 2009; Patel, 2009).

Three questions that can improve siting decisions include:

1. Can the risk of triggered seismicity be lessened by locating EGS projects away from population centers?
2. How much of the potential recoverable resource of EGS is lost with such remote siting?
3. How can the industry's external stakeholders (the public, politicians, local economic interests) be engaged so as to ensure a fair evaluation of EGS? (Hoşgör, Apt, & Fischhoff, 2013)

The risk for a number of geothermal areas in the United States was examined using risk assessment methods based on both worst-case assumptions and historical data. Figure 4.9 provides an analysis of the risk of seismic activity for El Centro,

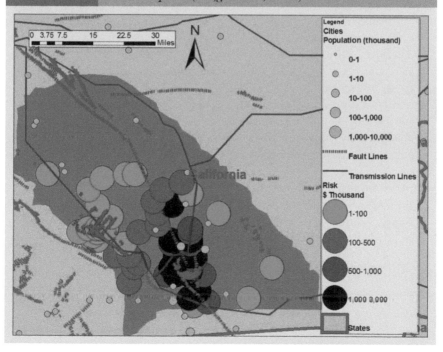

FIGURE 4.9   *Relative risk map for the area near El Centro, California, where the risk is computed by multiplying the calculated relative direct economic loss by the historical probability of an earthquake (Hoşgör et al., 2013).*

California. As shown here, there are many potential location options for EGS where the risk is low (green) for areas with low population and infrastructure density. The analysis for the entire country found that most of the EGS resources in the United States are located in areas with low seismic risk based on both worst-case and historic seismic activity data.

Appropriate EGS site selection can reduce the risk of induced seismic activity, and this technology is an option that should continue to be included in the portfolio of renewable electric generation resources. The research found that early involvement in the planning project is essential for public acceptance and has been successful in attaining trust for past energy projects. Public involvement and trust development along with restriction of site selection to those with a limited chance of seismic activity should be sufficient to support the continued use of this technology. Similar lessons can be applicable to siting other renewable resources.

4.2 Summary

As discussed in this chapter, planning is an essential component when expanding renewable energy facilities to reach economic, environmental, and security policy goals. While there are clear criteria, such as the optimal location for wind, solar, and geothermal facilities, there are other less obvious factors that should be considered. Among these are:

- Currently, the United States does not have any offshore wind turbines. The DOE has estimated that tens of gigawatts of capacity from offshore turbines will be essential if the United States is to achieve a 20% renewable electricity goal. Although hurricanes could be a threat to offshore wind turbines, the risk can be substantially reduced by locating the turbines in less hurricane-prone areas and incorporating an emergency power supply so that turbines can be yawed into hurricane-force winds even without external power.
- Renewable energy sources that are located in areas with heavy use of traditional fuel sources such as coal can lead to greater reductions in criteria pollutant and greenhouse gas emissions.
- Hybrid renewable-natural gas plants in a few favorable areas can lead to greater pollution reduction cost effectiveness than the exclusive use of renewable energy plants.
- Before siting a geothermal plant, a historical analysis of the seismic activity in the area should be conducted to identify areas that have a low risk of seismic activity. Similar considerations should be taken into account when assessing siting of other renewable projects in areas prone to natural risks.

References

C2ES. (2013). *Carbon pricing proposals of the 113th Congress.* Center for Climate and Energy Solutions. www.c2es.org/publications/carbon-pricing-proposals-113th-congress

Glanz, J. (2009). Geothermal project in California is shut down. *New York Times*, December 11, 2009. www.nytimes.com/2009/12/12/science/earth/12quake.html?_r=0

Hoşgör, E., Apt, J., & Fischhoff, B. (2013). Incorporating seismic concerns in site selection for enhanced geothermal power generation. *Journal of Risk Research, 16*(8), 1021–1036.

Katzenstein, W., & Apt, J. (2009). Air emissions due to wind and solar power. *Environmental Science and Technology, 43*(2), 253–258.

Moore, J., & Apt, J. (2013). Can hybrid solar-fossil power plants mitigate CO_2 at lower cost than PV or CSP? *Environmental Science and Technology, 47*(6), 2487–2493.

Muller, N.Z., & Mendelsohn, R. (2007). Measuring the damages of air pollution in the United States. *Journal of Environmental Economics and Management, 54,* 1–14.

Oates, D.L., & Jaramillo, P. (2013). Production cost and air emissions impacts of coal cycling in power systems with large-scale wind penetration. *Environmental Research Letters, 8.* DOI:10.1088/1748-9326/8/2/024022

Patel, S. (2009, December 1). Assessing the earthquake risk of enhanced geothermal systems. *Power.* www.powermag.com/assessing-the-earthquake-risk-of-enhanced-geothermal-systems

Rose, S., Jaramillo, P., Small, M.J., Grossmann, I., & Apt, J. (2012). Quantifying the hurricane risk to offshore wind turbines. *Proceedings of the National Academy of Sciences, 109*(9), 3247–3252.

Rose, S., Jaramillo, P., Small, M.J., & Apt, J. (2013). Quantifying the hurricane catastrophe risk to offshore wind power. *Risk Analysis, 33*(12), 2126–2141. DOI:10.1111/risa.12085

Siler-Evans, K., Azevedo, I., & Morgan, M. G. (2012). Marginal emissions factors for the U.S. electricity system. *Environmental Science and Technology, 46*(9), 4742–4748.

Siler-Evans, K., Azevedo, I., Morgan, M.G., & Apt, J. (2013). Regional variations in the health, environmental, and climate benefits of wind and solar generation. *Proceedings of the National Academy of Sciences, 110*(29), 11768–11773.

U.S. DOE. (2008). *20% wind energy by 2030.* DOE/GO-102008-2567. S. Lindenberg, B. Smith, K. O'Dell, E. DeMeo, & B. Ram (eds.). Washington, DC: U.S. Department of Energy.

U.S. DOE. (2009). *Geothermal resource of the United States.* U.S. Department of Energy. www.nrel.gov/gis/images/geothermal_resource2009-final.jpg

U.S. DOE. (2013a). *Resource assessment and characterization.* U.S. Department of Energy. www1.eere.energy.gov/wind/resource_assessment_characterization.html

U.S. DOE. (2013b). *Solar maps.* U.S. Department of Energy. www.nrel.gov/gis/solar.html

Valentino, L., Valenzuela, V., Botterud, A., Zhou, Z., & Conzelmann, G. (2012). System-wide emissions implications of increased wind power penetration. *Environmental Science and Technology, 46*(7), 4200–4206.

5

NEW REGULATIONS, RATE STRUCTURES, AND STANDARDS TO SUPPORT VARIABLE ENERGY RESOURCES

We have focused thus far primarily on the engineering aspects of integrating variable energy resources into the grid, but policies are important as well. State policies, such as the renewable portfolio standards discussed in Chapter 1, set goals and many impose penalties if the goals are not met. Both national and state policies provide incentives that encourage the construction or use of renewable energy. This chapter concentrates on how national and regional regulatory, rate structure, and industry standard policies influence the integration of renewable energy resources into the United States grid.

5.1 Current regulations, rate structures, and standards

In the absence of a national low-carbon policy, nonhydroelectric renewable power in the United States has been made economically feasible at its current scale by the combination of state renewable portfolio standards, the federal production and investment tax credits, and state tax incentives in certain states. There are a number of other policies that have contributed and are important considerations in the feasibility of a greatly expanded role for variable renewables.

At the national level, the Federal Energy Regulatory Commission (FERC) is the independent agency within the U.S. Department of Energy (DOE) that regulates the interstate transmission of electricity, as well as the transportation of natural gas and oil through interstate pipelines. FERC and the states share regulatory responsibility over renewable energy projects. FERC, in particular, regulates the terms under which renewable generators are interconnected with the transmission grid as well as the terms and conditions of transmission service. The Energy Policy Act of 2005 gave FERC additional responsibilities related to the integration of renewable energy and enforcement. States have exclusive authority to regulate

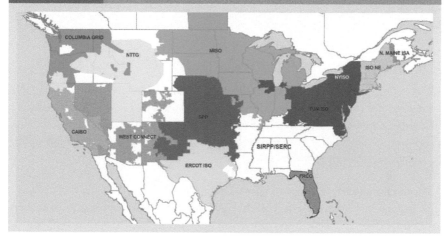

FIGURE 5.1   *Current transmission planning regions (data source: Ventyx). CAISO = California ISO; FRCC = Florida Reliability Coordinating Council; MISO = Midcontinent ISO; ISO-NE = New England ISO; NTTG = Northern Transmission Tier Group; NYISO = New York ISO; PJM = Pennsylvania, Jersey, Maryland Interconnection; SERC = Southeast Reliability Council; SIRPP = Southeast Inter-regional Participation Process; SPP = Southwestern Power Pool.*

many policies affecting the development of renewable energy, including siting of generation and transmission facilities other than on federal lands, interconnection with local distribution, net metering, and the rates, terms, and conditions of retail power sales on a bundled and, to a lesser extent, unbundled basis.

FERC is taking actions to integrate renewables at large scale. Among these are:

- Modifications to FERC's *pro forma* Open Access Transmission Tariff (OATT) to better accommodate variable and intermittent renewable generation, including wind- (and solar-) related ancillary and load-following resources on the inter-state transmission grid
- Revising rules applicable to energy, capacity, and ancillary services markets in centralized power markets
- Transmission planning, siting, cost allocation, and rate incentives
- Facilitating use of demand response and energy storage policies
- Interconnection procedures and agreements for renewable generators
- General NERC reliability standards; and
- Coordination of data availability and transmission operation between balancing authorities (Nordhaus, Yaffe, & Stern, 2010).

FERC works with regional transmission organizations (RTOs) and independent transmission system operators (ISOs) "to identify possible reforms to [RTO/ISO] market rules related to market access that, if adopted, can improve the competitiveness of wholesale energy markets" (FERC, 2013; see Figure 5.1 for the current transmission regions in the United States). The current emphasis of these reforms is on the incorporation of "new or emerging services and technologies, such as demand response, renewable energy, and electric energy storage" (FERC, 2013).

FERC has issued two major orders related to renewable energy resources: Orders 764 and 1000. In this chapter, we discuss these two orders and selected other policy issues related to integration of variable energy resources into the grid, including balancing area consolidation and decommissioning requirements.

5.1.1 FERC Order 764: Integration of variable energy resources

FERC Order 764, issued on June 22, 2012, focused on integration of variable energy resources (VER) into the grid. It defined VER as energy sources that are renewable, cannot be stored by the facility owner or operator, and have variability that is beyond the control of the facility owner or operator. The order requires public utility transmission providers to:

1. offer intrahourly transmission scheduling; and
2. require new (after the effective date of Order 764) interconnection customers under large generator interconnection agreements (LGIAs) whose generating facilities are variable energy resources to provide meteorological and forced outage data to the public utility transmission provider for the purpose of power production forecasting (FERC, 2012b).

The first requirement addresses the schedule on which a power generator must commit its generation output to the grid. The order reduces the interval in which a generator must forecast its output to the grid from 1 hour to 15 minutes. While all ISOs/RTOs have already moved to 5-minute scheduling, most of the balancing areas in the west and the southeast still use hourly scheduling. As illustrated in Figure 5.2, intrahour scheduling enhances the ability to incorporate variable energy resources, as it provides a much closer matching with real outputs; it also benefits nonvariable sources. These benefits are not without additional cost, and FERC indicated that transmission organizations could incorporate this cost into their rates.

The second requirement considers the need of grid operators for more meteorological and forced outage data so that they can improve their power production forecasts given the increased presence of VER sources. FERC 764 requires new interconnecting renewable generators to submit this data to the transmission provider and allows the provider to charge its customers through rate changes for the increased cost it incurs in making power production forecasts.

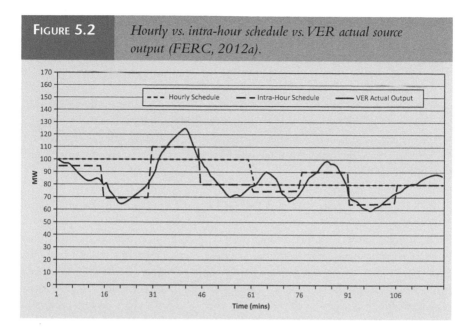

| FIGURE 5.2 | *Hourly vs. intra-hour schedule vs. VER actual source output (FERC, 2012a).* |

5.1.2 FERC Order 1000: Transmission planning and cost allocation

FERC Order 1000, finalized on July 21, 2011, addresses transmission planning and cost allocation requirements for transmission grid expansion. This order, a follow-up to Order 890, requires each transmission provider to participate in designing regional transmission plans. It also mandates incorporation of public policy requirements (for example, an RPS) into those plans, increased involvement of new transmission project developers into the planning process, and coordination between transmission planning regions (FERC, 2011a).

From the perspective of variable energy resources, these new requirements are important for two main reasons. They further the public policy goal of integrating renewable generators into the grid. And they require transmission system operators to develop in advance a transparent method consistent with six principles for allocating the costs of new transmission projects consistent with the benefits received from them. These principles, which apply to the allocation of costs both within a planning region and between planning regions for interregional projects, are:

1. The costs of transmission facilities are to be allocated "in a manner that is at least roughly commensurate with estimated benefits."
2. Those not benefiting from transmission facilities "must not be involuntarily allocated the costs of those facilities."
3. If a cost ratio threshold is used, it must not be so high "that facilities with significant positive net benefits are excluded from cost allocation."

4. Costs for a regional transmission facility must be allocated only within that particular transmission region, unless an outside entity "voluntarily agrees to assume a portion of those costs."
5. Cost allocation methods for determining benefits and beneficiaries must be transparent.
6. "A transmission planning region may choose to use a different cost allocation method for different types of transmission facilities in the regional plan, such as transmission facilities needed for reliability, congestion relief, or to achieve public policy requirements established by state or federal laws or regulations" (FERC, 2011b).

FERC did not require that each ISO/RTO have the same allocation method, only that each apply the general principles listed above. We examined the plans from different ISOs/RTOs and categorized the different methods used by transmission organizations. For example, CAISO uses a "postage stamp" rate that apportions costs among all market participants in a region. CAISO has taken the position that this method is appropriate because the entire region benefits from the development of renewable resources and enhanced reliability and economics.

Other ISOs use a "license plate" rate that allocates costs on a proportional basis among various rate zones within the region based on the transmission facility benefits from each zone. For example, the Northern Tier Transmission Group (NTTG) has a cost allocation committee of state regulatory and consumer protection representatives that uses the "beneficiary pays" model. This committee reviews and makes nonbinding recommendations based on a proposal from each market participant. Each proposal provides a cost allocation method based on four principles. Two of these are that "[c]ost causers should be cost bearers," and "beneficiaries should pay" in amounts reflective of benefits received; and projects should be consistent with renewable portfolio standards (NTTG, 2012).

5.2 The effectiveness of FERC Orders 764 and 1000

One model of widespread adoption of renewable electricity is to locate wind in the windiest areas and require an extensive transmission infrastructure to carry wind resources from their geographically isolated locations to major load centers. Earlier regulations governing transmission did not provide measures to facilitate adequate development of the expansive transmission infrastructure that might be needed to support the development of the full potential of renewable energy from prime areas of wind (or solar) resources to major load centers. FERC Order 1000 is expected to improve on the previous regulations by explicitly requiring consideration of the transmission needs driven by public policy requirements to develop renewable energy in each region's transmission planning and cost allocation processes.

FERC Order 764 could be improved. One difficult problem in the VER integration is how to refine the type of data to be provided by generators for use in forecasting models that are still undergoing substantial development. These forecasting results must then be coupled with the risks faced by transmission providers and balancing authorities in the minute-to-minute operations of their systems. The following changes to Order 764 should be considered:

- **Scheduling:** Require 5-minute scheduling in all areas with significant VER integration needs, improving on the current 15 minutes. Five-minute intervals are already in use in organized market regions, are technically feasible, and will ensure that VER dispatch is conducted at the maximum achievable accuracy and efficiency.
- **Power production forecasting techniques:** Express a preference for common forecasting methods but avoid prematurely locking in existing and still-developing methods, and develop a requirement to retain data so that researchers can develop better forecasts. At present, forecasting for wind and solar power production continues to have high error rates, particularly in balancing areas where data are limited. It is thus appropriate to be cautious against locking in forecasting methods based on insufficient data and flawed methods. This could have unintended negative consequences for VER integration and the cost of overall system management.
- **Application of the data reporting requirements to existing large generator interconnection agreements (LGIAs):** Apply any data reporting requirements to existing VERs by requiring retroactive changes to LGIAs already in effect, and request existing VER operators make existing data collections available. These data would allow public utility transmission providers (and other interested parties) to gain a better understanding of the variability of wind and solar generation. Existing wind plants are farther along on the operating learning curve, and there may be lessons to be learned by collecting their data in the future.
- **Application of the data reporting requirements through the small generator interconnection agreement (SGIA):** Apply data reporting requirements to generators with capacities of less than 20 MW, including existing customers as well as those interconnection customers whose generating facilities are solar VERs with a capacity of 1 MW or greater. Collecting these data will ensure that transmission providers can develop VER power production forecasts that account for a small but growing portion of generation in their balancing areas.
- **Definition of "variable energy resource":** Clarify the VER definition so that it applies to all VERs regardless of the use of energy storage and so that the use of energy storage or other methods to mitigate variability by VER owners and operators does not undercut the quality and quantity of data provided by VERs.

5.3 Other regulatory, rate structure, and rule strategies to improve the integration of renewable energy into the grid

Beyond FERC orders, other actions influence the regulatory, rate structure, and rule strategies when incorporating variable energy resources. Two examples are balancing area consolidation and decommissioning requirements.

5.3.1 Balancing area consolidation

Balancing authorities maintain the balance of generation and load within a geographic area so that the grid frequency is stable. The California ISO is the largest of 10 balancing authorities in California (see Figure 5.3) and 40 in the Western

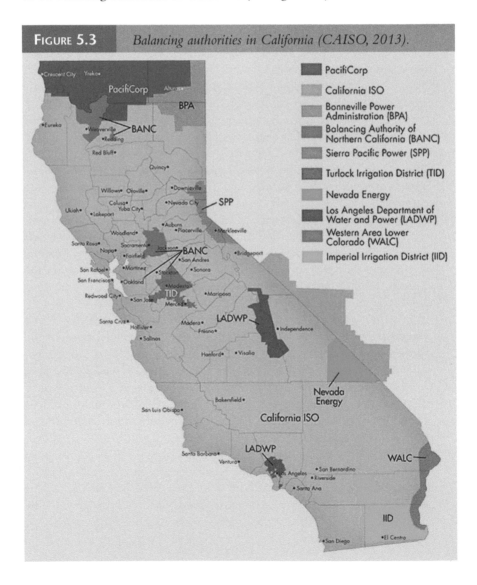

| FIGURE 5.3 | *Balancing authorities in California (CAISO, 2013).* |

Interconnection. There are currently 103 balancing areas in North America; these vary in size from 40 MW to 153 GW. Balancing area consolidation might take a variety of forms, including sharing power and resources with a neighboring area or reorganizing to create an RTO/ISO. It has been suggested that consolidating balancing areas would provide economic benefits and is required in order to integrate large amounts of variable renewables (Milligan, Kirby, & Beuning, 2010). A RenewElec case study (Ryan, Jaramillo, & Hug, 2013) investigates this issue by modeling a large-scale consolidation with a near-term wind penetration (~20% by energy) and estimating the associated economic benefits.

Milligan and colleagues (2010) propose that balancing area (BA) consolidation might help manage renewable energy variability in two ways. First, enlarging the BA allows more cost-effective resources to be used to compensate for variability. Second, variability is somewhat reduced when wind is averaged over a large area. This leads to cost reductions by reducing the quantity of balancing services required. These economic gains can be quantified in terms of the change in total, producer, and consumer surplus; each of these is a measure of economic benefits that go to society, producers, and consumers.

The RenewElec case study (Ryan et al., 2013) modeled these economic gains for both the energy and frequency regulation[1] markets, such as the amount of additional energy and frequency regulation services foregone through balancing area consolidation. Two wind penetration levels were modeled, 0 and 20% by energy. The results showed that economic surplus might increase as a result of balancing area consolidation, equivalent to a total cost reduction of $0.02 to $0.20/MWh, but it could be as high as $1.70/MWh, with more than 90% of the benefits occurring in the energy market. This large uncertainty is driven by the study's counterfactual assumption − that all balancing area needs are met by resources internal to the balancing area. Most balancing areas today do not strictly meet this assumption; instead, they cooperate with their neighbors without consolidation (e.g., imports, exports, and power purchase agreements). If balancing areas do not import or export electricity, then all reductions in energy and frequency regulation service accrue to the area and the actual results could be significant (~$1.70/MWh or 9% of modeled system cost).

It is much more likely that balancing areas do import and export power to take advantage of cheaper resources and thus the "starting" base costs of integration for purposes of analysis may be less than computed on a single balancing area basis. This would indicate that a large portion of the economic gains associated with consolidating balancing areas may already be realized. Therefore the actual near-term benefits of balancing area consolidation are likely to be modest ($0.02–$0.20/MWh or 0.1%–1.2% of the modeled system cost). Additionally, the presence of wind in the system (~20% by energy) did not increase the benefits of balancing area consolidation in the model; it increased only the variance of these benefits.

---

1. Frequency regulation is a type of ancillary service that is responsible for the short-term balancing of electric power.

These two results suggest that there is little economic motivation to consolidate balancing areas today or at the scale of renewables mandated by current RPS legislation.

5.3.2 Decommissioning requirements

All energy generation facilities have a limited lifetime, and the cost of decommissioning these facilities at the end of their economic life must be taken into consideration while they are still providing an economic benefit. This has long been an issue for oil and gas wells and is now increasingly important as the number of wind (and solar) plants increases throughout the United States. Decommissioning requirements are determined by federal authorities for projects that occur in lands managed by the Bureau of Land Management. State, county, and municipal authorities determine the decommissioning requirements for projects developed outside of federal lands. State and local bonding requirements for energy projects are generally greater than federal requirements.

Decommissioning bonds or financial assurances are important due to the economic cost of managing retired or abandoned infrastructure, which may occur decades after project development. Decommissioning requirements can also help to ensure that all energy sources are on a level playing field. A RenewElec study (Changala, Dworkin, Apt, & Jaramillo, 2012) examined this issue for oil and gas wells and for wind. The comparison was chosen because oil and gas production activities and wind power activities are similarly distributed among large areas and can also be similar in size.

For both wind and the oil and gas wells, the study found that the existing funds are insufficient to meet the actual costs of decommissioning (See Figure 5.4).

5.3 Summary

Federal, state, and RTO/ISO policies play a significant role in influencing the degree to which renewable energy is encouraged or discouraged to be part of the nation's energy supply. Although they can be improved, recent regulatory actions such as FERC Orders 764 and 1000 are major steps in advancing the potential of renewable electricity, along with state renewable energy portfolio standards.

FERC's Order 1000 requires transmission providers to participate in a regional transmission planning process that considers renewable electricity projects as mandated by public policies (for example, state RPS requirements). The regional planning must also include allocation of transmission costs in a way that satisfies specific regional cost allocation principles; these must include allocation of costs that satisfy RPS requirements. This 2010 order has led to increased activity in planning transmission for renewables. Lines such as a 3,500 MW high-voltage direct current transmission line from Iowa to Chicago are being proposed under this order, highlighting the influence that regulation can have on the large-scale deployment of variable energy resources like wind and solar.

A comparison of federal, state, and county decommissioning regulations for oil and gas extraction sites and wind energy projects reveals that, generally, regulatory

FIGURE 5.4	*Minimum bonding requirements and range of reclamation/ decommissioning costs (modified from Changala et al., 2012). The vertical axis shows the cost of reclamation/ decommissioning normalized by the area affected by the projects.*

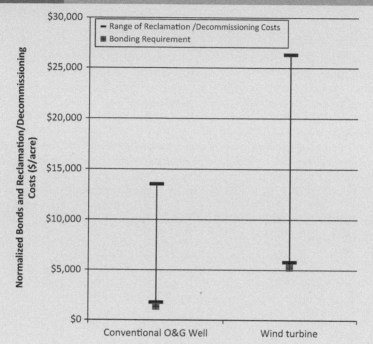

requirements are insufficient to adequately secure the costs of decommissioning. The states that have established decommissioning regulations for wind projects require bonding amounts on a case-by-case basis. Evaluating decommissioning bonding requirements on a case-by-case basis will most likely produce more accurate bond amounts. Consequently, state decommissioning requirements may potentially serve as a useful regulatory model for establishing adequate decommissioning requirements for wind and solar plants.

References

CAISO. (2013). *The ISO grid.* www.caiso.com/about/Pages/OurBusiness/Understand ingtheISO/The-ISO-grid.aspx

Changala, D., Dworkin, M., Apt, J., & Jaramillo, P. (2012). Comparative analysis of conventional oil and gas and wind project decommissioning regulations on federal, state, and county lands. *The Electricity Journal, 25*(1), 29–45

FERC. (2011a). *FERC Order 1000 fact sheet.* Federal Energy Regulatory Commission. www. ferc.gov/media/news-releases/2011/2011-3/07-21-11-E-6-factsheet.pdf

FERC. (2011b). *Order 1000 transmission planning and cost allocation by transmission owning and operating public utilities*. Federal Energy Regulatory Commission. www.ferc.gov/whats new/comm–meet/2011/072111/E-6.pdf

FERC.(2012a).Integration of variable energy resources,final rule. *Federal Register.* www.federal register.gov/articles/2012/07/13/2012-15762/integration-of-variable-energy-resources

FERC. (2012b). *Integration of variable energy resources*. Federal Energy Regulatory Commission. www.ferc.gov/whats-new/comm-meet/2012/062112/E-3.pdf

FERC. (2013). *FY 2014 Congressional performance budget request*. Federal Energy Regulatory Commission. www.ferc.gov/about/strat-docs/fy14-budg.pdf

Milligan, M.R., Kirby, B., & Beuning, S. (2010). *Combining balancing areas' variability: Impacts on wind integration in the Western Interconnection*. National Renewable Energy Laboratory paper, NREL/CP-550-48249. www.nrel.gov/docs/fy10osti/48249.pdf

Nordhaus,R.,Yaffe,D.,& Stern,H.(2010). *Preliminary survey of FERC Renewable Electricity Regulatory Initiatives*. https://wpweb2.tepper.cmu.edu/rlang/RenewElec/PRELIMINARY% 20SURVEY%20OF%20%20FERC%20RENEWABLE%20ELECTRICITY%20REG ULATORY%20INITIATIVES%20.pdf

NTTG. (2012). *Regional planning and cost allocation practice*. Northern Tier Transmission Group. http://nttg.biz/site/index.php?option=com_docman&task=doc_download&gid=1871 &Itemid=31

Ryan, T., Jaramillo, P., & Hug, G. (2013). *Near-term economics and equity of balancing area consolidation to support wind integration*. Carnegie Mellon Electricity Industry Center working paper, CEIC-13-03. https://wpweb2.tepper.cmu.edu/ceic/PDFS/CEIC_/electricity/ papers/ceic_13_03_EEB.pdf

6

POLICIES TO MANAGE VARIABLE GENERATION AT INCREASED RENEWABLE CONTRIBUTION

Throughout this first part of the book, we have discussed the importance of managing variability so that we can increase the contribution of variable renewable energy resources to the nation's electricity supply. Managing that variability requires actions that begin with predicting variability, understanding the technical options available to reduce or manage that variability, and planning for expanding renewable energy's contribution to the grid.

In this chapter, we identify policies that can be adopted by key decision makers to increase renewable energy's contribution to the grid. Our definition of decision makers is broad and includes:

- grid operators
- renewable power forecasters
- analysts performing integration studies for variable renewable generation
- independent system operators and regional transmission organizations
- electric power markets
- project finance firms
- standards organizations and
- federal, state, and local legislative bodies, governments, and regulatory agencies.

Accurately assessing and preparing for the operational effects of renewable generation can ensure that renewables can play a much larger role in the U.S. power system.

Figure 6.1 provides an overview of the short- and long-term strategies our research has identified as helpful in reaching these renewable energy goals. Short-term strategies are those that can be taken today with an immediate effect. Long-term strategies are those that can be taken today but that focus on long-term planning of the grid. For example, transmission lines are planned long before they

FIGURE 6.1

Short- and long-term strategies to increase renewable energy's contribution to the grid.

Short-Term Strategies

Grid operators should incorporate wind forecast uncertainty in unit commitment and dispatch decisions. Forecast uncertainty is strongly dependent on whether the forecast is for low or high wind. Load errors are also less accurate at high load.

Forecasters and grid operators should correct for wind forecast biases before using the data for unit commitment and dispatch. Present methods tend to over predict wind power output when the wind blows strongly and under predict in light winds.

Analysts conducting large-scale integration studies should recognize that wind and load are correlated and that they do not fit Gaussian (bell-shaped) curves. Analysts should also base capacity credit calculations on at least 5 years of wind data and consider increased capacity credit for plants that have low correlation with other plants.

Large ISO/RTOs should recognize that dynamically limiting all turbines within a wind plant below maximum generation to provide regulation capacity is typically more costly than other options available to enhance reliability. It is not costly to give wind plants the ability to dynamically limit, but its use should be very infrequent.

Electricity markets should allow fossil-fueled plant operators to incorporate cycling costs in their bids to limit excessive cycling of units when they are used to compensate for variability.

Electricity markets should strongly consider implementing a market for ramping services provided by generators, storage, and demand response to facilitate large-scale integration of variable renewable generators.

The financial community should allow developers to combine projects to reduce year-to-year variability's effect on debt-to-equity ratios.

FERC should require 5-minute scheduling in all areas with significant variable energy resource penetration. FERC should also increase data reporting requirements for variable generators. FERC should clarify the definition of variable energy resources to apply to facilities that make use of energy storage or other variability mitigation.

Long-Term Strategies

ISO/RTOs and planning agencies should recognize that large-scale geographic aggregation is not necessary to mitigate the variability of wind. While variability slightly decreases when large areas are combined, and firm capacity slightly increases, using natural gas turbines and/or RTO-RTO coordination market products to counter variability is likely to be more cost effective.

ISO/RTOs should develop strategies to compensate energy storage operators for the benefits they provide to electricity consumers. Storage can significantly reduce high-frequency variability of wind and solar and better match these resources to load.

Local, state, and federal governments should establish appropriate decommissioning requirements for wind and solar power plants. **States** should set wind and solar plant decommissioning bonds to adequately cover end-of-life costs on a case-by-case basis.

Regulatory agencies and insurers should provide incentives for the development of renewable resources in areas with lowest risks of hazards like earthquakes and hurricanes.

Legislative bodies and regulatory agencies should provide incentives to site wind and solar power plants where benefits from displacing pollution are highest.

Standards organizations such as the IEC should consider requiring wind turbines in higher-risk hurricane areas to have the capability to yaw into the wind even without grid power.

are built, and energy storage technologies are still not fully mature, but actions taken today affect their ability to be developed and deployed.

6.1 Short-term strategies

Grid operators and forecasters should incorporate wind forecast uncertainties and biases before using the data for unit commitment and dispatch decisions.

Grid system operators manage many energy sources to meet customer demand for energy (load). Because electricity cannot be easily stored, system operators must instantaneously match supply and demand by instructing different generators to produce a given amount of electricity. Generators are constrained by technical limitations. For example, they cannot be turned on and off instantaneously. They are also limited in the amount of output that can be changed in a given period of time. Load forecasting, which takes place on time scales from a day ahead to minutes ahead of time, allows the system operators to prepare to instantaneously match generation and load. Unlike conventional generators, variable energy resources are not fully controllable, and their power production at any given moment depends on environmental and technical factors. Forecasting the availability of these variable renewable resources is thus becoming of increasing importance to grid operators, who have to schedule conventional resources according to their expectations for demand and for availability of renewable resources.

Forecasts are uncertain (see Chapters 2 and 9). This uncertainty, which is quantified as forecast errors, is not surprising given that forecasts are based on statistical and physical models that incorporate information on the season, time of day, and weather conditions – and we are all familiar with the challenges in weather forecasting. It is not just renewable generation whose forecasts are uncertain. For large areas, even those without renewables, electricity demand forecast errors are approximately 2 to 3% of the mean demand. Wind plant power forecasts rely on more than weather and incorporate issues such as wind plant size and local terrain. Understanding and incorporating the uncertainty of wind/solar and demand forecasts can reduce inefficiencies in the system. For example, if operators overpredict the amount of conventional generation that will be needed, valuable resources would be wasted and the cost of providing electricity would increase. If, on the other hand, operators underpredict the amount of conventional generation that will be needed, they would be unable to meet demand. We see this on occasion when utilities ask customers to reduce demand on a particularly hot day. Using forecast errors appropriately could reduce the number of these events.

One way to make better use of forecasts is to correct them to account for the forecast biases. A RenewElec analysis of load and wind forecasts in ERCOT and MISO found that although load forecasts errors were generally assumed in the published literature to have a normal (bell-shaped curve) distribution, this is not what is observed (see Figure 9.2). The same is true for wind forecasts (see Figure 9.3). In fact, the analysis showed that the forecast error distribution is strongly dependent on the forecast level of wind power: for low wind power forecast, the forecasts tend to underpredict the actual wind power produced; when the forecast is for high power, the forecast tends to overpredict the actual wind power. The researchers developed a mathematical method to correct for forecast biases. They also used mathematical tools to evaluate the amounts of generation that must be scheduled as reserves in order to prepare for forecast errors (additional details are in Chapters 9 and 10). This analysis showed that if wind capacity reaches 30 GW, 30% and 13% of the installed wind capacity must be matched with dispatchable capacity to counter day-ahead forecast errors in ERCOT and MISO, respectively (see Figure 10.7).

One of the limits to current forecast accuracy is that wind plants tend to be clustered in close proximity. If future wind plant development substantially broadens the areas in which wind plants are located, forecast accuracy may improve. If forecast accuracy improves through future spatial diversity in wind plant locations, the ERCOT and MISO day-ahead reserves could be reduced to 7% and 4%, respectively. Reserve requirements will differ from one system to the next, but the method used in this analysis can be used for other systems.

Analysts conducting large-scale integration studies should recognize that wind and load are correlated and that they do not fit Gaussian (bell-shaped) curves. Capacity credit calculations should incorporate year-to-year variation and should give credit for plants exhibiting low correlation with other plants.

Many organizations, including DOE, NREL, GE Energy, RTOs/ISOs, and consultants in the U.S. and Europe, have produced large-scale wind integration studies. These studies have typically examined a period 10 to 20 years in the future. Chapter 17 is a critical analysis of many of these studies.

Net load is the demand for electricity minus the renewable production. Almost all the studies reviewed used the standard deviation of net load step changes (the difference in the net load from one time period to the next) to estimate the need for regulation and load-following reserves. Doing so implicitly assumes that load and wind are uncorrelated and that the data fit Gaussian statistical models. Neither assumption is accurate. Low-probability events that have large consequences are the events that may cause power systems to become unreliable. Methods that use the magnitude of low-probability ramping events rather than standard deviations are likely to produce balancing resource estimates that more accurately predict what will be needed to maintain system reliability.

In future studies, it may be valuable to think more broadly about the ways in which different types of balancing services can be purchased from different types of power plants. For example, storage and demand response could provide highly responsive balancing services but will be deployed only if electricity markets reward market participants for their responsiveness.

There is generally lower year-to-year variability of individual wind plants at sites with better wind resource. Connecting wind plants with a transmission system reduces the year-to-year variability of their combined energy generation, depending on the correlation between them (see Chapter 11). When capacity credit is allocated for wind plants, consideration should be given to the degree to which plants exhibit correlation with other variable plants in the system, just as some current wind and solar capacity credit allocations give consideration of correlation with load.

Large ISO/RTOs should recognize that dynamically limiting all turbines within a wind plant below maximum generation to provide regulation capacity is typically more costly than other options available to enhance reliability.

Maintaining an alternating current frequency (60 Hz in North America, 50 Hz in much of the rest of the world) on the grid is important for maintaining grid stability. Grid operators maintain frequency in part by adjusting the amount of power generated to match constantly changing load within rather tight limits.

In a system with a large penetration of wind power, wind turbines can be used to help maintain the frequency of the grid. Because wind turbines running at full capacity cannot be adjusted to provide additional power, grid operators in some countries require that wind plant operators continuously operate all their turbines below maximum power so they can adjust their frequency within seconds and thus provide frequency support. This is typically done by changing the blade angle of the turbine.

A RenewElec analysis found that this mechanism (used for "up-regulation") almost always costs more than the market price of purchasing this service elsewhere from conventional sources (see Chapter 15). A wind plant must sacrifice more energy production than would a conventional generator to supply the same amount of capacity. In addition, the wind plant loses the production subsidies provided by governments to encourage wind production. These two factors constitute the opportunity cost (see Figure 15.4). This observation does not hold true, however, when conventional sources are particularly expensive or wind power producers are on long-term low-cost contracts or are not subsidized.

Although it is expensive to limit a wind plant below maximum generation to regulate the grid frequency, it is relatively inexpensive to install the hardware and algorithms to give a wind plant the *capability* to curtail. For that reason, ISO/RTOs should not routinely dynamically limit wind power to provide regulation capacity, but it is reasonable to require them to be able to dynamically limit on rare occasions when it may be needed for grid reliability.

Note that this capability, done by changing the angle at which the wind turbine blades meet the wind, is not the same as dispatchable intermittent resources programs, which require wind turbines to eliminate their power output when commanded to do so by the ISO/RTO. Such programs have no opportunity cost when the wind plant is being operated normally.

Electricity markets should allow fossil-fuel plant operators to incorporate cycling costs in their bids to limit excessive cycling units. Markets for ramping services should be implemented to facilitate large-scale integration of variable renewables.

Increasingly, we ask the machines that generate our electric energy to change the amount of power they produce; the industry calls this cycling. The growing market penetration of variable renewable generators, such as wind, is causing existing generators to cycle in order to compensate for wind's variability. For large generators, cycling means operating in a less efficient way, leading to higher fuel costs and air emissions. It also means an increased risk of creep and fatigue damage, which leads to higher maintenance costs and more outages.

A model of the PJM region developed by researchers in the RenewElec project shows that 20% wind penetration would result in increased coal cycling and reduced coal capacity factors (see Chapter 13). The presence of large-scale wind power would reduce the revenue of coal power plants and could increase their operation costs. However, if the costs of cycling were incorporated in the decision to dispatch coal power plants, the burden of balancing system variability would be transferred to more flexible units, which would also result in further reductions of air emissions. Current rules are largely silent on how to incorporate cycling costs in electricity markets. The results of the RenewElec work suggest that market rules should allow coal operators to incorporate cycling costs into their bids to the market.

Cycling is turning a plant on and off. Ramping is changing the output of a plant once it is running. The presence of variable renewables on the system will require that other plants be ramped. The Midcontinent ISO and a few other markets are considering market products for ramping services. The implementation might take the form of a constraint to the dispatch model that requires the ISO/RTO to have enough ramping capacity every 15 minutes. The ramping resources should not be limited to generators but should also include storage and verifiable demand response.

The financial community should allow developers to combine projects to reduce year-to-year variability's effect on debt-to-equity ratios.

The most important consequence of year-to-year variability of wind power is that it determines the debt-to-equity ratio of wind plant projects, which affects the profitability for their owners. The lower the year-to-year variability, the more the wind plant developer can leverage its equity. Some sites in the U.S. Great Plains have nearly twice the year-to-year variability in annual energy as others, which puts them at a financial disadvantage. However, if two or more distant (poorly correlated) wind plants can be financed together (instead of the current practice of a single-purpose LLC for each wind plant), they may be able to get better financing terms than if they are considered separately. For example, if developers financially combine a site in Iowa with a coefficient of variation (COV, the standard deviation of several years of wind power production normalized by the mean) of 6% and a site with the same mean generation in Illinois with a COV of 11%, the combined COV of annual energy is 7.5% (see Chapter 11). Separate ownership makes combined financing difficult, but the potential savings should incent investment banks to develop appropriate investment vehicles to capture this potential.

FERC should require 5-minute scheduling in all areas with significant variable energy resource penetration. FERC should also increase data reporting requirements for variable generators. FERC should clarify the definition of variable energy resources to apply to facilities that make use of energy storage or other variability mitigation.

Five-minute scheduling intervals will help integrate variable generation; they are already in use in organized market regions, are technically feasible, and will ensure that VER dispatch is conducted at the maximum achievable accuracy and efficiency (see Chapter 5). FERC should require 5-minute scheduling in all areas with significant VER integration needs, improving on the current 15 minutes.

Both large variable generators and those as small as 3 MW should be required to collect and report generation data. Such data would allow public utility transmission providers (and other interested parties) to gain a better understanding of the variability of wind and solar generation. Such data would also ensure that transmission providers can develop VER power production forecasts that account for a small but growing portion of generation in their balancing areas (see Chapter 5).

FERC should clarify the VER definition so that it applies to all VERs regardless of the use of energy storage; it should also clarify that the use of energy storage or other methods to mitigate variability by VER owners and operators not undercut the quality and quantity of data provided by VERs (see Chapter 5).

6.2 Long-term strategies

ISO/RTOs and planning agencies should recognize that large-scale geographic aggregation is not necessary to mitigate the variability of wind.

As discussed in Chapters 3 and 12, wind plants within a region or large geographical area can be interconnected with transmission lines with the goal of mitigating the variability of added wind sources. Electrically combining the output of several wind plants in a region can reduce variability in the aggregate power output, but the amount of reduction is dependent on the time scale involved. The small variability at intervals of an hour or shorter can be reduced by 95%, but the large variability at 12 hours and longer is reduced by only 50%. It is the strong fluctuations at long time scales that require the most compensating resources. These fluctuations at long intervals mean that slow-ramping plants such as natural gas combined-cycle and coal can be scheduled to compensate for the variability.

Large-scale interconnection of regions results in diminishing returns in the reduction of variability beyond interconnecting about five wind plants and is more expensive than other strategies. We note that interconnection can somewhat increase the firm power output compared with that available within a single region.

While variability slightly decreases when large areas are combined, and firm capacity slightly increases, using fossil-fueled plants and/or RTO-RTO coordination market products to counter variability is likely to be more cost effective than building transmission just to reduce variability. Thus, large new investments in transmission systems designed to interconnect large areas of the country are neither required nor desirable to integrate wind. Decreased transmission construction costs could change this conclusion (see Chapter 3).

ISO/RTOs should develop strategies to compensate energy storage operators for the benefits they provide to electricity customers.

As discussed in Chapters 3 and 14, energy storage technologies can help integrate variable renewables and can provide a number of benefits to electricity consumers by allowing the ability to save and then use electricity when we most need it. For example, the optimal time for solar energy facilities to produce power might be 10 a.m. to 2 p.m., but we need electricity the most from 4 to 6 p.m. when we are active in both offices and homes. In the United States, this period coincides in some regions with the gap between the ramp-down of solar and the evening onset of wind's ramp-up. Energy storage can enhance the ability to balance power when there are variable renewable energy sources, optimize the ability to incorporate bulk power production into the grid, encourage integration of plug-in hybrid electric vehicles, defer infrastructure investments in transmission and distribution to respond to peak loads, and provide general services to grid and market operators.

Storage provides substantial benefits to consumers in two ways: by reducing prices in the wholesale energy market and by reducing the need for peaking generators. Economic benefits rise rapidly with increasing large-scale storage up to approximately 20 GW in PJM. The benefits include savings to consumers of roughly 10% of total energy market costs in PJM at a level of 20 GW of storage (see Chapter 14). This is a transfer of producer surplus to consumers, reducing the revenues to companies that generate power.

ISO/RTOs should develop strategies to compensate energy storage operators for the benefits they provide to electricity customers. Storage can significantly reduce the effects of high-frequency variability of wind and solar power on the grid and better match these resources to load. Possible policies include those that provide the ability for storage activities to cover fixed costs, including an appropriate return, such as the ability to be eligible for capacity payments and the ability to participate in ancillary service markets. Storage providers could also be compensated for reducing the need for new transmission infrastructure.

Local, state, and federal governments should establish appropriate decommissioning requirements for wind and solar power plants.

As discussed in Chapter 5, wind and solar power plants, like all energy-generation facilities, will reach a time, normally decades after the initial installation, when they are no longer economically viable for use and must be decommissioned. Because decommissioning costs can be significant, the responsible government authorities should ensure that developers have made adequate arrangements to cover those costs at the time the facilities are installed. Although local, state, and federal requirements factor into decommissioning and each may have some oversight over the decommissioning process, the experience and regulations of state and local governments make them the likely places to enforce such requirements. Indeed, state and local bonding requirements are generally higher than those of the federal government. A comparison of federal, state, and county decommissioning regulations for oil and gas extraction sites and wind energy projects reveals that, generally, regulatory requirements are insufficient to adequately secure the costs of decommissioning. It is important that the growth of all energy sources, including wind and solar, be accompanied with the appropriate regulatory structure that guarantees the ability of sustainably retiring installations at the end of their useful life. States should set wind and solar plant decommissioning bonds to adequately cover end-of-life costs on a case-by-case basis. The states that have established decommissioning regulations for wind projects require bonding amounts on a case-by-case basis (see Chapter 5). Evaluating decommissioning bonding requirements on a case-by-case basis will most likely produce more accurate bond amounts. Consequently, state decommissioning requirements may potentially serve as a useful regulatory model for establishing adequate decommissioning requirements for wind and solar plants.

Regulatory agencies and insurers should provide incentives for the development of renewable resources in areas with lowest risks of hazards like hurricanes, and standards organizations such as the IEC should consider hurricane risks in wind turbine design standards.

In 2008, the U.S. Department of Energy envisioned development of 54 GW of shallow offshore wind capacity by 2030. Today in 2014, the United States currently has none – although several such facilities are in the planning stages and offshore wind development in the Atlantic, Pacific, Gulf of Mexico, and Great Lake Coasts is possible. In the Atlantic and Gulf of Mexico, hurricane wind speeds can exceed present wind turbine design limits, posing a challenge to those hoping to install wind turbines in this region. A RenewElec analysis found that the risk of hurricanes to wind plants is manageable off the coasts of the Mid-Atlantic and New England regions (see Chapter 16). There is higher risk in the Gulf of Mexico and the Southeast (see Figure 16.3). Identified changes to the design and operation of turbines could help reduce these risks. Efforts are underway to determine design standards for offshore wind turbines in hurricane-prone areas. To mitigate the risks of hurricanes, offshore wind turbines can be designed for higher maximum wind speeds, designed to track the wind direction (yaw) quickly enough to match wind changes in a hurricane even if grid power is cut off, or placed in areas with lower hurricane risk.

Regulatory agencies and insurers, who can provide incentives to wind developers through insurance rates, should provide incentives for the development of renewable resources in areas with lowest risks of hazards like hurricanes. If riskier areas are to be developed, appropriate engineering design codes should be established to build sufficiently strong turbines. Standards organizations should consider requiring wind turbines in higher-risk hurricane areas to have the capability to yaw into the wind even without grid power.

Legislative bodies and regulatory agencies should provide incentives to site wind and solar power plants where benefits from displacing pollution are highest.

When wind or solar energy displaces conventional generation like coal- and gas-powered electricity, the reduction in air emissions and thus the social benefits vary dramatically across the United States (see Chapter 4). While on an energy basis the southwest United States is more suitable for solar power, greater environmental benefits are realized in the Ohio Valley and the largest CO_2 reduction occurs in the Great Plains. Similarly, wind power provides the largest reductions in damages associated with air pollutants in the Ohio Valley, even though this region does not have the best wind resources.

Production-based subsidies like the PTC encourage developers to seek sites with high energy output, although electricity production may not be the goal of taxpayers and policy makers. Larger pollution reduction benefits can be obtained by tuning incentives to encourage wind and solar plant development where they will displace higher pollution generation.

PART II
Scientific findings

7

THE SOCIAL COSTS AND BENEFITS OF WIND ENERGY

Case study in the PJM Interconnection

7.1 Introduction

In the United States, a variety of government incentives have resulted in nationwide deployments of more than 60 GW of wind capacity since 2002 (Wiser & Bolinger, 2013). Wind introduces system-level costs into the management of the electricity grid that arise from the need to manage wind's inherent variability and unpredictability. Although forecasting tools are in use, no forecast is perfect (see Chapter 9), and the unforecast fluctuations of renewable energy can cause the grid to be operated suboptimally and require additional reserves (see Chapter 10 and Chapter 17). Wind benefits society by reducing emissions of pollutants from the electric power sector. Large deployments of wind create social costs and benefits that accrue to entities other than the investor. These social costs and benefits are not captured by traditional levelized cost of electricity (LCOE) or calculations of project revenue. This chapter considers whether the benefits from deploying wind exceed the costs of integrating this variable generation source into the power system.

Social costs and benefits (SCBs) are highly uncertain and affected by factors such as the quality of wind resources, the makeup of the underlying grid, and the capacity of wind installed. However, an evaluation of wind's SCBs is useful in determining if additional deployments of wind are beneficial or harmful to society. Table 7.1 divides wind's SCBs into six categories and provides definitions for each. These categories are consistent with existing terminology used in the academic literature and by industry.

The unpriced social costs of wind, sometimes called integration costs, are due to wind's variability and partial unpredictability, which create difficulties for system

TABLE 7.1	Definitions of wind's socialized cost and benefit categories.
Cost and benefit categories	**Definition**
Operational costs	The cost of ensuring stable grid operations, distributed across different markets (unit commitment, load following, regulation, and reserves)
Transmission costs	The cost of connecting electricity produced by distant and variable renewables to load
Curtailment costs	Intentionally reducing the power produced by wind turbines to ensure grid stability
Capacity costs	The cost of building backup generation (natural gas combustion turbines) to provide grid reliability similar to dispatchable generators
Greenhouse gas reduction benefits	The societal benefit of reducing CO_2 and other greenhouse gas pollutants by displacing fossil-fueled generation with wind
Criteria pollutant reduction benefits	The societal benefit of reducing criteria pollutant emissions (NO_x, SO_2, particulate matter) that harm human health and the environment by displacing fossil-fueled generation with wind

operators managing the grid. These costs occur at timescales that range from seconds to decades in the future and are generally socialized among all market participants. Several analyses estimate the social costs for different electrical systems (Acker, 2007; Electrotek Concepts, 2003; EnerNex, 2007a, 2007b, 2011; Lueken et al., 2012; PacifiCorp, 2012). These studies vary in terms of completeness and complexity (see Chapter 17).

The primary social benefit of wind is its low emissions of greenhouse gases (GHGs) and criteria pollutants (CPs) relative to fossil-fueled generators. CP reductions are valued by estimating how wind power reduces harm to human health and the environment (EPA, 2005; Fann, Fulcher, & Hubbell, 2009; Muller, Mendelsohn, Nordhaus, 2011). Researchers have found that the benefits of CP reductions due to wind range from $3/MWh to $82/MWh in the United States depending on the location and thus what type of generators are displaced by wind (Siler-Evans et al., 2013). The social cost of carbon (SCC) is difficult to quantify (see, for example, Pindyck, 2013), so we use the estimates developed by the U.S. government (U.S. IWGSCC, 2013).

Few studies have attempted to comprehensively measure wind's social costs and benefits. OECD (2012) analyzes the comprehensive costs of wind for several developed countries. However, the study is limited by a focus on country-level and not subnational, regional costs. The study also excludes the benefits of wind and relies heavily on a limited number of existing studies.

In this chapter, we estimate the social costs and benefits of wind in the PJM Interconnection. We analyze two scenarios: a low-wind scenario designed to model PJM as it was in 2010 with 1.5% of energy from wind and ample reserve margins

(20% in 2010; Monitoring Analytics, 2011) and a high-wind scenario with 20% of energy from wind. These two scenarios can be viewed as lower and upper bounds of wind's SCBs in PJM.

Because both the benefits and the costs are highly uncertain, we present the results as probability density functions (PDFs). Our estimates of each cost and benefit category were derived from both our modeling and existing literature. A comprehensive analysis of wind's SCBs should allow market operators and policymakers to establish rules and incentives that encourage the socially optimal level of wind to be built.

7.2 Methods

Costs and benefits vary regionally based on grid topology and wind resources. Estimates are highly dependent on assumptions of the makeup of the electricity grid, generator technologies, and fuel costs. Most importantly, estimates vary due to differences in methods among studies.

To gain a more complete picture of wind's SCBs, we treated them probabilistically with Monte Carlo simulation (Weber & Clavin, 2012). For each SCB category, we developed a triangular probability distribution. We then used Monte Carlo simulation to calculate the probability density function of the net SCB. We did this for both a low-wind scenario designed to mimic PJM as it was in 2010 with 1.5% of electric energy from wind and ample reserve margins and a high-wind scenario with 20% of energy from wind.

Studies modeling the social costs of wind typically involve both statistical models of wind generation and unit commitment and economic dispatch (UCED) models to simulate grid operations. We used the results of existing UCED analyses as well as our own reduced-form, open-source UCED of the PJM Interconnection's day-ahead market, using 2010 data. Our UCED, the PHORUM model, was used to estimate operational costs, criteria pollution reduction benefits, and greenhouse gas reduction benefits (Lueken & Apt, 2013).[1] PHORUM uses mixed-integer linear optimization to find the least-cost combination of generators to meet load at each hour of the year. The low-wind scenario was modeled with 1.5% of energy from wind, as PJM was in 2010. The high-wind scenario, with 20% of energy from wind, used data from the Eastern Wind Integration and Transmission Study (EWITS) to characterize likely locations for new wind plants in PJM states (EnerNex Corporation, 2011).

We separately analyzed the value of six categories of costs and benefits: operational costs, transmission costs, curtailment costs, capacity costs, GHG reduction benefits, and criteria pollutant reduction benefits. For each category, we estimated a lower bound, upper bound, and expected value for triangular distributions in the low-wind and high-wind scenarios. These analyses are described in detail in what follows. We then used Monte Carlo simulation to calculate probabilistic estimates of total social costs and total social benefits of wind in each scenario.

---

1. PHORUM is available at https://github.com/rlueken/PHORUM.

7.2.1 Operational costs

Operational costs are the costs of maintaining grid stability by continuously balancing total generation with total load, given the variability and unpredictability of renewable energy. Operational costs occur from the next 48 hours to real time (NERC, 2011). The net effect of these costs is increased prices in several formal markets run by the independent system operator (ISO), including the unit commitment/day-ahead market, load following/real-time market, regulation market, and reserve markets. Compensating for wind variability requires ramping other generators in the system, which in turn can cause generators to operate suboptimally and increase the frequency of generator cycling. The variability of wind leads to forecasting errors that increase reserve requirements and, when realized, may force system operators to use inefficient but fast-ramping generation instead of more cost-effective generators. Day-ahead wind forecast errors are typically 8% to 14% (root mean squared error; Mauch et al., 2012).

Calculating increases in operational costs requires both a statistical model of wind generation and a model of the electricity grid. Wind models use either measured or simulated wind speed data. Grid simulations vary in complexity from simple unit commitment models to more sophisticated models that capture forecast uncertainty and electrical dynamics of the grid (Chapter 17). To isolate the costs of wind variability and unpredictability, researchers can use the flat-block approach, in which a scenario with wind is compared not to a scenario without wind but to a scenario in which the wind generation is constant and perfectly known (Milligan & Kirby, 2009).

Figure 7.1 shows operational cost estimates of several published studies, as well as our internal modeling with PHORUM. The studies vary in the cost categories they include (Table 7.2). Lueken and colleagues (2012) used historical price data instead of simulation techniques to estimate operational costs; the high resulting cost suggests simulation methods may be biased to underpredict operational costs or that the observed California price data may be unrepresentative of areas used in simulations.

The low-wind scenario costs range from $0 to $4.3/MWh, with an expected value of $1.2/MWh, and high-wind scenario costs range from $1.9 to $9.7/MWh, with an expected value of $4.0. For both scenarios, bounds were derived from existing literature and most likely values from PHORUM simulations.

7.2.2 Transmission costs

The cost of connecting electricity produced by distant and variable renewables to load is an appreciable and unique cost for wind energy. Typically, transmission costs are omitted from estimates of wind LCOE because they are very site specific (Borenstein, 2011). Since transmission costs will either be socialized among ratepayers or paid by developers, we include them as a category of SCBs. Studies of wind transmission costs can be bottom up, in which costs are estimated for connecting

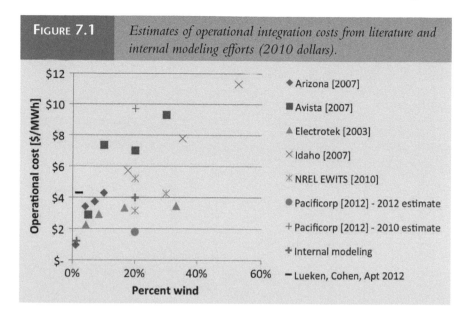

FIGURE 7.1 *Estimates of operational integration costs from literature and internal modeling efforts (2010 dollars).*

TABLE 7.2 *Cost categories included in each wind integration study.*

	Unit commitment	Load following	Forecast error	Reserves	Regulation
Arizona (2007)					
Avista (2007)					
Electrotek (2003)					
Idaho (2007)					
NREL EWITS (2009)					
PacifiCorp (2012)					
Internal modeling					
Lueken et al. (2012)					

Source: Arizona (2007), see Acker (2007); Avista (2007), see EnerNex Corporation (2007a); Electrotek (2003); Idaho (2007), see EnerNex Corporation. (2007b); NREL EWITS (2009), see EnerNex Corporation (2011); PacifiCorp (2012); Lueken et al. (2012).

individual wind plants, or top down, in which a significant expansion of the transmission grid is designed to integrate very high wind penetrations.

A bottom-up cost study by the Lawrence Berkeley National Laboratory (LBNL) study reviewed a sample of 40 transmission planning studies to assess the range of costs allocated to wind for transmission (Mills et al., 2009). LBNL found that the cost of transmission has varied for different wind plant projects, with a median

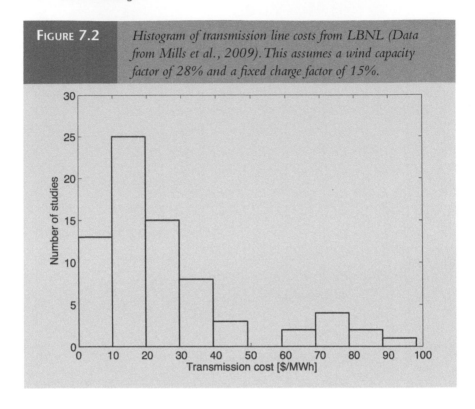

FIGURE 7.2   *Histogram of transmission line costs from LBNL (Data from Mills et al., 2009). This assumes a wind capacity factor of 28% and a fixed charge factor of 15%.*

cost of $300/kW of wind capacity. We converted these numbers to a levelized cost ($/MWh of wind) assuming a 28% capacity factor for PJM wind projects (U.S. DOE, 2011) and a fixed-charged factor of 15% as assumed by the LBNL authors. A histogram of the costs is shown below in Figure 7.2.

Costs varied from $0/MWh to $98/MWh with a median of $18/MWh. Cost estimates at the high end are due to projects with transmission oversized for future plant development. Ignoring these projects, we assume transmission costs range from $0/MWh to $48/MWh with a median of $16/MWh. The study also examined the case in which costs were not allocated to fossil plants on the same transmission line. This made a small difference, increasing the cost allocated to wind by $30/kW ($2/MWh). However, it should be noted that new transmission may have other cobenefits such as easing transmission congestion or increasing grid reliability by connecting dispatchable generators. New transmission can also increase system congestion (Blumsack, Lave, & Ilic, 2007). We do not include these cobenefits or costs in this analysis but note that they may be appreciable (EnerNex Corporation, 2011; Renewable Energy Transmission Initiative, 2010).

In order to realize 20% penetration of renewable energy, a significant top-down expansion of the transmission grid may be necessary (EnerNex Corporation, 2011). Table 7.3 shows capital cost estimates from studies of very large transmission expansions in order to incorporate high wind penetrations.

TABLE 7.3	Capital costs from various large transmission studies and estimated cost per MWh of wind using financial assumptions described in the text.	
Cost per kW of Wind [$/kW]	Cost Per MWh of Wind [$/MWh]	Study
150–300	9–18	AEP (Mills et al., 2009)
207	13	NREL EWITS (EnerNex Corporation, 2011)
316	19	NEMS (Mills et al., 2009)
67–367	4–22	Holttinen (Holttinen et al., 2011)
350–570	21–35	ERCOT (U.S. DOE, 2008)
—	9	Dobesova, Apt, & Lave, et al. (2005)

For the low-wind scenario, we assumed bounds of $0/MWh and $48/MWh with a most likely value of $16/MWh per the distribution in Figure 7.2. We excluded estimates above $45/MWh, as the authors noted these high costs were due to intentional overbuilding of transmission lines for these projects. For the high-wind scenario, we assumed bounds of $4 to $35/MWh of wind, with a most likely value of $15/MWh based on the findings of top-down transmission studies (Table 7.2).

7.2.3 Curtailment costs

Curtailment costs occur when wind plants intentionally reduce power output due to grid or market conditions. Curtailment costs are generally socialized via make-whole payments to wind generators for the energy they were forced to curtail, although market rules vary by region (Rogers, Fink, & Porter, 2010). To date, wind curtailment has been insignificant in PJM. However, regions with significant wind have experienced appreciable curtailment; annual wind curtailment in ERCOT was 17% in 2009 before declining to 3.7% in 2012 after new transmission lines and nodal pricing eased transmission congestion (Chesnez, 2012; Fink, Mudd, Porter, & Morgenstern, 2009).

To estimate curtailment rates in the low-wind scenario, we used data from the Midwest Independent System Operator (MISO), which has similar generator characteristics to PJM. Curtailment in MISO has varied from 2.0% to 4.2% from 2009 to 2012; these are the bounds we use for our distribution. We used as the median value the midpoint of 3.1% (Wiser & Bolinger, 2013). For the high-wind scenario, we used simulated curtailment estimates from the EWITS study, as historic curtailment data do not exist for wind penetrations of 20% in the United States. EWITS estimated curtailments would range from 3.6% to 10%, assuming a large expansion of the transmission grid necessary to support 20% wind (EnerNex Corporation, 2011); these are the bounds we used for the high-wind scenario. Our median value is the midpoint, 6.8%.

We quantified the societal costs of curtailment as the make-whole revenues paid to wind producers for energy they curtail. As a first order approximation, we assumed these revenues equal the levelized power purchase agreement (PPA) price ($/MWh) multiplied by the percentage of power curtailed. Average levelized PPA price in 2012 was approximately $50/MWh in the PJM region (Wiser & Bolinger, 2013). With our assumed percent curtailment, this translated to curtailment costs in the low-wind scenario of $1/MWh to $2.1/MWh (median value $1.6/MWh) and curtailment costs for the high-wind scenario of $1.8/MWh to $5.0/MWh (median value $3.4/MWh).

7.2.4 Capacity costs

The North American Electric Reliability Corporation (NERC) requires system operators to ensure that sufficient capacity exists in their system to reliably meet system load. In 2007, PJM created the Reliability Pricing Model to ensure that adequate capacity is available to meet peak demand. Reliability can be met through a variety of technologies such as generation capacity, demand-side management, energy efficiency, and imports. PJM has a need for approximately 170 GW of capacity; approximately 15 GW will be met in 2015 through demand-side management (PJM, 2013b).

A generator's net expected contribution to reliability in terms of capacity is defined as the equivalent load-carrying capability (ELCC; Milligan & Porter, 2008). An ELCC of 90% means that a generator can be expected to provide 90% of its nameplate capacity during peak hours. The ELCC provided by onshore wind in the United States is much lower than that of conventional generators. At low penetrations of wind, the ELCC of wind could be roughly approximated by the capacity factor (28% in PJM; Milligan & Kirby, 2009). In PJM, wind receives a capacity credit of only 13%, because wind output does not coincide well with peak demand periods (PJM, 2013b). Typically, as penetration of onshore wind or solar increases, the value of ELCC diminishes (Milligan & Kirby, 2009; Mills et al., 2009).

In order for wind plants to provide the same capacity as dispatchable power plants, some form of capacity must be added to offer the same ELCC as dispatchable plants. For the low-wind scenario, we assumed that the cost of capacity is the marginal cost of capacity from PJM's Reliability Pricing Model. Capacity prices have varied from $16 to $174/MW-day with a median value of $106/MW-day since the RPM began in 2007 (PJM, 2013b). This provides our range and expected value for capacity costs in the low-penetration scenario.

To find the range for the high-penetration scenario, we broadened our range. For the lower bound, we assumed that PJM has excess capacity resulting in no capacity costs. For the upper bound, we relied on PJM estimates of the cost of new entry (CONE) of new simple-cycle power plants. The CONE is used to construct the demand curve for capacity for the RPM. PJM estimates the CONE of new simple-cycle power plants to be approximately $382/MW-day (PJM, 2013a).

For the expected capacity cost in a high-penetration scenario, we assumed that the marginal capacity cost is $200/MW-day. This figure is based on two estimates.

TABLE 7.4	Social cost and benefit parameters used in Monte Carlo simulation.					
Cost and benefit categories	Low-wind scenario ($/MWh)			High-wind scenario ($/MWh)		
	Lower bound	Median	Upper bound	Lower bound	Median	Upper bound
Operational costs	$0	$1	$4	$2	$4	$10
Transmission costs	$0	$16	$48	$4	$15	$35
Curtailment costs	$1	$2	$2	$2	$3	$5
Capacity costs	$2	$14	$23	$0	$26	$49
Greenhouse gas reduction benefits	$10	$35	$110	$10	$36	$110
Criteria pollutant reduction benefits★	$64	$64	$64	$77	$77	$77

★ Point estimates used for low-wind and high-wind scenarios.

The first estimate is the amount of "missing-money" needed for adequate capacity in ERCOT's energy-only market ($190/MW-day) as calculated by Spees and colleagues (2013). The second estimate is the avoidable cost rate of subcritical coal-fired power plants in PJM ($210/MW-day; PJM, 2013c). We assumed that if low-cost gas and high penetrations of wind undercut energy profit revenues of coal-fired power plants in PJM, capacity costs would be set by the avoidable cost rate of coal-fired power plants. For both the low and high penetrations, we converted capacity costs from $/MW-day to $/MWh wind assuming a 28% wind capacity factor and an ELCC of 13% for wind in PJM (see Table 7.4).

7.2.5 Pollution reduction benefits

A primary benefit of wind energy is pollution reduction. Because wind has very low short-run marginal costs, system operators dispatch it before expensive generators. If wind displaces fossil-fueled generators, it reduces net grid emissions. Emission reductions are typically stated as pounds of emissions avoided per MWh of electricity produced by wind. We monetized the benefit of pollution reductions with the estimated social cost of each pollutant.

We modeled pollution reduction benefits in the low-wind and high-wind scenarios as triangular distributions. We used PHORUM to simulate how adding wind to PJM in 2010 would have changed each plant's annual power generation and emissions. CO_2 emission reductions are valued with a social cost of carbon (SCC) of $13 to $136/ton, with a mode of $45/ton (2010 dollars) as valued by the U.S. government (U.S. IWGSCC, 2013). We valued criteria pollutant reductions (NO_x, SO_2, 2.5 micrometer particulate matter [$PM_{2.5}$]) with the APEEP model, a reduced-form, integrated assessment model that links emissions of criteria pollutants to human health and environmental damages for all U.S. counties (Muller, 2011). We assigned location-specific damage rates to each plant. Because APEEP does not provide uncertainty estimates for the correlated damages between plants in different locations, we used a point estimate of damages for each plant rather than a triangular distribution. In the high-wind scenario, it can be argued that APEEP's baseline emissions are affected enough so that the human health effects are no longer accurate. In the case of SO_2, there is clear evidence that the effects are linear with reduced SO_2 emissions, at least to the levels that are foreseeable if wind reaches 20 to 30% of electric production and displaces fossil generation (Koo et al., 2009). Thus, for our high-wind case at 20% wind, the APEEP model predictions are justified.

Since 2010, the year simulated, emissions of CO_2 and criteria pollutants have dropped significantly in PJM due to lower natural gas prices, the Clean Air Interstate Rule (CAIR; EPA, 2005), and the Mercury and Air Toxics Standard (MATS; EPA, 2011a). In 2012, emissions of SO_2 were 42% lower than 2010 levels in PJM states, and NO_x and CO_2 emissions have both dropped 15% (EPA, 2013). To compensate for these reductions, we reduced the simulated 2010 emissions and associated damages from each plant by 42% for SO_2, 15% for NO_x, and 15% for CO_2. This adjustment ignores any changes to the dispatch order that may have occurred since 2010. We have applied this adjustment in the results that follow.

7.3 Results and discussion

Table 7.4 summarizes the parameters used in our Monte Carlo analysis of social costs and benefits in PJM. Social costs are significant when compared to private costs – the average PPA price in 2012 was approximately $50/MWh in the PJM region (Wiser & Bolinger, 2013). However, social costs are much smaller than both GHG emission reduction benefits and criteria pollutant emission reduction benefits (Figure 7.3). Emission reduction benefits are higher in PJM than in other ISOs due to the combination of PJM's reliance on high-emitting fossil-fueled generators and high population, resulting in increased pollution exposure compared to other ISOs.

Monte Carlo simulation results are shown in Figure 7.3. Total social benefits are highly uncertain but with very high probability exceed total social costs for both the low-wind and high-wind scenarios. Total costs in the high-wind scenario likely exceed those in the low-wind scenario and are more uncertain. Total benefits are higher in the high-wind scenario than in the low-wind scenario, as wind offsets a greater proportion of coal generation.

We next calculated the net social benefit of wind power, or total social benefit minus total social cost (Figure 7.4). The net benefit is positive for both scenarios; the median expected net benefit is $74/MWh for both the low-wind scenario and high-wind scenario.

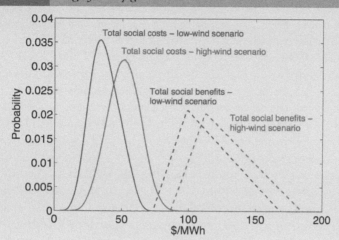

FIGURE 7.3 *Distribution of total social costs and benefits. Social costs and benefits are larger in the high-wind scenario than in the low-wind scenario. Total social benefits are highly uncertain but have a very high probability of being significantly greater than costs.*

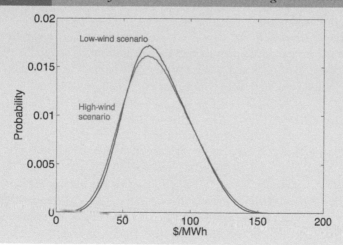

FIGURE 7.4 *Net social benefit (total social benefit minus total social cost). Expected net benefits are $74/MWh in each scenario; net benefits are more uncertain in the high-wind scenario.*

7.3.1 Net social benefits under a future, cleaner grid

Over the next decade, several rules by the U.S. Environmental Protection Agency are expected to force many of PJM's coal generators to either retire or retrofit with improved emission control technologies. Rules include the Clean Air Interstate Rule (CAIR), which limits emissions of NO_x and SO_2 (EPA, 2005); the Mercury and Air Toxics Standard (MATS), which limits emissions of mercury and primary particulate matter (EPA, 2011a); and the US Environmental Protection Agency's forthcoming rules regulating CO_2 on existing power plants. EPA has proposed the Cross-State Air Pollution Rule (CSAPR) to replace CAIR (EPA, 2011b). Although CSAPR was voided by the D.C. Circuit Court of Appeals (EME Homer City Generation, L.P. v EPA, 2012) and (as of this writing) the case is pending before the U.S. Supreme Court, CAIR remains in effect. PJM anticipates as much as 20 GW of coal capacity is placed at risk of retirement by CAIR/CSAPR and MATS, or 25% of total coal capacity. An additional 29 GW of capacity may need at least two retrofits to comply with the rules (PJM, 2011).

Two future scenarios are possible under the EPA regulations. The first scenario is that the emission caps established by CAIR bind. In this case, total emissions of NO_x and SO_2 will be fixed at the emissions cap and new additions of wind will not result in a net reduction in emissions. Rather, the displaced NO_x and SO_2 emissions will be valued at the market emission permit price, anticipated by the EPA to be $1,300/ton for SO_2 and $2,100/ton for NO_x in 2015 (2010 dollars; EPA, 2005). These anticipated permit prices are much lower than the health damages caused by emissions from PJM plants. According to the APEEP model, expected damages are as high as $71,000/ton for SO_2 and $13,000/ton for NO_x for PJM plants, depending on plant location. If CAIR emission caps bind, the effect of additional wind would be downward pressure on permit prices and minimal reductions in criteria pollutant emissions. This suggests that for the socially optimal amount of wind to be deployed under a cap system, the permit price would need to be closer to the estimated health damages.

The second scenario is that emission caps do not bind due to significant wind deployment, low natural gas prices, or tightened regulations under MATS (Paul et al., 2013). In this scenario, new additions of wind would reduce criteria pollutant emissions and should be valued by the human health benefits they induce.

7.4 Summary

The addition of wind to electric power systems creates social costs and benefits that are not priced in today's markets. These social costs and benefits (SCBs) are highly uncertain and vary between markets. In PJM, our median estimate of total social costs is $36/MWh in a low-wind scenario and $51/MWh in a high-wind scenario with 20% of energy from wind. The social benefits wind creates by reducing GHG and criteria pollutant emissions are expected to exceed total social costs. The median expected net societal benefit of wind in PJM is $74/MWh for both the low-wind and high-wind scenarios. If CAIR results in binding emission caps, additional wind will not reduce criteria pollutant emissions. If caps do not bind, additional wind will reduce criteria pollutant emissions and human health damages. Policy makers and market operators should establish rules that correctly price the social costs and benefits of wind and therefore encourage that the socially optimal amount of wind be deployed.

References

Acker, T. (2007). *Final report: Arizona public service wind integration cost impact study.* Flagstaff, AZ: Northern Arizona University.

Avista 2007: EnerNex Corporation. (2007). *Final report: Avista Corporation wind integration study.* www.uwig.org/avistawindintegrationstudy.pdf

Blumsack, S., Lave, L.B., & Ilic, M. (2007). A quantitative analysis of the relationship between congestion and reliability in electric power networks. *Energy Journal, 28*(4), 73–100.

Borenstein, S. (2011). *The private and public economics of renewable electricity generation.* WP 221R. http://ei.haas.berkeley.edu/pdf/working_papers/WP221.pdf

Chesnez, J. (2012). *Impact of curtailment on wind economics.* www.renewableenergyworld.com/rea/news/article/2012/03/impact-of-curtailment-on-wind-economics

Dobesova, K., Apt, J., & Lave, L. (2005). Are renewables portfolio standards cost effective emission abatement policy? *Environmental Science & Technology, 39*(22), 8578–8583.

Electrotek Concepts. (2003). *WE energies energy: System operations impacts of wind generation integration study.* www.uwig.org/weenergieswindimpacts_finalreport.pdf

EME Homer City Generation, L.P. v. EPA. (2012). 696 F.3d 7, (D.C. Cir. 2012). www.cadc.uscourts.gov/internet/opinions.nsf/19346B280C78405C85257A61004DC0E5/$file/11-1302-1390314.pdf

EnerNex Corporation. (2007a). *Final report: Avista Corporation wind integration study.* www.uwig.org/avistawindintegrationstudy.pdf

EnerNex Corporation. (2007b). *Operational impacts of integrating wind generation into Idaho Power's existing resource portfolio.* www.idahopower.com/pdfs/AboutUs/PlanningForFuture/wind/Petition_ReviseAvoidedCostRates1.pdf?id=238&.pdf

EnerNex Corporation. (2011). *Eastern wind integration and transmission study.* National Renewable Energy Laboratory report, NREL/SR-5500-47078. www.nrel.gov/docs/fy11osti/47078.pdf

EPA. (2005). *Regulatory impact analysis for the final Clean Air Interstate Rule.* www.epa.gov/cair/pdfs/finaltech08.pdf

EPA. (2011a). *Regulatory impact analysis for the final Mercury and Air Toxics Standard.* www.epa.gov/mats/pdfs/20111221MATSfinalRIA.pdf

EPA. (2011b). *Regulatory impact analysis for the federal implementation plans to reduce interstate transport of fine particulate matter and ozone in 27 states.* www.epa.gov/airtransport/pdfs/FinalRIA.pdf

EPA. (2013). *Air markets program data.* http://ampd.epa.gov/ampd/

Fann, N., Fulcher, C., & Hubbell, B. (2009). The influence of location, source, and emission type in estimates of the human health benefits of reducing a ton of air pollution. *Air Quality, Atmosphere and Health, 2*(3), 169–176.

Fink, S., Mudd, C., Porter, K., & Morgenstern, B. (2009). *Wind energy curtailment case studies.* NREL/SR-550-46716, www.nrel.gov/docs/fy10osti/46716.pdf

Holttinen, H., Meibom, P., Orths, A., Lange, B., O'Malley, M., Tande, J.O., et al. (2011). Impacts of large amounts of wind power on design and operation of power systems, results of IEA collaboration. *Wind Energy, 14*(2), 179–192.

Koo, B., Wilson, G., Morris, R., Dunker, A., & Yarwood, G. (2009). Comparison of source apportionment and sensitivity analysis in a particulate matter air quality model. *Environmental Science and Technology, 43*(17), 6669–6675.

Lueken, C., Cohen, G., & Apt, J. (2012). Costs of solar and wind power variability for reducing CO2 emissions. *Environmental Science and Technology, 46*(17), 9761–9767.

Lueken, R., & Apt, J. (2013). *The effects of bulk electricity storage on the PJM market.* Working paper. http://wpweb2.tepper.cmu.edu/electricity/papers/ceic-13-05.asp

Mauch, B., Carvalho, P.M.S., & Apt, J. (2012). Can a wind farm with CAES survive in the day-ahead market? *Energy Policy, 48,* 584–593.

Milligan, M., & Kirby, B. (2009). *Calculating wind integration costs: separating wind energy value from integration cost impacts.* NREL/TP-550-46275. www.nrel.gov/docs/fy09osti/46275.pdf

Milligan, M., & Porter, K. (2008). *Determining the capacity value of wind: An updated survey of methods and implementation.* NREL/CP-500-43433. www.nrel.gov/docs/fy08osti/43433.pdf

Mills, A., Wiser, R., & Porter, K. (2009). *The cost of transmission for wind energy: A review of transmission planning studies.* Berkeley, CA: Lawrence Berkeley National Laboratory.

Monitoring Analytics. (2011). *State of the market report for PJM – 2010.* www.monitoringanalytics.com/reports/PJM_State_of_the_Market/2010.shtml

Muller, N. (2011). Linking policy to statistical uncertainty in air pollution damages. *BE Journal of Economic Analysis & Policy, 11*(1), 1–29.

Muller, N., Mendelsohn, R., & Nordhaus, W. (2011). Environmental accounting for pollution in the United States economy. *American Economic Review, 101*(5), 1649–1675.

NERC. (2011). *Task 2.4 report: Operating practices, procedures, and tools.* www.nerc.com/files/ivgtf2-4.pdf

OECD. (2012). *Nuclear energy and renewables: System effects in low-carbon electricity systems.* www.oecd-nea.org/ndd/reports/2012/system-effects-exec-sum.pdf

PacifiCorp. (2012). *PacifiCorp 2012 wind integration resource study – DRAFT for IRP public participants review.* www.pacificorp.com/content/dam/pacificorp/doc/Energy_Sources/Integrated_Resource_Plan/Wind_Integration/2012WIS/2013IRP_2012WindIntegration-DRAFTReport-11-15-12.pdf

Paul, A., Blair, B., & Palmer, K. (2013). *Taxing electricity sector carbon emissions at social cost.* Resources for the Future. www.rff.org/RFF/Documents/RFF-DP-13-23-REV.pdf

Pindyck, R. S. (2013). Climate change policy: What do the models tell us? *Journal of Economic Literature, 51*(3), 860–872.

PJM. (2011). *Coal capacity at risk of retirement in PJM: Potential impacts of the finalized EPA Cross State Air Pollution Rule and proposed national emissions standards for hazardous air pollutants.* Valley Forge, PA: PJM Interconnection. http://pjm.com/~/media/documents/reports/20110826-coal-capacity-at-risk-for-retirement.ashx

PJM. (2013a). *2016/2017 RPM base residual auction planning period parameters.* Valley Forge, PA: PJM Interconnection.

PJM. (2013b). *2016/2017 RPM base residual auction results.* Valley Forge, PA: PJM Interconnection.

PJM. (2013c). *PJM RPM default avoidable cost rates for the 2015/2016 delivery year.* Valley Forge, PA: PJM Interconnection.

Renewable Energy Transmission Initiative. (2010). *Renewable energy transmission initiative, phase 2b final report.* www.energy.ca.gov/2010publications/RETI-1000-2010-002/RETI-1000-2010-002-F.PDF

Rogers, J., Fink, S., & Porter, K. (2010). *Examples of wind energy curtailment practices*. NREL/SR-550-48737. www.nrel.gov/docs/fy10osti/48737.pdf

Siler-Evans, K., Azevedo, I., Morgan, M.G., & Apt, J. (2013). Regional variations in the health, environmental, and climate benefits of wind and solar generation. *Proceedings of the National Academy of Sciences, 110*(29), 11768–11773.

Spees, K., Newell S., & Pfeifenberger, J. (2013). Capacity markets – lessons learned from the first decade. *Economics of Energy & Environmental Policy, 2*(2), 1–26.

U.S. DOE. (2008). *20% wind energy by 2030: Increasing wind energy's contribution to U.S. electricity supply*. www.nrel.gov/docs/fy08osti/41869.pdf

U.S. DOE. (2011). *2011 wind technologies market report*. www1.eere.energy.gov/wind/pdfs/2011_wind_technologies_market_report.pdf

U.S. Executive Office of the President. (2013). *The president's climate action plan*. www.white house.gov/sites/default/files/image/president27sclimateactionplan.pdf

U.S. IWGSCC. (2013). *Technical support document: Technical update of the social cost of carbon for regulatory impact analysis under Executive Order 12866*. www.whitehouse.gov/sites/default/files/omb/inforeg/social_cost_of_carbon_for_ria_2013_update.pdf

Weber, C., & Clavin, C. (2012). Life cycle carbon footprint of shale gas: Review of evidence and implications. *Environmental Science and Technology, 46*(11), 5688–5695.

Wiser, R., & Bolinger, M. (2013). *2012 Wind technologies market report*. Washington, DC: Lawrence Berkeley National Laboratory.

8

CHARACTERISTICS OF WIND, SOLAR PHOTOVOLTAIC, AND SOLAR THERMAL POWER

8.1 Introduction

Wind, solar photovoltaic, and solar thermal electric power generation exhibit variability at all time scales. When examined closely, the variability has characteristics that arise from the nature of the physical processes that produce turbulence and clouds. Here we expand on the discussion in Part I to show the details of the analysis.

One commonly used method is to construct a histogram of the changes in power produced by the renewable source. For example, constructing a histogram of the 15-minute step changes in wind or solar power is a common technique (Wan, 2004; Wan & Bucaneg, 2002). As an illustration, consider the summed output of seven wind turbines in Texas taken at 10-second time resolution for a bit more than 2 weeks, during which the capacity factor[1] was 42% (Figure 8.1).

A histogram of the change in the summed power output of the seven turbines from one 10-second period to the next is shown in Figure 8.2, where the step changes are shown as a fraction of the maximum power observed during the 2 weeks. Most of the step changes are very small, but even in 10 seconds, the output of the wind plant occasionally exhibits large changes.

In order to examine the character of the largest step changes (these are infrequent but of most interest to power system operators), the same histogram can be plotted with the vertical axis (the frequency of occurrence of the step change) as a logarithmic scale (Figure 8.3).

---

1. Capacity factor is the actual energy (in units, for example, of kWh) produced by a generator in a particular time period (for example, a year) divided by the product of the nameplate capacity of that generator (in kW) times the number of hours in the period.

FIGURE 8.1

The output of seven 1.5 MW turbines from one Texas wind plant taken at 10-second time resolution over a period of 15.4 days.

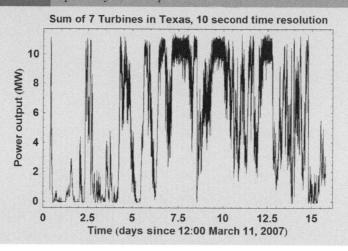

FIGURE 8.2

Sample-to-sample wind power step changes for the data in Figure 8.1. The step changes are shown as a fraction of the maximum power observed during the 15.4 days, 11.434 MW.

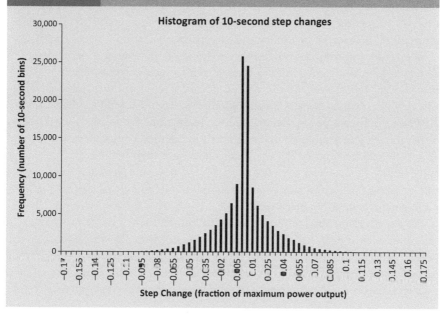

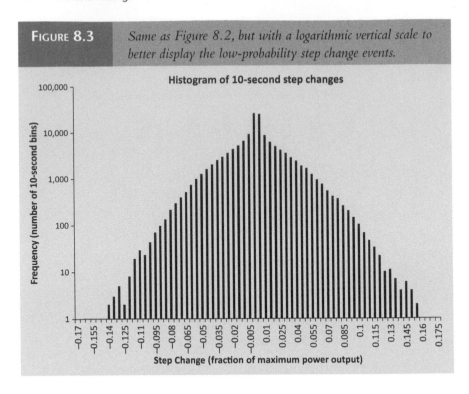

| FIGURE 8.3 | *Same as Figure 8.2, but with a logarithmic vertical scale to better display the low-probability step change events.* |

This allows us to observe, for example, that a step change of more than 5% of the maximum power observed during the 2 weeks of data occurred in 1 of about every 11 10-second periods. Similarly, a step change of more than 10% of the maximum power is observed in 1 of every 300 periods.

We could construct similar histograms for each time period of interest (say, 5-minute, 15-minute, or 1-hour step changes). However, it becomes tedious to examine every period of interest in this way. Fortunately, there is another technique available that also provides physical insight into the processes that produce the variability.

This technique is power spectral analysis. To examine the relative contributions of bass drums and piccolos to the sound of a symphony orchestra, power spectral analysis would decompose the music into low- and high-frequency components and compare them. A plot with frequency on the horizontal axis and amplitude of the music on the vertical axis would show quite different characteristics for symphonic music and heavy-metal rock music.

As described in Apt (2007), we examine the frequency domain behavior of the time series of power output data from the generation plants by estimating the power spectrum (sometimes termed the power spectral density or PSD). To estimate the power spectrum of the real power output of a wind plant, the discrete Fourier transform of the time series of output measurements is computed.

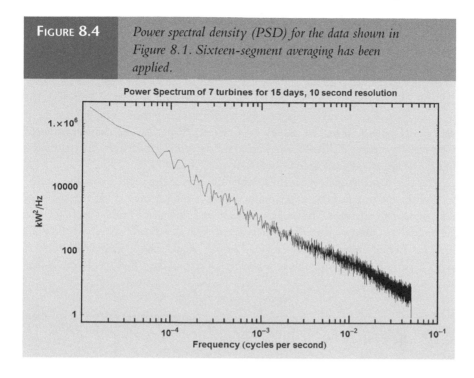

FIGURE 8.4   *Power spectral density (PSD) for the data shown in Figure 8.1. Sixteen-segment averaging has been applied.*

One of the attributes of power spectrum estimation is that increasing the number of time samples (N) does not decrease the noise at any given frequency f_k. In order to take advantage of a large number of data points in a data set to reduce the variance at f_k, the data set may be partitioned into several time segments. The Fourier transform of each segment is then taken. The Fourier transforms are then averaged at each frequency, reducing the variance of the final estimate by the number of segments (and reducing the noise by the reciprocal of the square root of the number of segments). This has no effect on f_{max} but increases the lowest nonzero frequency (f_1) by a factor equal to the number of segments.

The data of Figure 8.1 can be examined in this way, and the power spectral density is shown in Figure 8.4.

The data set in Figure 8.1 spans 1.3×10^6 seconds. That long data set has been separated into 16 segments (each of 8.3×10^4 seconds) that have been averaged as described to increase the signal-to-noise ratio. Thus, the lowest frequency in Figure 8.4 is 1.2×10^{-5} Hz (the inverse of the length of each segment). The data were sampled at 10-second intervals. Roughly speaking, two samples are required to define the amplitudes of the sine or cosine used to decompose the signal into frequency components (the Nyquist sampling theorem). Thus, the highest frequency corresponds to 20 seconds, or 0.05 Hz.

8.2 Information from the power spectrum

In order to examine the power spectrum over the widest range of frequencies, it would be desirable to have high-time-resolution wind turbine power output data (say, sampled at 1-second intervals) for many months. While these data are generally difficult to obtain, we have data for a wind plant sampled at 1-second intervals for short periods of time and at 1-hour intervals for longer periods. Combining these data sets (Figure 8.5), we can see several features of interest. The characteristics of the power spectra in Figure 8.4 and Figure 8.5 are similar for wind plants throughout the world and do not depend on a particular location.

The first observation we can make from the power spectrum is that the amplitude of fluctuations from wind generators is very much smaller at high frequencies than at low frequencies. This characteristic has important consequences for the generators that must be used to compensate for wind's variability. Suppose the PSD were flat (white noise). In this case, wind power would have equally large fluctuations at periods of, say, a minute as at periods of several days. This would mean that

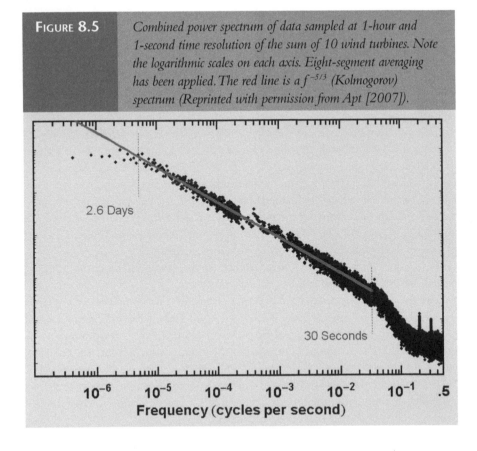

| FIGURE 8.5 | *Combined power spectrum of data sampled at 1-hour and 1-second time resolution of the sum of 10 wind turbines. Note the logarithmic scales on each axis. Eight-segment averaging has been applied. The red line is a $f^{-5/3}$ (Kolmogorov) spectrum (Reprinted with permission from Apt [2007]).* |

2.6 Days

30 Seconds

10^{-6} 10^{-5} 10^{-4} 10^{-3} 10^{-2} 10^{-1} .5

Frequency (cycles per second)

large numbers of very fast generators or energy storage devices would be required. Nature has kindly arranged things so that the largest variability is at low frequency (the power variations at time scales of a minute are a thousand times smaller than those at time scales of a day).[2] Thus, slow-ramping plants (coal or combined-cycle natural gas, for example) can compensate for most of wind's variability.

Second, there is a remarkably linear region of the power spectrum (exponential in frequency), covering four orders of magnitude in frequency, where the amplitude of the variations in output power of the wind plant decreases with increasing frequency f, as $f^{-5/3}$ (red line in Figure 8.5). This characteristic is observed in data taken from wind plants in many regions of the world. Its origin is that the atmosphere dissipates energy in turbulent eddies; these cascade in scale from large eddies to tiny ones where viscous dissipation finally turns the energy into heat. Kolmogorov (1941) recognized that dimensional analysis of this sort of energy cascade leads to a $f^{-5/3}$ dependence. Kolmogorov's frequency dependence of turbulence is a characteristic of many fluids; the first experimental verification of Kolmogorov's spectrum was in an ocean tidal channel near Vancouver (Grant, Stewart, & Moilliet, 1961).

At frequencies above roughly 0.05 Hz, the inertia of the wind turbines (both physical and electrical) act as low-pass filters, reducing the high-frequency variability below that predicted by the Kolmogorov spectrum. Frequencies above about 0.1 Hz show the minimum resolution (termed the noise floor) of the output power sensors. The two peaks at about 0.2 and 0.3 Hz are likely due to the turbine blades passing the support tower.

At frequencies below ~5×10^{-6} Hz (corresponding to ~2.5 days), the power spectrum of output from wind plants becomes quite flat. The basic physics is that the turbulence that dominates the higher frequencies is replaced at low frequencies with mesoscale phenomena such as the approximately 5-day planetary wave. The low- and mid-frequency behaviors are well fit by a model of the form shown in Figure 8.6.

The strong peak at a frequency corresponding to 24 hours that appears in the power spectrum of most large wind plants is due to the repeatable pattern of when in the day the wind blows most strongly. Figure 8.7 shows the average pattern for the sum of all wind plants in the California ISO during 2010.

It is time to put to rest a persistent piece of misinformation that continues to be reprinted. A 1957 wind speed PSD covering the frequency region from 1.9×10^{-7} to 0.25 Hz appears to show a region with diminished wind variability between roughly 3×10^{-5} and 7×10^{-3} Hz, termed by several authors the "spectral gap" (Van der Hoven, 1957). Despite the frequent repetition of this particular PSD in modern handbooks and review articles (Burton, Sharpe, Jenkins, & Bossanyi, 2001; Fordham, 1985), there is no evidence that such a gap exists.

---

2. The amplitude fluctuations are the square root of the power fluctuations, or about 30 times smaller at high frequencies than at low frequencies.

PSD of 1 year of data from a 65 MW wind plant in Texas taken at 15-minute resolution on a log-log scale. The highest frequency corresponds to time scales of 30 minutes. Four-segment averaging has been applied, so the lowest frequency plotted corresponds to 3 months. The red line is a best 2-parameter fit to the function $A/(1 + Bf^{5/3})$. The spectral peak at a frequency corresponding to 24 hours is due to the diurnal repeatability of wind power (in most locations in the United States, the wind blows more strongly at night than during the day; at some European sites that is not the case).

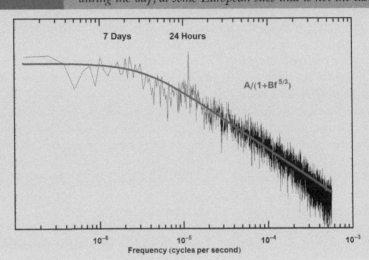

Hourly average wind data for all wind plants in the California ISO during 2010; data were sampled at 10-minute intervals.

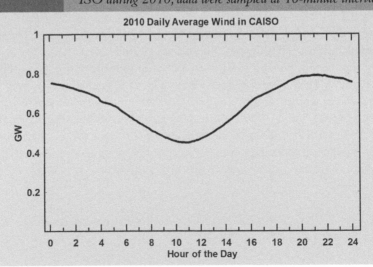

8.3 Power spectra of wind compared to that of load

System operators have sometimes treated wind as negative load, using the same techniques honed over many years of dealing with load variations to compensate for variability in net load, the load less the wind generation.

Apt (2007) noted that 14-second time resolution load data for 6 months obtained from a control area in the Mid-Atlantic region has a power spectrum that, in the region corresponding to times between about 1 hour and 2 minutes, is fit well by a Kolmogorov spectrum. He wrote,

> In this interval, the practice of treating wind power fluctuations as negative load appears to be justified. Note that this is not the same thing as saying that load fluctuations cancel wind fluctuations, as is sometimes stated, since the two would have to be both of the same magnitude and anti-correlated for the assertion to be valid.
>
> (Apt, 2007, p. 373)

A similar analysis using 10-minute time resolution data for the California ISO for data spanning the entire year of 2010 shows that treating wind as negative load over a large region is justified (Figure 8.8).

FIGURE 8.8	*The power spectrum of wind in CAISO (red) and load in CAISO (purple) for the year 2010 at 10-minute resolution, shown on a log-log scale. The strong peaks observed in the load spectrum are largely due to the 24-hour repeatability; the harmonics of this frequency are prominent in the PSD.*

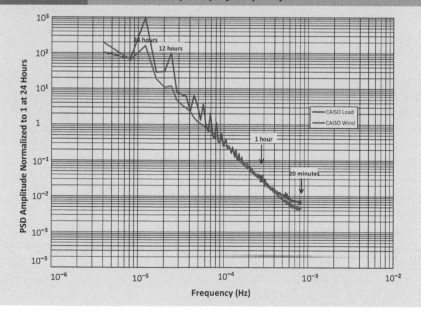

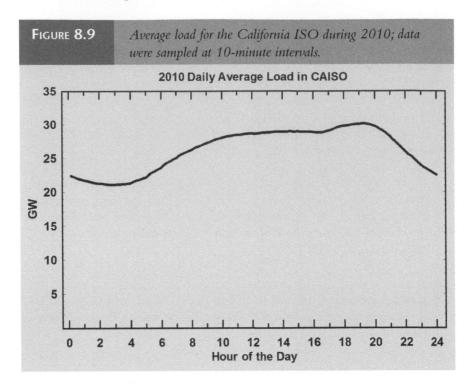

FIGURE 8.9   *Average load for the California ISO during 2010; data were sampled at 10-minute intervals.*

Load exhibits a well-known daily pattern that causes the strong peaks observed in its power spectrum. Figure 8.9 shows the average pattern for CAISO load for 2010, using the same 10-minute data whose PSD is shown in Figure 8.8.

8.4 Power spectra of solar photovoltaic and solar thermal generation compared to wind

Displaying the PSD of several renewable generation sources together and normalizing the spectra at a frequency corresponding to a range near 24 hours reveals a difference in the variability of each source at high frequencies (Figure 8.10). In contrast to the $f^{-5/3}$ spectrum exhibited by wind plants, the power spectrum of solar PV power output has a significantly flatter spectrum, approximated by $f^{-1.3}$. The power spectral analysis shows that the power fluctuations of solar photovoltaic electricity generation have approximately 100 times larger amplitude of variations at frequencies near 10^{-3} Hz than solar thermal electricity generation (this frequency corresponds to a time scale of ~15 minutes).[3]

It is possible that the flatter solar PV spectrum compared to wind's PSD is due to one or more of the following factors. First, sunlight is focused by common puffy

---

3. The amplitude fluctuations are the square root of the power fluctuations, or about a factor of 10 larger for PV than for solar thermal.

FIGURE **8.10** *Power spectra of solar PV, wind, and solar thermal generation plants, shown on a log-log scale. The spectra have been normalized to one at a frequency corresponding to approximately 24 hours. There is very little difference between the 5 MW Springerville, Arizona, PV spectrum and that of a much larger PV array. Reprinted with permission from Lueken and colleagues (2012) © American Chemical Society.*

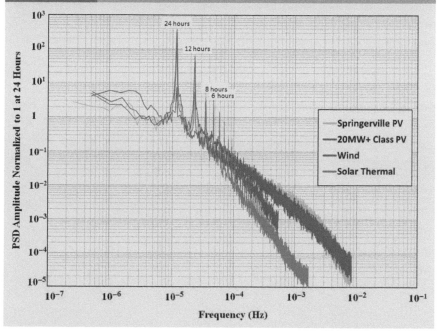

cumulus clouds so that there is a bright band around the shadow of the cloud. This causes the spikes above the nameplate capacity observed in PV data. It also introduces high-frequency variability. Second, the edges of such clouds are fractal, so their shadows exhibit high-frequency variability. Third, thin clouds are brighter than the sky due to enhanced diffuse sunlight, also increasing variability.

Both types of solar generation exhibit strong peaks corresponding to a 24-hour period and its higher harmonics, as expected from the cessation of generation each night. Wind power exhibits this property to a lesser extent (in the continental U.S., wind tends to have a diurnal variation, blowing more strongly at night).

Electricity from wind plants is intermediate between solar PV and solar thermal in terms of variability in this frequency range. The power spectra are similar for the three generation types at frequencies lower than approximately 4×10^{-5} Hz (corresponding to periods greater than 6 hours).

High variability at frequencies corresponding to less than 1 hour creates the need for more ancillary energy services to avoid quality problems or interruptions in electricity service to customers. Lueken and colleagues (2012) have analyzed the costs of variability for wind, solar photovoltaic, and solar thermal power by using ancillary service and energy costs observed in the California ISO. Their analysis used the load-following ancillary services of up- and down-regulation to compensate for variability. The authors note that the ISO would also use frequency response ancillary services to mitigate very short-term (1- to 10-second) variability, but their data set cannot include those costs because the highest-time-resolution data were sampled at 1 minute.

As reported by Lueken and colleagues (2012), the average price of power in the southern CAISO region in 2010 was $42/MWh. Variability cost as a percentage of the price of power varies significantly across power sources (Table 8.1). The average cost of variability per megawatt of installed capacity (Table 8.1) is consistent with the observed variability characteristics (Figure 8.10).

The majority of the cost of variability consists of charges for balancing energy for all plants considered (Table 8.2). The average energy costs in 2010 were higher than the average regulation costs by nearly a factor of eight (Table 8.1).

TABLE 8.1	*Cost of variability of solar PV, solar thermal, and wind and the average price of electricity in the CAISO zone or region.*			
	Solar thermal (NSO)	ERCOT wind	Solar PV (Springerville, AZ)	Solar PV (20 MW+ class)
Average cost of variability per MWh (2010)	$5.2	$4.3	$11.0	$7.9
Median cost of variability per MWh (2010)	$0.0	$2.2	$0.3	$0.2
Variability cost as a percentage of total cost of power (2010)	11.9%	10.2%	26.5%	18.9%
Capacity factor (or average capacity factor)	23%	34%	19%	25%

Source: Modified with permission from Lueken and colleagues (2012). Original table © American Chemical Society.

TABLE 8.2	*Cost of variability breakdown between energy and regulation charges.*	
	Energy costs	**Regulation costs**
Springerville solar PV	69%	31%
20 MW+ solar PV	65%	35%
NSO solar thermal	69%	31%
Wind (average)	73%	27%

Source: Reprinted with permission from Lueken and colleagues (2012) © American Chemical Society.

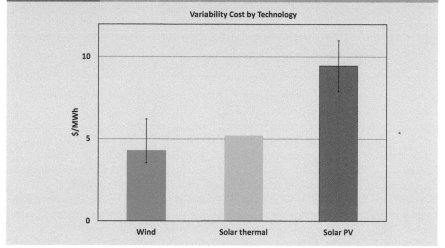

FIGURE **8.11**   *Cost of variability of solar thermal and solar PV and wind estimated by using the costs of ancillary services in the California ISO and the energy required to compensate for its variability. Reprinted with permission from Lueken and colleagues (2012) © American Chemical Society.*

The summary of the variability cost differences found by Lueken and colleagues (2012) is shown in Figure 8.11.

These results suggest that not all renewable technologies should be treated equally in terms of variability charges. To avoid creating market biases, utilities and regulators can use methods like these to determine how each variable generator contributes to total variability cost in its service area. Adopting proposals for intra-hourly scheduling would also help ISOs reduce the cost of renewable electricity's variability.

8.5 Summary

The variations in wind, solar photovoltaic, and solar thermal power generation are much stronger at low frequencies than at high frequencies. This characteristic means that a great deal of the variability can be compensated for by dispatchable generators with relatively slow ramp rates, including natural gas combined-cycle and coal plants. This greatly decreases the cost of integrating these variable sources into the power system below what they would have been if the variability was equally strong at all frequencies.

The frequency-dependence of the variations in wind turbine output power is a remarkable match to that of the observed variations in demand for electricity, as observed both in a small control area and in the large California ISO.

The variability of utility-scale solar PV generation has considerably larger amplitude than that of wind or solar thermal generation and thus incurs larger integration costs.

References

Apt. J. (2007). The spectrum of power from wind turbines. *Journal of Power Sources, 169*(2), 369–374.

Burton, T., Sharpe, D., Jenkins, N., & Bossanyi, E. (2001). *Wind energy handbook.* Chichester, UK: John Wiley & Sons.

Fordham, E.J. (1985). The spatial structure of turbulence in the atmospheric boundary layer. *Wind Engineering, 9*, 95–133.

Grant, H.L., Stewart, R.W., & Moilliet, A. (1961). Turbulence spectra from a tidal channel. *Journal of Fluid Mechanics, 12*, 241–263.

Kolmogorov, A.N. (1941). The local structure of turbulence in incompressible viscous fluid for very large Reynolds numbers. *Dokl. Akad. Nauk. SSSR, 30*, 301–305. Reprinted in *Proceedings of the Royal Society of London: Math. Phys. Sci.* (1991), *434*(1890), 9–13.

Lueken, C., Cohen, G.E., & Apt, J. (2012). Costs of solar and wind power variability for reducing CO_2 emissions. *Environmental Science & Technology, 46*, 9761–9767.

Van der Hoven, I. (1957). Power spectrum of horizontal wind speed in the frequency range from 0.0007 to 900 cycles per hour. *Journal of Meteorology, 14*, 160–164.

Wan, Y. (2004). *Wind power plant behaviors: Analyses of long-term wind power data.* National Renewables Energy Laboratory technical report, NREL/TP-500-36551. www.nrel.gov/docs/fy04osti/36551.pdf

Wan, Y., & Bucaneg, D. (2002). Short-term power fluctuations of large wind power plants. *Journal of Solar Energy Engineering, 124*(4), 427–431.

9

FORECAST ERROR CHARACTERISTICS OF WIND AND OF LOAD

9.1 Introduction

Electric grids consist of thousands of generators and loads. System operators for these large, complex systems strive to ensure that energy is delivered from generators to loads in a reliable and cost-effective manner. This process involves decisions about which generators should run and at what levels of output. Due to the long start-up times required for some baseload generators, planning decisions begin a day or two in advance, while shorter-term dispatch decisions take place on time scales ranging from hours to minutes.

In order to determine the optimal use of resources, system operators use forecasts of load and wind power as inputs for algorithms to determine the most cost-effective method to operate generators. Since planning decisions occur over multiple time scales, forecasts are used to predict load and wind days in advance as well as 5 minutes in advance. As more wind turbines are deployed and wind power provides a greater portion of the electricity generated, wind forecasts are becoming more important to system operators. Conventional generation is scheduled to supply the expected net load, which is the amount of forecast load above the amount of forecast wind energy.

Forecasts are also used in some electricity markets. These markets generally operate in two stages, day-ahead and real-time. The day-ahead market makers produce energy and operating reserve schedules covering the following day. The real-time market is used to make adjustments to the schedules as more information becomes known. In both time intervals, good estimates of load and wind in the grid will produce more efficient market results.

Load forecasts are produced with statistical tools that model load as a function of season, time of day, weather conditions, and other variables (Hahn, Meyer Nieberg, & Pickl, 2009). Model selection and variables used in the model depend on the geographic location of the grid. Load diversity also has a large effect on load forecast accuracy. When a grid covers a large area, electric demand at load centers located throughout the region will not correlate perfectly. The result is that variability of loads in different locations tends to smooth out fluctuations in the aggregate load, which makes the aggregate load easier to forecast.

Load forecasts for large areas at present tend to have errors that are 2 to 3% of the mean load (Table 9.1). However, if customer-side generation (for example, solar or natural gas combined heat and power) or aggressive customer load management become more prevalent, it is possible that there will be an increase in the uncertainty of load forecasts.

Wind forecasts are created using statistical or physical models. Statistical models rely on historical wind data to predict future values based on trends and generally employ time series or neural network analysis methods (Giebel, Brownsword, & Kariniotakis, 2003). Physical models, often referred to as numerical weather prediction (NWP) models, use meteorological data to predict wind speeds that are converted to wind turbine output electrical power with a transfer function (Landberg et al., 2003).

Wind forecast accuracy is influenced by factors such as wind plant size, local terrain, model selection, and weather fluctuations (Monteiro et al., 2009). Similar to load forecasting, the aggregation of wind plant forecasts over a large region tends to reduce forecast errors due to spatial smoothing of generation output (Focken et al., 2002; Giebel, Sørensen, & Holttinen, 2007). Wind forecast errors at different locations tend to cancel each other out to some extent.

Since forecasts are never exact, the uncertainty of predicted values must be understood in order to effectively use forecasts. When given a load and wind forecast, a system operator must be able to evaluate the amount of generation required to satisfy a reliability target in the event the forecasts are wrong. Scheduling too much generation is expensive, and too little generation leads to potential load curtailments. For this reason, there is a trade-off between the cost of scheduling more generation and the probability that it will be needed.

The most common definition of a forecast error e is the difference between the predicted value (F) and the actual value (A),

$$e = F - A \tag{9.1}$$

The most common metrics used to characterize forecast accuracy are mean average error (MAE) and root mean square error ($RMSE$):

$$MAE = \frac{\sum_{i=1}^{N} |e_i|}{N} \quad RMSE = \sqrt{\frac{\sum_{i=1}^{N} e_i^2}{N}}$$

The variable e_i indicates the forecast errors at time period i, and N is the number of observations. The most common way to express the MAE and $RMSE$ is to normalize them in a manner that allows one to compare forecasts from different regions. In the case of load forecasts, one could use the mean load or the peak load.

With wind forecasts, the installed capacity is almost always used. The result is an indication of forecast accuracy as a percentage of the mean load or installed capacity.

9.2 Analysis of forecasts in ERCOT and MISO

We (Mauch et al., 2013b) analyzed load and wind forecasts in two U.S. electric grids, the Electric Reliability Council of Texas (ERCOT) and the Midwest Independent System Operator (MISO), recently renamed the Midcontinent ISO. Data collected from each organization were from different time periods. Table 9.1 shows a summary of the data collected.

The map shown in Figure 9.1 shows both regions as they were when the data were collected (MISO's boundaries and name have changed since then). Dots on the map indicate where wind plants are located in each grid. Note that the MISO wind plants span much more area than those in ERCOT, and thus MISO benefits from greater geographical diversity in both wind and load forecasts. That is one reason the uncertainty metrics in Table 9.1 for MISO are smaller than for ERCOT.

9.2.1 Load forecast errors

Mauch and colleagues (2013b) analyzed day-ahead load forecast errors for ERCOT and MISO. Load forecast errors are generally assumed to have a normal (Gaussian) distribution, but this has been shown to be a poor assumption in many regions

TABLE 9.1	*Summary of wind and load in ERCOT and MISO during the time periods of the data used in this research.*	
	ERCOT	**MISO**
Date Range	Jan 2009–Dec 2010	Feb 2011–May 2012
Average load	36 GW	61 GW
Maximum load	66 GW	104 GW
Average nameplate wind capacity	8.8 GW	10 GW
Average wind capacity factor★	0.35	0.33
Percentage of load served by wind★	8.7%	5.5%
Mean absolute wind forecast error (percentage of installed wind capacity)	11%	6.1%
Root mean squared wind forecast error (percentage of installed wind capacity)	14%	8.1%
Mean absolute forecast error (percentage of mean load)	3.1%	1.7%
Root mean squared load forecast error (percentage of mean load)	4.2%	2.2%

★ ERCOT wind data were corrected for wind curtailments.

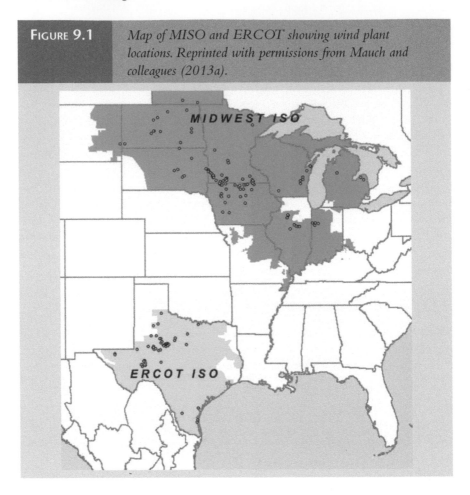

Map of MISO and ERCOT showing wind plant locations. Reprinted with permissions from Mauch and colleagues (2013a).

(Hodge et al., 2012). Our analysis of data from ERCOT and MISO further indicates that this is not a correct assumption. In fact, day-ahead load forecast errors tend to be more peaked than normal distributions and have fatter tails. Figure 9.2 shows the frequency distribution of day-ahead load forecast errors in both grids with a fitted normal distribution and logistic distribution overlaid. The logistic distribution has the density function shown in (9.2) and was used in our analysis since it provides a better fit to the data.

$$f(x) = \frac{e^{-\frac{x-\mu}{s}}}{s\left(1 + e^{-\frac{x-\mu}{s}}\right)^2} \tag{9.2}$$

While the distribution of day-ahead load forecast errors is leptokurtic in both regions, it is more peaked in the ERCOT grid than in MISO. The tails in the

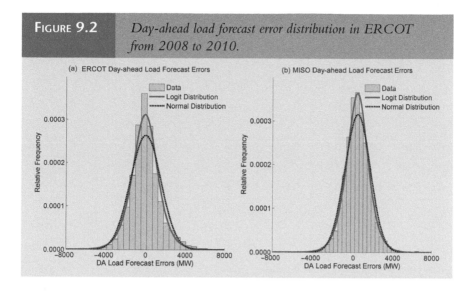

FIGURE 9.2 *Day-ahead load forecast error distribution in ERCOT from 2008 to 2010.*

ERCOT grid are also fatter than in MISO. This suggests that large forecast errors are more common in ERCOT. One potential reason for this is the higher level of load diversity in MISO. The large footprint in MISO ensures that unexpected weather events affect only a portion of the load and, therefore, do not result in extreme forecast errors. In both grids, load forecast errors span roughly the same range even though aggregate load in MISO is much larger than in ERCOT (Table 9.1).

When considering load forecast errors for different forecast levels, the distribution has a similar shape, but the parameters change. Figure 9.3 shows the ERCOT and MISO day-ahead load forecast errors plotted against the load forecasts. All data are normalized by the mean load value in order to compare the two grids. Note that the spread of the errors increases with the forecast values, indicating higher uncertainty for high forecast values. The data are divided into three ranges, low (less than 90% of mean load), medium (between 90% and 120% of mean load), and high (greater than 120% of mean load). This was done to show what the distribution looks like at different forecast levels in Figure 9.4, where the forecast errors for both ERCOT and MISO are displayed with a logistic distribution.

9.2.2 Wind power forecast errors

A normal distribution does not fit wind forecast errors either (Hodge et al., 2012). Figure 9.5 shows the distribution of day-ahead wind forecast errors in ERCOT and MISO. As with load forecast errors, the errors are more peaked in the center of the distribution, with fatter tails than what would be predicted with a normal distribution (leptokurtic). Note that the wind forecast errors shown in the plots are normalized by the installed wind capacity in each grid. This is necessary when displaying wind forecast data over a period of time to account for increases in the amount of installed wind capacity.

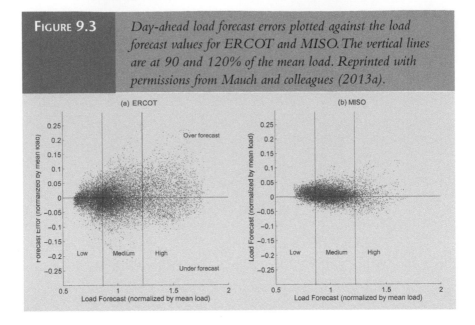

FIGURE 9.3 — Day-ahead load forecast errors plotted against the load forecast values for ERCOT and MISO. The vertical lines are at 90 and 120% of the mean load. Reprinted with permissions from Mauch and colleagues (2013a).

In a similar manner to load forecasts, the dependence of the uncertainty of wind forecasts on the forecast values can be examined. The graphs in Figure 9.6 show wind forecast values on the horizontal axis and wind forecast errors on the vertical axis. There are two interesting features to note in these graphs. First, since wind forecast errors are defined as the forecasted value minus the actual value, forecast values of zero cannot have positive errors while forecasts of max power cannot have negative errors. The error definition also means that forecasts in the lower half of the forecast range are more likely to produce negative errors, while high forecasts are more likely to produce positive errors. Second, the variance of the forecast errors is higher in the middle of the forecast range than near zero or max power. In ERCOT, the variance is a maximum for forecasts of 75% of the installed capacity and in MISO at 60% of the installed capacity. The implication of this is that forecast uncertainty is relatively low for forecasts near zero power and also near the maximum power.

One reason forecast uncertainty is dependent on the forecast level is the relationship between wind speed and wind power shown schematically in Figure 9.7 and for an operational wind turbine in Figure 2.2. Wind speed is predicted and converted to wind power using such a curve. Errors in wind speed predictions cause errors in the wind power predictions. As shown in the graph, the slope of the wind power curve is highest around the middle of the range of wind power output. In this region, wind speed prediction errors create larger wind power prediction errors.

The distribution of the wind forecast errors changes dramatically at different forecast values. For example, errors are skewed toward the negative end at low forecasts and toward the positive end at high forecasts. Near the center of the forecast range, the error distributions are fairly symmetrical. Error distributions based on

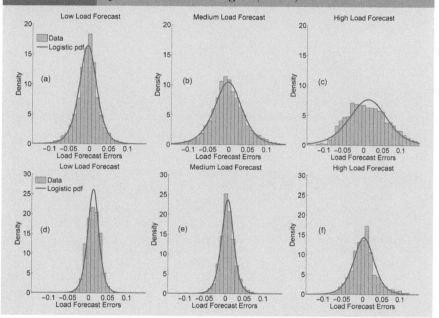

FIGURE 9.4	*Load forecast errors for ERCOT (top) and MISO (bottom) with data separated into three different load forecast classes. All values are normalized by the mean load forecast. Low forecasts include all forecasts less than 90% of the mean forecast; medium forecasts are between 90 and 120% of the mean forecast; and high forecasts are greater than 120% of the mean forecast. Histogram bars show the relative frequency of the actual data while solid lines show the fitted logistic distributions. Reprinted with permission from Mauch and colleagues (2013a).*

the data in Figure 9.6 are shown for ERCOT and MISO at three different forecast levels in Figure 9.8. Solid lines in the graphs show the results of a model based on the logistic-normal distribution for observed wind values as well as forecasted wind values. In order to develop the logistic-normal model, the variables were transformed according to (9.3).

$$F^{\star} = \ln\left(\frac{F}{1-F}\right) \quad W^{\star} = \ln\left(\frac{W}{1-W}\right) \tag{9.3}$$

where F and W are the forecasted and observed wind values respectively and $F^{\star}$ and $W^{\star}$ are the logit transformations of F and W. The transformed variables were then fit to a bivariate normal distribution shown in (9.4).

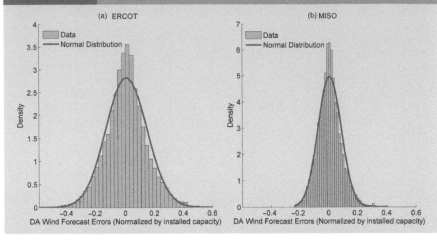

FIGURE 9.5   *Day-ahead wind forecast errors for ERCOT and MISO. See Table 9.1 for time periods when data were collected for the plots. Reprinted with permission from Mauch and colleagues (2013b).*

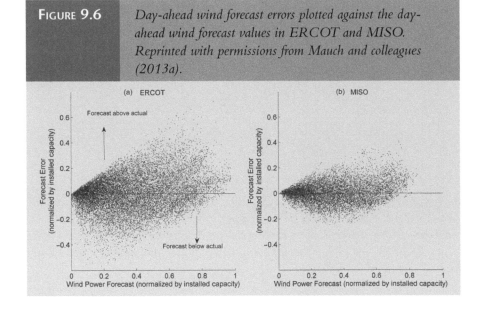

FIGURE 9.6   *Day-ahead wind forecast errors plotted against the day-ahead wind forecast values in ERCOT and MISO. Reprinted with permissions from Mauch and colleagues (2013a).*

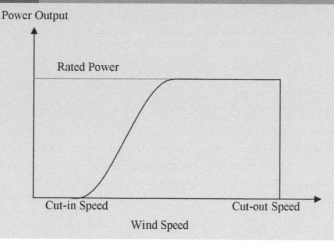

FIGURE 9.7 *Typical wind turbine power curve showing the relationship between wind speed and the power generated.*

$$f(F^{\star}, W^{\star}) = \frac{1}{2\pi\sigma_{F^{\star}}\sigma_{W^{\star}}\sqrt{1-\rho^2}} exp\left(-\frac{1}{2(1-\rho^2)}\left[\frac{(F^{\star}-\mu_{F^{\star}})^2}{\sigma_{F^{\star}}^2}\right.\right.$$
$$\left.\left. + \frac{(W^{\star}-\mu_{W^{\star}})^2}{\sigma_{W^{\star}}^2} - \frac{2\rho(F^{\star}-\mu_{F^{\star}})(W^{\star}-\mu_{W^{\star}})}{\sigma_{F^{\star}}\sigma_{W^{\star}}}\right]\right).$$

(9.4)

where $\rho = \dfrac{cov(W^{\star}, F^{\star})}{\sigma_{W^{\star}}\sigma_{F^{\star}}}$ and cov(,) is the covariance.

Using the joint distribution of the forecasts and observed wind power values, one can easily get conditional distributions for observed wind power given a forecasted value. For a particular forecast value, the transformed wind power is modeled with a normal distribution with the conditional probability density defined by:

$$f(W^{\star}\mid F^{\star}) = \frac{1}{\sqrt{2\pi}\sigma_{W^{\star}|F^{\star}}} exp\left(-\frac{(W^{\star}-\mu_{W^{\star}|F^{\star}})^2}{2\sigma_{W^{\star}|F^{\star}}^2}\right)$$

(9.5)

where $\mu_{W^{\star}|F^{\star}} = \mu_{W^{\star}} + \dfrac{\rho\sigma_{W^{\star}}}{\sigma_{F^{\star}}}(F^{\star}-\mu_{F^{\star}})$

(9.6)

$$\sigma_{W^*|F^*} = \sigma_{W^*} \sqrt{1 - \rho^2} \tag{9.7}$$

Using (9.5), one can determine confidence intervals for $W^\star$ as a function of $F^\star$. In order to provide useful values, the transformed values are converted back to the original space with (9.8).

$$F = \frac{1}{1 + e^{-F^*}} \qquad W = \frac{1}{1 + e^{-W^*}} \tag{9.8}$$

Using the notation for forecast errors in (9.1), confidence intervals can be expressed in terms of forecast errors by subtracting the wind value confidence intervals from the wind forecast values. The results in Figure 9.9 for MISO and ERCOT were calculated with the procedure outlined above. Due to a lack of high forecasts in the MISO data, the model did not provide as good a fit in the high forecast range.

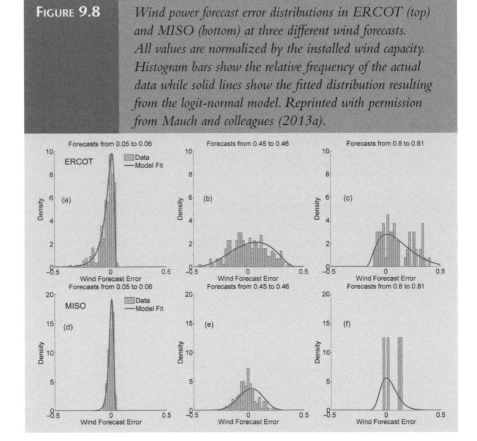

FIGURE 9.8 *Wind power forecast error distributions in ERCOT (top) and MISO (bottom) at three different wind forecasts. All values are normalized by the installed wind capacity. Histogram bars show the relative frequency of the actual data while solid lines show the fitted distribution resulting from the logit-normal model. Reprinted with permission from Mauch and colleagues (2013a).*

We also applied the model to hour-ahead wind forecast data. Results are shown in Figure 9.10. Since hour-ahead forecasts are much more accurate than the day-ahead forecasts, the confidence intervals are much tighter.

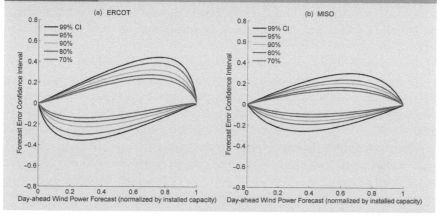

FIGURE 9.9 *Day-ahead wind forecast error confidence intervals as a function of the wind forecast value in ERCOT and MISO. The confidence intervals were created with the logit-normal model. Reprinted with permission from Mauch and colleagues (2013b).*

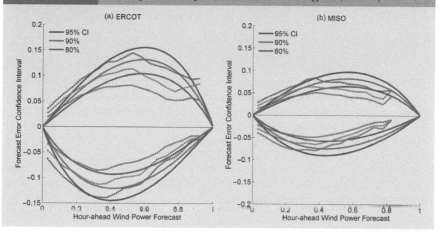

FIGURE 9.10 *Hour-ahead wind forecast error confidence intervals in ERCOT and MISO. Dashed lines are determined with direct calculations using a moving window of size 0.1. Solid lines are determined with the logit-normal model. Reprinted with permission from Mauch and colleagues (2013b).*

9.2.3 Summary

Normal distributions do not adequately model the distribution of forecast errors in load and wind forecasts. The day-ahead forecast error data from ERCOT and MISO indicate that normal distributions underpredict small forecast errors near zero and overpredict forecast errors slightly farther from zero. While not shown in detail here, normal distributions also underpredict extreme forecast errors in the tails of the distribution. These are the errors that cause the most problems.

A logistic distribution is a much better fit for load forecast errors than a normal distribution. However, the parameters change with the amount of load forecasted since the uncertainty of high load forecasts is greater than the uncertainty associated with low load forecasts. Analyzing load forecasts at different levels provides a system operator with more information on the level of uncertainty in the forecast.

Wind power forecast uncertainty has a much stronger dependence on the value of the wind forecast. Instead of increasing monotonically with the forecast values, wind forecast uncertainty reaches a maximum slightly above the center of the forecast range and becomes less at very high forecasts. This characteristic is due primarily to the way wind turbines operate. In the middle of the forecast range, small wind speed fluctuations lead to large changes in the wind power output. At the low and high ends of the forecast range, wind speed changes have little effect on the output of the wind turbines.

Fitting a logit-normal model to the wind forecast and actual wind power data provides a method to accurately model wind forecast errors.

References

Focken, U., Lange, M., Mönnich, K., Waldl, H.-P., Beyer, H.G., & Luig, A. (2002). Short-term prediction of the aggregated power output of wind farms—a statistical analysis of the reduction of the prediction error by spatial smoothing effects. *Journal of Wind Engineering and Industrial Aerodynamics, 90*, 231–246.

Giebel G., Brownsword, R., & Kariniotakis, G. (2003). *State of the art on short-term wind power prediction.* ANEMOS deliverable report, D1.1.

Giebel, G., Sørensen, P., & Holttinen, H. (2007). *Forecast error of aggregated wind power.* TradeWind deliverable report, Risø-I-2567(EN).

Hahn, H., Meyer-Nieberg, S., & Pickl, S. (2009). Electric load forecasting methods: Tools for decision making. *European Journal of Operational Research, 199*(3), 902–907.

Hodge, B., Florita, A., Orwig, K., Lew, D., & Milligan, M. (2012). A comparison of wind power and load forecasting distributions. *2012 World Renewable Energy Forum*, NREL/ CP-5500–54384. www.nrel.gov/docs/fy12osti/54384.pdf

Landberg, L., Giebel, G., Nielsen, H.A., Nielsen, T.S., & Madsen, H. (2003). Short-term prediction—an overview. *Wind Energy, 6*(3), 273–280.

Mauch, B., Apt, J., Carvalho, P.M.S., & Jaramillo, P. (2013a). What day-ahead reserves are needed in electric grids with high levels of wind power? *Environmental Research Letters, 8*(3). DOI: 10.1088/1748–9326/8/3/034013

Mauch, B., Apt, J., Carvalho, P.M.S., & Small, M.J. (2013b). An effective method for modeling wind power forecast uncertainty. *Energy Systems, 4*(4), 393–417.

Monteiro, C., Bessa, R., Miranda, V., Botterud, A., Wang, J., & Conzelmann, G. (2009). *Wind Power Forecasting: State-of-the-art 2009*. Argonne National Labs report, ANL/ DIS-10–1.

10

DAY-AHEAD WIND RESERVE REQUIREMENTS

10.1 Introduction

To ensure that all load is served, system operators of electric grids procure greater amounts of generation capacity than the expected requirement. Most short-term capacity is procured a day in advance when the day-ahead market determines which generators will be online the following day. Capacity procurement is done with a forecast of system conditions the following day. The excess capacity provides insurance against unexpected contingencies as well as forecast errors such as higher-than-expected load or lower-than-expected wind power (see Chapter 9).

The excess generation capacity procured in the day-ahead market is collectively referred to here as operational reserves. Since markets are structured differently among regional transmission organizations (RTOs), the use of operational reserves is also different. For example, the Midcontinent Independent System Operator (MISO) procures operating reserves to guard against contingencies in the grid (e.g., generator outages) but not to balance load forecast errors. The Electric Reliability Council of Texas (ERCOT), on the other hand, procures operating reserves to cover contingencies as well as load forecast errors. MISO relies on energy purchases in the real-time market to balance forecast errors. A significant difference between the two RTOs is the must-offer requirement for generators cleared in the MISO capacity market (Newell, Spees, & Hajos, 2010). ERCOT has no capacity market and, therefore, no must-offer requirement (Schubert, Hurlbut, Adib, & Oren, 2006).

Whether system operators rely on operating reserves for forecast errors or not, they still must ensure adequate capacity is available to protect the system against a capacity shortfall. The proper amount of excess capacity depends on the uncertainty of forecasts and the cost associated with procuring additional capacity.

We analyzed load and wind forecasts in ERCOT and MISO to determine day-ahead net load uncertainty (Mauch et al., 2013). The data used included hourly values of day-ahead load and wind forecast values along with actual load and wind values. ERCOT data covered the time period from January 2009 to December 2010. The MISO data were recorded over the period February 2011 to May 2012. Net load is defined as the load minus the available wind power. It represents the amount of load served with fully dispatchable (controllable) resources. Net load uncertainty represents only a portion of the total system uncertainty hedged against with excess capacity. Other sources of uncertainty include grid contingencies and interchange power transfers. However, the analysis focused on net load uncertainty to examine the effect that wind forecast uncertainty has on reserve requirements.

10.2 Net load forecast uncertainty

In order to procure adequate capacity to cover net load forecast errors, system operators of electric grids must have a good understanding of the net load forecast error characteristics. Net load forecast errors are defined as the load forecast error minus the wind forecast error,

$$e_N = e_L - e_W. \tag{10.1}$$

Forecast errors of load and wind are often assumed to be independent variables. This assumption is valid for the ERCOT and MISO data, with some exceptions. For this reason, Mauch and colleagues (2013) used the independence assumption to determine the distribution of net load errors and performed sensitivity analysis on this assumption. If the variables are uncorrelated, the distribution of net load errors is the convolution of the load and wind forecast error variables. Knowing the probability density function of the net load forecast errors, we can determine the probability that available capacity will not be adequate to serve the load. Figure 10.1 shows a sample cumulative distribution function of the load forecast errors and the net load forecast errors to illustrate the effect of adding wind to the system. The dashed lines indicate the 95% confidence interval for the forecast errors. The net load forecast distribution is broadened with respect to the load forecast distribution so that the confidence interval increases from the additional wind uncertainty.

Load and wind forecast uncertainty is dependent on the level of load or wind expected (Chapter 9). We are primarily interested in the errors that result from an under-forecast of net load since these errors are corrected by increasing generation or load curtailment from the dispatchable resources. Therefore, we calculated the confidence intervals for net load errors based only on the negative (underforecast) errors.

10.3 Day-ahead dispatchable capacity requirements

Using day-ahead load and wind forecast data for ERCOT and MISO, the net load forecast confidence intervals at the 95% level can be determined. The results indicate the amount of dispatch capacity required to compensate for 95% of the

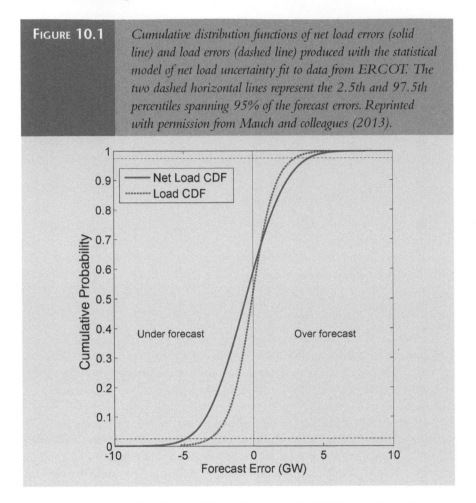

FIGURE **10.1** *Cumulative distribution functions of net load errors (solid line) and load errors (dashed line) produced with the statistical model of net load uncertainty fit to data from ERCOT. The two dashed horizontal lines represent the 2.5th and 97.5th percentiles spanning 95% of the forecast errors. Reprinted with permission from Mauch and colleagues (2013).*

underforecast errors. We select a 95% level because ERCOT uses this level to procure operating reserves (ERCOT, 2012).

During the time period in which the data were collected, ERCOT's installed wind capacity increased from 8.3 GW to 9.5 GW, and MISO's wind capacity increased from 9.1 GW to 10.8 GW. Mauch and colleagues (2013) scaled the wind data to simulate an installed wind capacity of 10 GW in both grids and compared dispatchable generation capacity requirements due to wind uncertainty for both grids.

Figure 10.2 shows the day-ahead 95% confidence level of underforecast net load errors as a function of the wind power forecast for ERCOT and MISO. The horizontal line in each figure indicates the amount of capacity required to cover the 95th percentile of load forecast errors if no wind power was present in the grid. The graphs show that the 95th percentile of load underforecast errors in ERCOT is 3,100 MW and 2,200 MW in MISO. Wind forecast uncertainty in ERCOT increases the required capacity to 5,300 MW at a wind power forecast of 7,500 MW. In MISO, the total capacity required to cover 95% of net load errors is 3,100

FIGURE **10.2**   *Dispatchable generation capacity required to cover 95% of day-ahead net load underforecast errors in (a) ERCOT and (b) MISO when load forecasts are in the range 90% to 120% of mean load forecast. Each graph shows the 95th percentile of load forecast errors as a horizontal line. Net load forecast uncertainty converges to load forecast uncertainty at zero wind forecast and also at the maximum wind forecast. Reprinted with permission from Mauch and colleagues (2013).*

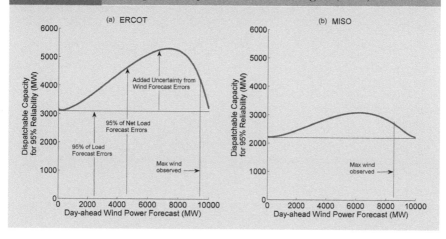

MW at a wind forecast of 6,000 MW. The greater accuracy in MISO wind and load forecasts leads to less net load uncertainty compared to ERCOT.

At zero wind forecasts, the net load uncertainty and load uncertainty are equal. Wind forecasts higher than zero increase the dispatchable capacity amount required to cover 95% of the net load underforecast errors. Also, for wind forecasts at the installed capacity (maximum wind), the net load uncertainty converges to the load uncertainty. Note that the maximum observed wind in each grid was below the installed wind capacity as indicated in the figure. Load forecast uncertainty is dependent on the load forecast values; higher load forecasts tend to be more uncertain. Therefore, the load forecasts used to create Figure 10.2 were constrained to be within 90% to 120% of the average load for the respective grids. This allowed the calculation of net load uncertainty for a range of wind forecast values without including the dependence of load forecast uncertainty on load forecasts.

Let us assume that different day-ahead reliability levels are used in different areas. The effect of higher reliability levels is shown in Figure 10.3. Increasing the level of reliability to 99% in ERCOT adds roughly 1,500 MW to the capacity requirement for day-ahead uncertainty coverage. In MISO, the requirement increases by only 1,000 MW.

The effect of load forecasts on the net load uncertainty is shown in Figure 10.4, where load forecast values are separated into three classes; high (greater than 120%

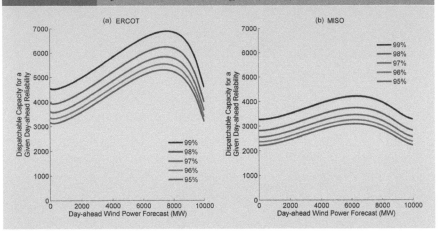

FIGURE 10.3 *Dispatchable generation capacity required to cover 95% to 99% of net load forecast errors in (a) ERCOT and (b) MISO for load forecasts in the range 90% and 120% of mean load forecast. Reprinted with permission from Mauch and colleagues (2013).*

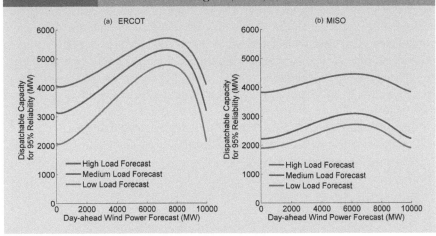

FIGURE 10.4 *Amount of dispatchable generation capacity required to cover 95% of net load underforecast errors as a function of wind forecast level for 10 GW of installed wind capacity for (a) ERCOT and (b) MISO. Backup requirements peak in ERCOT at a wind forecast of 7,500 MW and in MISO at 6,000 MW. Reprinted with permission from Mauch and colleagues (2013).*

of the mean load), low (less than 90% of the mean load), and medium (between 90% and 120% of the mean load). The ranges of forecast values to define the three categories were determined by inspecting the standard deviation of load forecast errors. Three categories were found to represent the different levels of forecast uncertainty well. At high load forecasts, the load forecast uncertainty increases, which increases the dispatchable capacity requirements.

10.4 Future day-ahead capacity requirements

We also considered how much dispatchable capacity will be required to compensate for day-ahead uncertainty in future grids. As wind capacity increases in ERCOT and MISO, future day-ahead uncertainty will change. If the added wind increases the geographic diversity of the wind locations, future uncertainty will decrease relative to the amount of wind in the system. However, if added wind is clustered around present locations, future wind forecast uncertainty will not benefit from additional geographic diversity. Scenarios were considered in which the wind uncertainty for a future grid is the same as today and when the future wind uncertainty decreases.

Previous work showed the standard deviation of wind forecast errors normalized by installed wind capacity decreases with additional wind capacity, largely due to spatial smoothing of the wind power (Focken et al., 2002). These results showed that the standard deviation of normalized forecast errors decreased 28% when the geographic area of the wind sites doubled. Mauch and colleagues (2013) used these results to model wind forecast uncertainty as a function of installed wind capacity, assuming that as wind capacity increases from 10 to 30 GW, the geographical area containing the wind sites doubles.

The ERCOT and MISO grids were modeled by scaling up wind forecast errors to simulate higher levels of wind power. High load forecast uncertainty was used without scaling the load data. We modeled future wind uncertainty with high load uncertainty to look at the maximum amount of generation capacity required beyond the net load forecast to compensate forecast uncertainty. A range of future capacity requirements is shown in Figure 10.5 as a function of installed wind capacity for ERCOT and MISO. The low end of the range was determined assuming improved (reduced) forecast uncertainty due to spatial smoothing. Worst-case scenarios were calculated assuming no change in wind forecast accuracy.

Capacity requirements in MISO are much lower than in ERCOT due to the higher accuracy of load and wind forecasts in MISO. As installed wind in ERCOT reaches 30 GW, dispatchable capacity requirements range from 9,000 MW to 13,000 MW. Since the capacity needed to cover load forecast errors is 4,000 MW, wind forecast uncertainty adds 5,000 MW to 9,000 MW to the day-ahead dispatchable capacity requirement. In MISO, wind adds 2,000 MW to

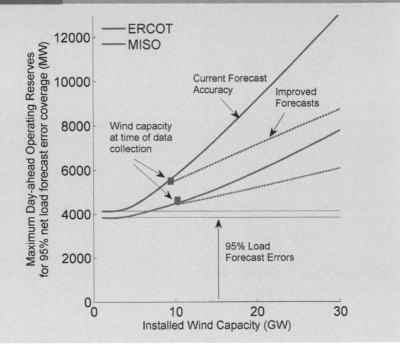

FIGURE **10.5** *Dispatchable generation capacity requirements for installed wind capacity up to 30 GW in ERCOT and MISO. Horizontal lines show the level of capacity required to cover load forecast errors at high load forecasts. Solid lines show operating reserves required for 95% of net load errors with current wind forecast accuracy. Dashed lines show the results with improved wind forecasts for wind capacity above 10 GW. Reprinted with permission from Mauch and colleagues (2013).*

4,000 MW of day-ahead dispatchable capacity requirements for net load forecast error balancing. However, when the two grids are compared based on the percentage of energy coming from wind, the capacity required to compensate wind forecast uncertainty looks similar. While ERCOT has more forecast uncertainty, it also has a higher penetration of wind energy. At the time these data were collected, 8.7% of ERCOT's load was served with wind power, while the percentage was 5.5% in MISO. Figure 10.6 shows the additional day-ahead dispatchable capacity requirements due to wind forecast uncertainty in ERCOT and MISO as a function of the share of energy from wind power in each grid. When viewed

FIGURE **10.6**   *Additional day-ahead reserve capacity for a range of wind penetration values in ERCOT and MISO. In this figure, the horizontal axis is the percentage of load served by wind power. Solid lines assume no change in wind forecast accuracy. Dashed lines show the effect of improved forecasts. Reprinted with permission from Mauch and colleagues (2013).*

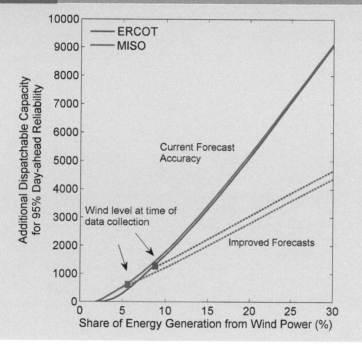

from this perspective, the additional capacity requirements are similar for both grids.

Another perspective that may be useful for system planners is to calculate the amount of additional reserves needed to balance wind forecast uncertainty as a percentage of peak load. In Figure 10.7, the additional capacity requirement as a percentage of peak load is displayed as a function of the percentage of energy generated from wind power.

As the share of energy generated from wind power increases, reserve capacity for day-ahead forecast error balancing will be required in greater quantities. The expansion of wind power in both grids should be accompanied with adequate dispatchable capacity to balance wind forecast uncertainty.

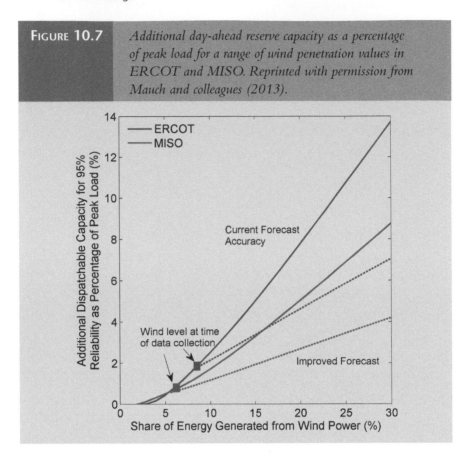

FIGURE 10.7	*Additional day-ahead reserve capacity as a percentage of peak load for a range of wind penetration values in ERCOT and MISO. Reprinted with permission from Mauch and colleagues (2013).*

10.5 Summary

At current levels of installed wind power capacity in the United States, grid operators have much more uncertainty in load forecasts than in wind forecasts. However, the rapid growth of wind power is likely to create a future in which day-ahead uncertainty is dominated by the wind forecast. With data from ERCOT and MISO, the day-ahead net load uncertainty has been modeled for different load and wind forecast levels. The procedure outlined here is applicable whether or not grid operators choose to cover 95% of the day-ahead net load underforecast errors with dispatchable capacity (as in ERCOT).

The proper level of dispatch capacity used to balance forecast errors depended on the level of load and wind forecasted. Day-ahead dispatchable capacity requirements for both grids was at a maximum when the wind forecast was in the upper half of the forecast range. In ERCOT, assuming 10 GW of installed wind capacity, the peak occurred at a 7,500 MW forecast, and in MISO it was at a 6,000 MW forecast. Load forecast levels also affect the net load uncertainty since high load forecasts were found to be less accurate than low load forecasts.

Ramp rate requirements for balancing forecast errors are an important aspect of wind power integration into electric grids and this aspect requires further research.

ERCOT requires operational reserves to cover 95% of the day-ahead forecast errors. We note that this may not be the most cost-effective method to handle day-ahead uncertainty. For example, in some regions it may be feasible to schedule reserves in the day-ahead market equal to the 50th percentile of day-ahead forecast errors in the day-ahead forecast and then reschedule during intraday commitment when uncertainties are reduced.

Wind is likely to play a larger role in electricity production in future grids. In ERCOT, as installed wind capacity reaches 30 GW, up to 30% of this capacity must be matched with dispatchable capacity to counter 95% of the day-ahead forecast errors. In MISO, up to 13% of the installed wind capacity must be matched with dispatchable capacity once wind reaches 30 GW. This result will have important policy implications in day-ahead unit commitment and operational reserve procurements. Other systems will have different results, but the method presented here can be applied to do system-specific analysis.

References

ERCOT. (2012). *ERCOT methodologies for determining ancillary service requirements, 2011–2012.* www.wecc.biz/committees/StandingCommittees/JGC/VGS/OWG/ActivityO1/Operating%20Reserve%20Requirement%20Methodologies/ERCOT%20Methodologies%20for%20Determining%20Ancillary%20Service%20Requirements.doc

Focken, U., Lange, M., Mönnich, K., Waldl, H.-P., Beyer, H.G., & Luig, A. (2002). Short-term prediction of the aggregated power output of wind farms—a statistical analysis of the reduction of the prediction error by spatial smoothing effects. *Journal of Wind Engineering and Industrial Aerodynamics, 90*(3), 231–246.

Mauch, B., Apt, J., Carvalho, P.M.S., & Jaramillo, P. (2013). What day-ahead reserves are needed in electric grids with high levels of wind power? *Environmental Research Letters, 8*(3). DOI: 10.1088/1748–9326/8/3/034013

Newell, S., Spees, K., & Hajos, A. (2010, Jauary 19). *Midwest ISO's resource adequacy construct: An evaluation of market design elements.* Report by The Brattle Group. www.brattle.com/system/publications/pdfs/000/004/844/original/Midwest_ISOs_Resource_Adequacy_Construct_Newell_et_al_Jan_19_2010.pdf?1378772134

Schubert, E.S., Hurlbut, D., Adib, P., & Oren, S. (2006). Texas energy-only resource adequacy mechanism. *Electricity Journal, 19*(10), 39–49.

11

YEAR-TO-YEAR VARIABILITY IN WIND POWER

11.1 Introduction

Researchers have been estimating the long-term variability of wind power since the 1940s, when large-scale wind power was first considered in detail. Thomas (1945), an early proponent of large-scale wind power, examined 25 years of wind speed data from seven weather stations around the United States and estimated that "the annual average velocity for any one year seldom drops more than 10 or 15 percent below the long-term average velocity" (p. 6). Around the same time, Golding and Stodhart (1949) analyzed wind speed records from weather stations in the UK and found that the annual average wind speed in a given year lies "between 65% and 130% of the long-term average" (p. 14). Later, when the energy crises of the 1970s renewed interest in wind power, Justus and colleagues (1979) used monthly average wind speeds from 40 sites around the United States to estimate variability and spatial correlation of monthly and annual wind speeds. Klink (1999) calculated the interannual variability of wind speed from historical weather data for 216 stations in the United States for the period 1961 through 1990. Wan (2012) did a similar analysis with 6 to 10 years of wind power data from four U.S. wind plants. Sinden (2007) analyzed wind speeds from weather stations around the UK to estimate the benefits of spatial diversity for wind power. Recently, Katzenstein and colleagues (2010) estimated the interannual variability of aggregate wind energy from 16 weather stations for 1973 through 2008.

Those studies based on historical data are limited by short periods of data, few sites, or sites that are not representative of wind plant locations. To address the limitations of the available historical data, several researchers have used reanalysis data, which interpolate meteorological observations in space and time using numerical weather prediction models. Giebel (2000) used 34 years of data from the NCEP/NCAR reanalysis to estimate correlated variability of wind power in northern Europe. Czisch and Ernst (2001) used 16 years of data from the European Centre for Medium-Range Weather Forecasts ERA-15 reanalysis to estimate variability and correlation of wind power on various time scales in Europe. Huang and colleagues (2014) used 5 years of data from the NASA Modern-Era Retrospective Analysis for Research and Applications (MERRA) reanalysis to estimate the smoothing effect of connecting distant wind plants in the United States.

A few researchers have investigated long-term trends in wind resources and the relationship of wind resources to long-term phenomena such as El Niño/La Niña. Palutikof and colleagues (1987) analyzed historical wind speed records from Britain to show that the relative frequency of westerly and anticyclonic conditions is related to the strength of the wind field. Klink (2002, 2007) analyzed data from sites in Minnesota to investigate long-term trends in wind speeds and relationships between periods of anomalous wind and large-scale circulation. Pryor and Barthelmie (2011) used regional climate models to assess the effects of climate change on wind resources in the United States.

The work in this chapter improves on earlier reanalysis studies in several ways. First, we use the Climate Forecast System (CFS) reanalysis, which has higher spatial and temporal resolution than most of the reanalyses used in previous papers (Saha et al., 2010). Second, we analyze locations of existing wind plants in the United States. Third, we more accurately estimate hub-height wind speeds by accounting for atmospheric stability and adjusting model outputs based on comparisons with measured data.

11.2 Method

We estimate hub-height wind power density at the location of every wind plant in the U.S. Great Plains for the period 1979 through 2010 using reanalysis data, which are meteorological data that have been interpolated in space and time from historical measurements using numerical weather-prediction models. Wind power density p (measured in W/m^2) is the power contained in the wind passing through one square meter of a turbine rotor; it is calculated with the following formula:

$$p = \frac{1}{2}\rho v^3 \tag{11.1}$$

where ρ is the air density (we use 1.225 kg/m^3) and v is the wind speed in m/s. A real wind turbine can capture only a fraction of the available power, determined by the power coefficient C_p. The power coefficient has a theoretical maximum value of 0.593, known as the Betz limit (Burton et al., 2001). We calculate average power density for each hour at each site and sum the power over longer periods (seasons or years) to get the energy density (measured in Wh/m^2).

11.2.1 Estimating wind speed at hub height

We estimate wind speed at 80-m height at each location using data from the Climate Forecast System (CFS) reanalysis (Saha et al., 2010) and the following formula for a logarithmic vertical wind profile given by Stull (1988):

$$u = \frac{u_*}{k}\left(\log\left(\frac{z}{z_0}\right) + \Psi \right)$$

(11.2)

where:

u_* = friction velocity [m/s]
k = 0.4 (von Karman constant)
z = hub height [m]
z_0 = surface roughness length [m]
Ψ = correction for atmospheric stability

The friction velocity and surface roughness values are taken directly from the CFS reanalysis data. Note that we extrapolate hub-height wind speed from friction velocity instead of 10-m wind speed; both variables are available in the reanalysis data. The term Ψ is calculated with another expression given by Stull (1988) using sensible and latent heat flux, surface air temperature, and surface air pressure values from the CFS reanalysis data. The advantage of reanalysis data is that they give accurate meteorological data in places that do not have weather stations and at times when measurements were not taken. The reanalysis data are given for points on a 0.5° grid, so we calculate the corresponding values at wind plant locations by linearly interpolating the values at the four nearest grid points.

The predicted wind power densities, calculated from wind speeds predicted with (11. 2), have biases compared to actual measurements at the same height. The predicted power density in unstable and neutral conditions tends to be higher than measured power density, so we fit a model to that error as a function of z/L, a measure of atmospheric stability, and use that model to correct the predictions. This process is known as Model Output Statistics (Potter, Gil, & McCaa, 2007). In stable atmospheric conditions ($z/L > 2$), the predicted power densities are unusable, so we replace them with power densities linearly interpolated from the preceding and following nonstable hours at the same site.

11.2.2 Location of wind plants

We simulate the wind power density at the locations of all commercial-scale wind plants in the U.S. Great Plains that are also in the Eastern Interconnect electrical system. The locations and installed capacities are taken from the OpenEI database

(U.S. National Renewable Energy Laboratory, 2013). We filter the data to use only sites operating by the end of 2012 with capacities greater than 10 MW in the following states: North Dakota, South Dakota, Nebraska, Kansas, Oklahoma, Texas, New Mexico, Minnesota, Michigan, Illinois, Indiana, and Ohio. We do not include wind plants in Eastern Interconnect located in the Appalachian Mountains because we have not validated wind speeds predicted by the CFS reanalysis data in mountainous terrain. When two or more sites are listed for the same location, we consolidate them into a single site with the combined capacity. This leaves 260 wind plants with a capacity of 27 GW, shown in Figure 11.1.

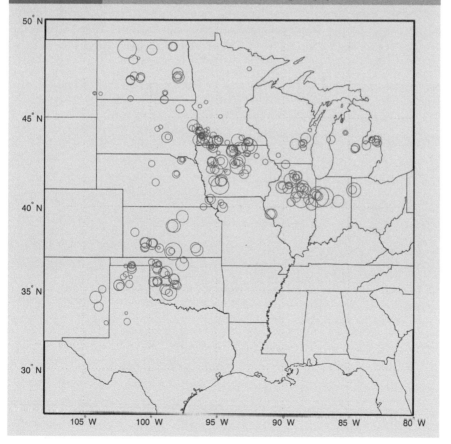

FIGURE 11.1 *Locations of the commercial (> 10 MW) wind plants in the U.S. Great Plains and Eastern Interconnect as of December 2012. The diameter of the circles is proportional to wind plant capacity. The largest site in this group is Fowler Ridge, Indiana, with a capacity of 600 MW.*

11.2.3 Measure of variability

In this chapter we use as a metric for the variability of energy density the coefficient of variation (COV), which is the standard deviation normalized by the mean. The COV does not assume energy density is normally (Gaussian) distributed, though we find that a normal distribution fits the annual energy density well. Both a Lilliefors test (1967) and a Jarque–Bera test (1987) find that the aggregate energy density of all wind plants combined is normally distributed at a 5% significance level. The same tests find that the annual energy densities of 88 to 92% of the individual sites are normally distributed. Given that annual energy densities can be modeled as normal distributions, approximately two thirds of the years will have annual densities within x of the long-term average, where x is the COV. Similarly, 99% of the years will have annual densities within $2.3x$ of the long-term average.

11.3 Results

11.3.1 Variability of aggregate annual energy generation

We use the wind speed data described previously to estimate the aggregate annual energy density for the Great Plains locations shown in Figure 11.1, weighted by the capacity of each wind plant. Figure 11.2 plots the annual energy density for all those

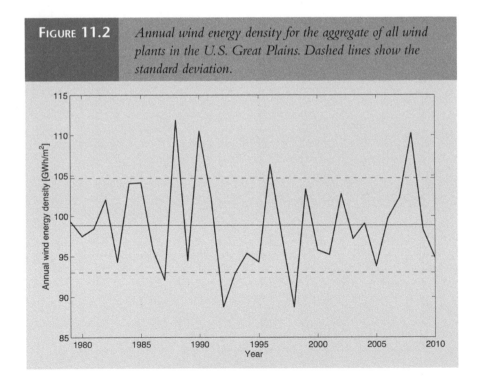

FIGURE 11.2   *Annual wind energy density for the aggregate of all wind plants in the U.S. Great Plains. Dashed lines show the standard deviation.*

sites combined for the period 1979 through 2010, overlaid with lines for the mean and standard deviation. This figure illustrates that using the COV as the measure of variability does not capture some significant aspects of the variation. For example, the highest energy generation is 13% above the mean and the largest year-to-year difference is 20% for the years 1987 to 1988.

We estimate that the COV of annual wind energy production in the U.S. Great Plains is 6%. This is similar to estimates of the COV of wind power by Katzenstein and colleagues (2010) and variability of wind speed by Thomas (1945) and Justus and colleagues (1979). We also estimate the COV of energy changes from one year to the next to be 9%, which is similar to the 7.4% Sinden (2007) estimated for the UK but smaller than the 15% estimated by Giebel (2000) for Europe.

11.3.2 Variability of single-site annual energy generation

The variability of annual energy generation for individual wind plants is larger than the variability of their aggregate. Figure 11.3 plots the COV of annual energy for each of the operating wind plants estimated from the 32 years of wind speed data. The COVs for these sites range from 5.7 to 11.2%. This variability is important to wind plant developers because it quantifies the uncertainty of future revenues from the wind plant and determines the size of loan a bank is willing to make. In the United States, banks typically make loans to a wind plant with annual payments that are less than the wind plant's revenue in 99 out of 100 of years. This is referred to as a debt service coverage ratio (DSCR) of 1.0 for the 1-year P99 revenue. If two identical wind plants have the same mean annual energy generation but different interannual variability, the wind plant with less variability will be able to borrow more money because its P99 revenue will be larger. It will have a higher debt-to-equity ratio because it can borrow more of the project cost but has the same average revenue as the other wind plant, so its internal rate of return (IRR) will be higher. A wind plant with a COV of annual energy generation of 6% will be able to borrow about 20% more than a plant with a COV of 11%.

The COVs of sites shown in Figure 11.3 appear to follow a pattern that is the inverse of the pattern of wind resource in the Great Plains. To investigate that pattern, we plot the COV against the long-term average wind power density for each site in Figure 11.4 and find a moderate correlation, $R^2 = 0.36$.

11.3.3 The smoothing effect of aggregating wind plants

The interannual variability of all wind plants combined is lower than the variability of nearly every individual wind plant. To analyze the smoothing effect of aggregating wind plants on their interannual variability, we aggregate each wind plant with all others within a given radius and calculate the COV of annual energy density. The results for radii of aggregation ranging from 0 to 2,200 km are shown in Figure 11.5,

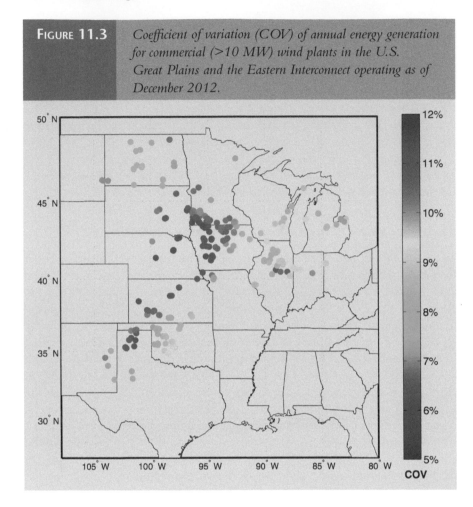

FIGURE 11.3   *Coefficient of variation (COV) of annual energy generation for commercial (>10 MW) wind plants in the U.S. Great Plains and the Eastern Interconnect operating as of December 2012.*

where the solid line plots the mean COV and the dashed lines plot the 5th and 95th percentile values.

The COV of annual energy density decreases steadily as larger and larger areas of wind plants are aggregated together. However, interannual variability is not reduced any farther by aggregating areas with a radius larger than 1,000 to 1,500 km. It is likely that this is not a fundamental limit of wind power; it is a limit of the correlations between the wind plants we analyze. The variance of the sum of correlated wind plants is the sum of their covariance. This means that variability of aggregated wind plants can be reduced by reducing the variability of some of the individual plants or reducing their correlation with each other. There is no easy way to reduce variability of annual energy generation, so we investigate changing the correlation between wind plants.

Figure 11.6 plots the correlation between annual energy density of pairs of wind plants against the distance between them. The correlation decreases with distance,

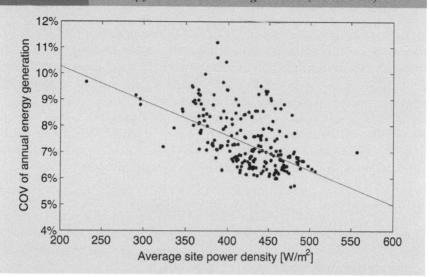

FIGURE 11.4 *COV of annual energy generation vs. average power density for sites shown in Figure 11.1 ($R^2 = 0.36$).*

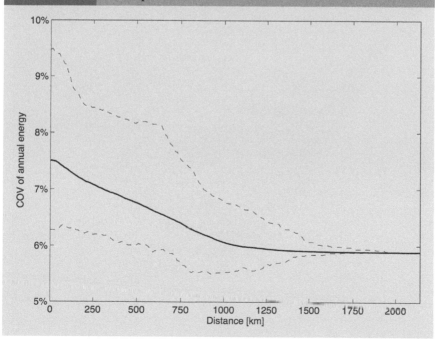

FIGURE 11.5 *COV of energy density of the aggregate of all wind plants within a given distance. Dashed lines plot the 5th and 95th percentile values.*

so we expect that adding distant wind plants will reduce the variability of the aggregate energy density. Klink (2007) and Kahn (1979) found a similar decrease in correlation of mean monthly wind speed with distance. Czisch and Ernst (2001) found that the correlation of mean monthly wind power does not decrease below zero even at distances greater than 3,000 km. Other authors have found that correlation of hourly or subhourly wind speed decreases exponentially with distance (Giebel, 2000; Katzenstein et al., 2010; Kempton et al., 2010; Sinden 2007). Robeson and Shein (1997) estimate spatial autocorrelation, different from the correlation versus distance we calculate here, and find no spatial autocorrelation of annual average wind speed at any distance. The correlation we and the other authors calculate shows that wind speed variations are more similar for nearby sites than for distant ones; the spatial autocorrelation Robeson and Shein calculate shows that wind speeds are more similar for nearby sites than for more distant ones.

11.3.4 Variability of seasonal energy generation

There are significant seasonal patterns in wind energy generation and its variability. Table 11.1 lists the seasonal COV, standard deviation, and mean of annual energy density for the wind plants shown in Figure 11.1. In the U.S. Great Plains, wind

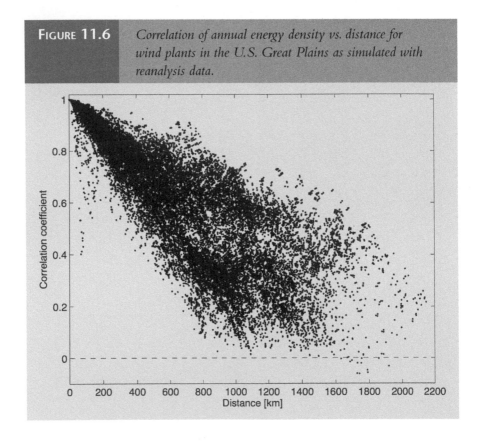

FIGURE 11.6 *Correlation of annual energy density vs. distance for wind plants in the U.S. Great Plains as simulated with reanalysis data.*

TABLE 11.1	*Seasonal statistics for wind energy density in the U.S. Great Plains.*			
	Winter	**Spring**	**Summer**	**Fall**
Energy density COV	12.6%	8.8%	10.8%	8.4%
Energy density std. dev.	3.9×10^9	2.5×10^9	1.6×10^9	2.1×10^9
Energy density mean (Wh/m²)	3.1×10^{10}	2.8×10^{10}	1.5×10^{10}	2.5×10^{10}

power peaks in the winter and is lowest in the summer. Variability also peaks in the winter but is nearly as large in the summer.

These findings match results from Klink (1999), who analyzed wind speeds rather than wind power. Klink suggests that the variance increases in the winter because the jet stream moves farther south and synoptic-scale weather systems become more frequent.

These seasonal COVs are not directly comparable to the annual COVs because they are calculated from energy summed over different periods. The sums of annual energy will have less variance than the sums of seasonal energy.

11.4 Summary

Annual wind energy density varies significantly from year to year, similar to many other meteorological quantities such as rainfall. We find that the coefficient of variation (the standard deviation normalized by the mean) of the aggregate annual wind energy for all wind plants in the U.S. Great Plains is approximately 6%. The COV of the individual wind plants in that area ranges from 6 to 11%, with generally lower variability at sites with better wind resource. Connecting wind plants together reduces the interannual variability of their combined energy generation, depending on the correlation between them. The correlation between sites decreases with the distance between them, so connecting more distant wind plants will generally reduce variability more than connecting nearby ones will.

We cannot estimate the interannual variability of wind energy from the entire Eastern Interconnect or the entire United States because we have not validated the CFS reanalysis data for inhomogeneous terrain such as the Appalachian Mountains, where many wind plants are installed. However, we expect the COV of annual energy generation for the entire Eastern Interconnect is smaller than the 6% we found for only the Great Plains because the correlation between Great Plains sites and Appalachian sites is likely to be lower than the correlations within either region. Fertig and colleagues (2012) show that the variability of hourly wind power from four connected regions in the western and central United States (BPA, CAISO, ERCOT, and MISO) is lower than the variability of any individual site. Czisch and Ernst (2001) use reanalysis data for Europe and Asia to show that the correlation of monthly-average wind speed falls to zero at distances of approximately 3,000 km.

The most important consequence of interannual variability is that it determines the debt-to-equity ratio of wind power projects, which affects the profitability for their owners. The lower the interannual variability, the more a wind plant developer can leverage his equity. Some sites in the U.S. Great Plains have nearly twice the interannual variability in annual energy as others, which puts them at a financial disadvantage. However, if two or more distant (poorly correlated) wind plants can be financed together, they may be able to get better financing terms than if they are considered separately. For example, if developers combine a site in Iowa with a COV of 6% and a site with the same mean generation in Illinois with a COV of 11%, the combined COV of annual energy is 7.5%.

Grid operators already manage the interannual variability of hydroelectric generation, which Katzenstein (2010) estimates to have a COV of 12%, twice as large as the variability we estimate for wind generation. Long-term planning for generation capacity may benefit from understanding the size of this interannual variability. However, grid operators typically estimate the contribution of wind power to peak generation capacity based on the correlation between hourly wind power and electricity demand.

References

Burton, T., Sharpe, D., Jenkins, N., & Bossanyi, E. (2001). *Wind energy handbook*. Chichester, UK: John Wiley & Sons.

Czisch, G., & Ernst, B. (2001). High wind power penetration by the systematic use of smoothing effects within huge catchment areas shown in a European example. Paper presented at Windpower 2001, Washington, DC, June 4–7.

Fertig, E., Apt, J., Jaramillo, P., & Katzenstein, W. (2012). The effect of long-distance interconnection on wind power variability. *Environmental Research Letters*, 7(3), 034017.

Giebel, G. (2000). *Equalizing effects of the wind energy production in northern Europe determined from reanalysis data*. Roskilde, Denmark: Risø National Laboratory.

Golding, E.W., & Stodhart, A.H. (1949). *The potentialities of wind power for electricity generation (with special reference to small-scale operation)*. Leatherhead, Surrey, United Kingdom: British Electrical and Allied Industries Research Association.

Huang, J., Lu, X., & McElroy, M.B. (2014). Meteorologically defined limits to reduction in the variability of outputs from a coupled wind farm system in the Central US. *Renewable Energy*, 62, 331–340.

Jarque, C.M., & Bera, A.K. (1987). A test for normality of observations and regression residuals. *International Statistical Review/Revue Internationale de Statistique*, 55(2), 163–172.

Justus, C.G., Mani, K., & Mikhail, A. (1979). Interannual and month-to-month variations of wind speed. *Journal of Applied Meteorology and Climatology*, 18(7), 913–920.

Kahn, E. (1979). The reliability of distributed wind generators. *Electric Power Systems Research*, 2(1), 1–14.

Katzenstein, W., Fertig, E., & Apt, J. (2010). The variability of interconnected wind plants. *Energy Policy*, 38(8), 4400–4410.

Kempton, W., Pimenta, F.M., Veron, D.E., & Colle, B.A. (2010). Electric power from offshore wind via synoptic-scale interconnection. *Proceedings of the National Academy of Sciences*, 107(16), 7240–7245.

Klink, K. (1999). Climatological mean and interannual variance of United States surface wind speed, direction and velocity. *International Journal of Climatology*, 19(5), 471–488.

Klink, K. (2002). Trends and interannual variability of wind speed distributions in Minnesota. *Journal of Climate*, *15*(22), 3311–3317.

Klink, K. (2007). Atmospheric circulation effects on wind speed variability at turbine height. *Journal of Applied Meteorology and Climatology*, *46*(4), 445–456.

Lilliefors, H.W. (1967). On the Kolmogorov-Smirnov test for normality with mean and variance unknown. *Journal of the American Statistical Association*, *62*(318), 399–402.

Palutikof, J.P., Kelly, P.M., & Davies, T.D. (1987). Impacts of spatial and temporal wind-speed variability on wind energy output. *Journal of Climate and Applied Meteorology*, *26*(9), 1124–1133.

Potter, C.W., Gil, H.A., & McCaa, J. (2007). Wind power data for grid integration studies. Paper presented at the IEEE Power Engineering Society General Meeting, Tampa, FL, June 24–28.

Pryor, S.C., & Barthelmie, R.J. (2011). Assessing climate change impacts on the near-term stability of the wind energy resource over the United States. *Proceedings of the National Academy of Sciences*, *108*(20), 8167–8171.

Robeson, S.M., & Shein, K.A. (1997). Spatial coherence and decay of wind speed and power in the north-central United States. *Physical Geography*, *18*(6), 479–495.

Saha, S., Moorthi, S., Pan, H.L., Wu, X., Wang, J., et al. (2010). The NCEP climate forecast system reanalysis. *Bulletin of the American Meteorological Society*, *91*(8), 1015–1057.

Sinden, G. (2007). Characteristics of the UK wind resource: Long-term patterns and relationship to electricity demand. *Energy Policy*, *35*(1), 112–127.

Stull, R.B. (1988). *An introduction to boundary layer meteorology*. Dordrecht, the Netherlands: Kluwer Academic Publishers.

Thomas, P.H. (1945). *Electric power from the wind*. Washington, DC: Federal Power Commission.

U.S. National Renewable Energy Laboratory. (2013). *Map of wind farms*. http://en.openei.org/w/index.php?title=Map_of_Wind_Farms&oldid=593614

Wan, Y.-H. (2012). *Long-term wind power variability*. National Renewables Energy Laboratory, NREL/TP-5500–53637. www.nrel.gov/docs/fy12osti/53637.pdf

12

REDUCTION OF WIND POWER VARIABILITY THROUGH GEOGRAPHIC DIVERSITY

12.1 Introduction

The variability of wind-generated electricity can be reduced by aggregating the outputs of wind generation plants spread over a large geographic area (see Chapter 3). In this chapter, we utilize Monte Carlo simulations to investigate upper bounds on the degree of achievable smoothing and clarify how the degree of smoothing depends on the *number* of wind plants and on the size of the geographic *area* over which they are spread.

We model two distinct benefits of geographic diversity that have different behaviors: (1) increased tendency of generation level to lie near its mean and (2) decreased tendency for it to lie near its extremes. Gaussian or normal probability distributions give accurate estimates of the first but underestimate the second. The second benefit, which has particular importance for electric grid reliability, has not been widely treated before.

The effect of geographic diversity on wind generation variability has been investigated by many, starting with Thomas (1945). Significant early work was carried out by Molly (1977), Justus and Mikhail (1978), Kahn (1979), Farmer and colleagues (1980), and Carlin and Haslett (1982). Some researchers have focused on particular geographic regions, including the U.S. Midwest (Archer & Jacobson, 2003, 2007; Fisher et al., 2013) and the Nordic countries (Holttinen, 2005). Others have investigated the effect on the frequency spectrum of the generated power (Beyer et al., 1993; Katzenstein et al., 2010; McNerney & Richardson, 1992; Nanahara et al., 2004; Tarroja et al., 2011). Hasche (2010) has modeled how the smoothing benefit saturates as the number of generator sites within a region increases. Some recent investigations have focused on the effects of spreading arrays of wind generators over especially large distances. Kempton and colleagues (2010) and Dvorak and colleagues (2012) considered an array of wind plants distributed along the entire

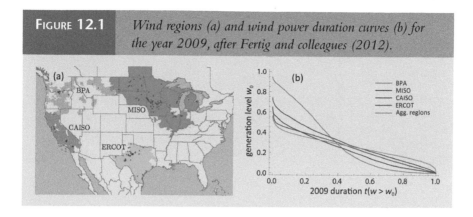

FIGURE **12.1** *Wind regions (a) and wind power duration curves (b) for the year 2009, after Fertig and colleagues (2012).*

extent of the U.S. East Coast, while Fertig and colleagues (2012) and Louie (2013) evaluated the smoothing effect on wind generation by interconnections between independent system operators (ISOs) across the United States, and Huang and colleagues (2014) considered wind plants spread over the Great Plains of the United States from Montana to Texas. These previous studies largely model the generation of an array of hypothetical wind plants by simulating the output of each individual plant from a few-year historical weather record, with some additionally fitting the empirical distribution of modeled array generation levels to some chosen standard parametric probability distribution.

As an example of that work, consider the four U.S. regions examined by Fertig and colleagues (2012). Figure 12.1 shows the regions and a generation duration curve for each region and for a sum of all regions using hourly data. The sum simulates the generation duration curve for the year 2009 if all four regions had been interconnected. The duration curve for the Bonneville Power Administration (BPA) region shows the effect of high correlation between turbines that are mostly located in the Columbia River gorge. During the year shown by the data in Figure 12.1(b), there is no hour in which the sum of the power from all four regions drops to zero. If data were available for a long enough interval, perhaps many decades, we would expect that in at least one hour, there would be no wind generation at all; the duration curves would always go to zero at 100% of the hours.

For probabilistic consideration of long-term capacity planning, the interesting question is: for a given percentage of hours (e.g., 99.97%), what is the minimum power output that can be counted upon from widely distributed wind turbines?

12.2 Monte Carlo model

To gain better insight into the implications of geographic diversity for low-probability events that influence grid reliability and to more clearly articulate the dependence of the different smoothing benefits on the number of included wind plants, we created a Monte Carlo model of an array with an adjustable number of independent wind plants.

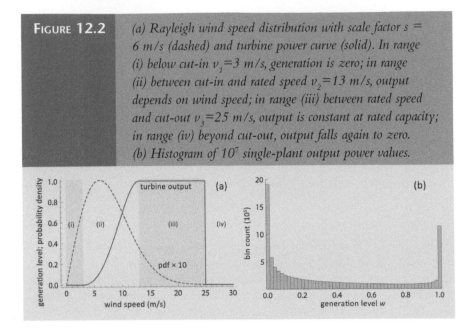

FIGURE **12.2** *(a) Rayleigh wind speed distribution with scale factor s = 6 m/s (dashed) and turbine power curve (solid). In range (i) below cut-in v_1=3 m/s, generation is zero; in range (ii) between cut-in and rated speed v_2=13 m/s, output depends on wind speed; in range (iii) between rated speed and cut-out v_3=25 m/s, output is constant at rated capacity; in range (iv) beyond cut-out, output falls again to zero. (b) Histogram of 10^7 single-plant output power values.*

The model treats the generation level w of an individual wind plant as a random variable ranging from 0 to 1, where $w = W/W_{cap}$ represents the plant's output power W normalized by its nameplate capacity W_{cap}. Transforming 10^7 wind speed samples drawn from a Rayleigh distribution by the turbine power curve depicted in Figure 12.2(a) (more details are in the Appendix to this chapter) yields a "base" sequence of 10^7 independent samples of generation level w having the histogram shown in Figure 12.2(b). For the chosen parameters, the capacity factor or average generation level $\mu = 0.373$. The large count or spike in the histogram bin centered at $w = 0$ arises from ranges (i) and (iv) in Figure 12.2(a), where wind speeds below cut-in and above cut-out give zero generation. Similarly, the spike in the bin centered at $w = 1$ arises from range (iii) where wind speeds drive the turbines to rated capacity.

While the histogram of Figure 12.2(b) is intended to be representative of the distribution of generation levels for a typical single turbine, we would expect the reduced variability of arrays of geographically diverse wind plants to be manifest in array histograms with rather different characteristics. We create a model of this behavior by supposing that all the wind plants in an array are identical, each having the same generation-level probability distribution as the single turbine of Figure 12.2, but with generation levels independent of the other plants in the array. By taking "diversity" to the extreme of "complete independence," our model clearly shows the effects due to diversity. We treat the degree to which independence overstates the diversity of realistic geographically distributed wind plants by later introducing the concept of effective sample size.

To model the output of an array of N independent wind plants, we draw N independent samples from the single-turbine distribution, add them together, and normalize them to the total array capacity. With the single-turbine levels already normalized, the array generation level is just the arithmetic average of the N independent samples. For computational efficiency, we create sequences for many different array sizes from a single $N = 1$ base sequence. Table 12.1 shows as an example a 5-element 3-sequence created from a 15-element base sequence.

12.3 Variability-reduction results

At small N values, the modeled array generation values cluster noticeably closer to the mean ($\mu = 0.373$ for all sequences here), as seen in Figure 12.3(a) for $N = 4$. Unlike the single-generator histogram of Figure 12.2(b), which has its most frequent values at the $w = 0$ and $w = 1$ extremes, the four-generator-array histogram has a central mode but still an appreciable fraction of zero-output generation levels. Increasing N further drives the histograms toward the bell shape characteristic of the Gaussian or normal distribution, as expected according to the Central Limit Theorem and as seen in Figure 12.3(b) for $N = 16$.

To quantify variability reduction, we define two metrics shown in Figure 12.4. The amount of time T the array of generators would spend in a particular state over a sufficiently long interval T_{tot} gives a fractional "duration" $t = T/T_{tot}$ we associate with the frequency at which that particular state occurs in the N-sequence. Duration $t_c(\delta)$ defined in Figure 12.4(a) measures the tendency of generation level w to lie within a range $\pm \delta/2$ of central mean μ; here δ is 0.15 and the corresponding t_c value is 0.38, meaning generation is within $\pm 7.5\%$ of the mean 38% of the time. Duration $t_e(\varepsilon)$ defined in Figure 12.4(b) measures the tendency of generation to lie within a range ε above its low extreme level (zero); here ε is 0.10 and the corresponding t_e value is 0.011, meaning that generation is less than 10% about 1.1% of the time. We have reversed the duration axis in Figure 12.4(b) relative to that in Figure 12.4(a) so small values of t_e can be clearly visualized on a conventional logarithmic axis. The two measures of reduced variability, larger t_c and smaller t_e, depend differently on N, as we show.

First, consider the occurrence of near-mean generation levels. Figure 12.5(a) shows generation duration curves for arrays of different N. By "N" we do not mean individual wind turbines or individual wind plants; rather, N is the number of statistically independent wind generators in our Monte Carlo model. In a later section, we estimate the effective value of N in various regions by using observed data. Increasing N progressively flattens the curves around the mean, increasing duration $t_c \equiv t(|w - \mu| < \delta)$. Note the similarity of the $N = 1$ curve to that of BPA's 2009 output in Figure 12.1.

Since the generation levels should become approximately normally distributed as N increases, we would expect duration t_c to behave approximately as $t_c \approx 2\Phi[\delta/\sigma_N] - 1$, where Φ is the unit-normal cumulative distribution function (CDF). The variance

TABLE 12.1 *Base sequence of random single-plant outputs (N = 1) and new 3-sequence of random outputs from an array of three independent plants (N = 3), generated by nonoverlapping arithmetic averages of base-sequence triples.*

base (N = 1)	0.7362	0.1890	0.0989	0.0573	0.0298	0.7022	0.0882	0.2247	0.1531	0.1332	0.1049	1.0000	0.1356	0.2763	0.0803
(N = 3)		0.3414			0.2631			0.1553			0.4127			0.1640	

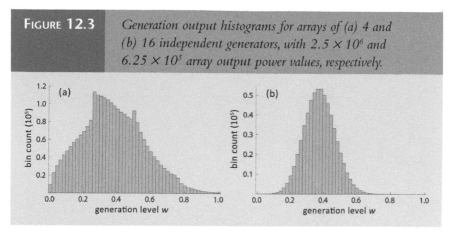

FIGURE **12.3** — *Generation output histograms for arrays of (a) 4 and (b) 16 independent generators, with 2.5×10^6 and 6.25×10^5 array output power values, respectively.*

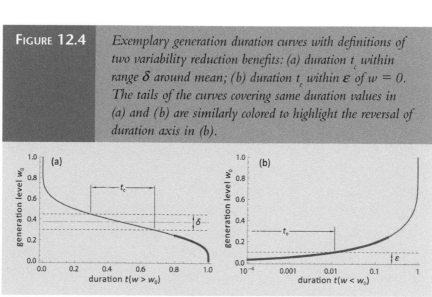

FIGURE **12.4** — *Exemplary generation duration curves with definitions of two variability reduction benefits: (a) duration t_c within range δ around mean; (b) duration t_e within ε of $w = 0$. The tails of the curves covering same duration values in (a) and (b) are similarly colored to highlight the reversal of duration axis in (b).*

is given by the Bienaymé formula as $\sigma_N^2 = \sigma_1^2 / N$, where $\sigma_1^2 = 0.134$ is the single-plant variance. The normal approximation for t_c, plotted as solid curves in Figure 12.5(b), is quite accurate for N larger than 5 or so. With near-normal behavior, t_c increases in proportion to $\sqrt{N}$ until it begins to saturation at $t = 1$.

The normal approximation, however, does not provide good estimates for the occurrence of near-zero generation levels, which we consider next. Figure 12.6 shows reversed generation duration curves for arrays of the same N values used in Figure 12.5. The increasingly square "toe" of the duration curve seen in Figure 12.5(a) correlates to the vanishing amounts of time spent at low generation levels observed in Figure 12.6(a). As can be seen by the dashed curves, the normal

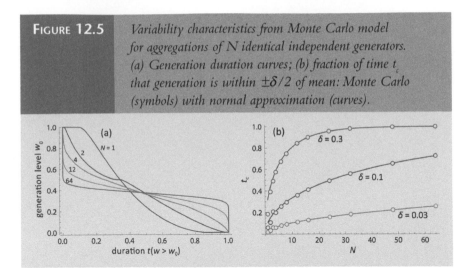

FIGURE 12.5 — Variability characteristics from Monte Carlo model for aggregations of N identical independent generators. (a) Generation duration curves; (b) fraction of time t_c that generation is within $\pm\delta/2$ of mean: Monte Carlo (symbols) with normal approximation (curves).

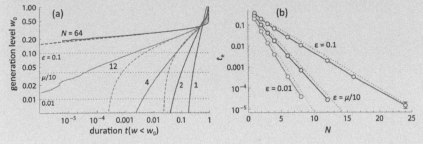

FIGURE 12.6 — Variability at low generation levels. (a) Reversed generation duration curves: Monte Carlo (solid curves) and normal approximation (dashed curves); (b) fraction of time t_e that generation is less than ε: Monte Carlo (symbols, solid curves), guide lines (dotted) of the form be^{-aN}, with $a = 0.43, 0.80,$ and 1.15. Error bar on last $\varepsilon = 0.1$ point spans the range of t_e values from 10 Monte Carlo runs each with a different starting seed.

estimate $t \approx \Phi[(w - \mu)/\sigma_N]$ gives only a poor approximation to the Monte Carlo results for low generation levels. Variability metric $t_e \equiv t(w < \varepsilon)$ measuring the fraction of time that generation drops below threshold ε falls approximately exponentially with N as shown in Figure 12.6(b), in accordance with the theory of large deviations (Lewis & Russell, 1997). For our presumably typical turbine and wind parameters and over the range of ε values down to at least 0.01, the rate of decline of t_e with N becomes larger the smaller ε gets.

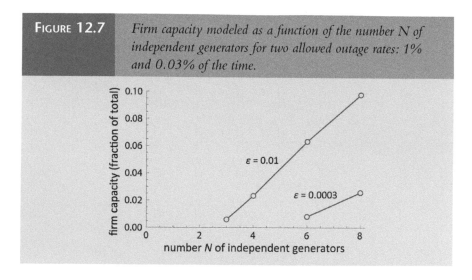

FIGURE 12.7 *Firm capacity modeled as a function of the number N of independent generators for two allowed outage rates: 1% and 0.03% of the time.*

In evaluating electric-grid reliability, a useful concept is that of *firm* capacity: capacity that is expected to be available all of the time except for infrequent outages. Figure 12.7 translates the duration versus capacity methods of Figure 12.6 into the form of firm capacities. If an outage rate of 1% were tolerable (as perhaps for an individual generator in a regional utility), a wind array could begin providing firm capacity with three or more independent plants. If outages were allowed only 0.03% of the time, a wind array would begin providing firm capacity at $N = 6$. These model results ignore diurnal and seasonal variations in average wind speed; these variations have significant correlations with electric load that make meaningful capacity calculations more complex.

12.4 Comparison with observations

Our model for the variability-reducing benefits arising from aggregation of independent wind generators compares well with observations based on real wind data, as we show by examining results previously reported by Kempton and colleagues (2010). Using wind-speed data from meteorological buoys, Kempton and colleagues created generation duration curves for 11 hypothetical offshore wind plants spread over the entire length of the U.S. Atlantic coast and for the aggregate of all 11, called P_{grid} (see their Figure S3). In Figure 12.8 (a), we reproduce two of their 12 curves: one for single Station S7 (blue) that falls near the middle of the various capacity factors and one for P_{grid} (red). Small adjustments to our turbine power curve parameters gave Monte Carlo model base sequence results ($N = 1$, blue dashed curve) approximating the single-station duration curve. Then, without further adjustments, we compared model results for various array sizes to the P_{grid} curve, selecting $N = 4$ as the best model match (red dashed). Figure 12.8 (b) shows the sensitivity of modeled low generation levels to choice of N: clearly $N = 3$ is too small and $N = 5$ is too large. (The error bar on the second point shows the width of Kempton's

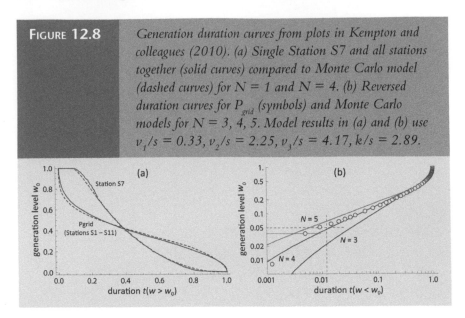

| FIGURE 12.8 | *Generation duration curves from plots in Kempton and colleagues (2010). (a) Single Station S7 and all stations together (solid curves) compared to Monte Carlo model (dashed curves) for N = 1 and N = 4. (b) Reversed duration curves for P_{grid} (symbols) and Monte Carlo models for N = 3, 4, 5. Model results in (a) and (b) use $v_1/s = 0.33, v_2/s = 2.25, v_3/s = 4.17, k/s = 2.89$.* |

printed P_{grid} curve, from which we digitized the points; the dashed crosshairs slightly above the curve but below the digitization symbols mark the 1.2% duration they explicitly report for $P_{grid} < 5\%$.) As nearly as can be determined from the published results, the variability-reducing benefits found by Kempton and colleagues for their geographically diverse Atlantic Transmission Grid are identical to the benefits of aggregating four statistically independent generators of similar characteristics.

12.5 Discussion

Our Monte Carlo model, along with well-defined quantitative metrics for variability reduction, provides a basis for understanding the benefits of geographically diverse wind arrays. Model calculations of the increase with N of the duration t_c of near-mean generation agree well with approximations made by assuming that aggregated generation levels are normally distributed. However, the same assumption does not suffice for estimating the duration t_e of near-zero generation levels, for reasons that can be readily appreciated from the characteristics of the underlying PDF (probability density function) shown in Figure 12.9(a).

While the domain of the normal PDF extends over $(-\infty, +\infty)$, any probability density representing generation normalized by nameplate capacity must necessarily equal zero outside of the interval [0,1], as shown by the red curve. Even though small, the area under the tails of the normal PDF outside the [0,1] interval is nonnegligible and will likely confound any attempted estimation of the probability (duration) of small generation levels. For example, Figure 12.9(b) compares the Monte Carlo model results previously presented in Figure 12.6(b) for the duration

FIGURE **12.9**   *Normal approximations. (a) Comparison of normal PDF (blue) with PDF of same mean (0.5) and standard deviation (0.15) but bounded to domain [0,1]: normal PDF tails extend significantly beyond domain boundaries (inset: linear axes). (b) Duration t_e of generation less than 10% of mean: Monte Carlo (symbols, solid curve), normal approximation (dashed curve).*

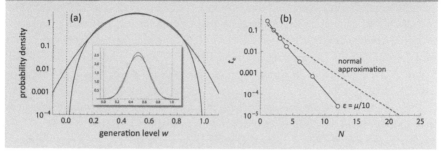

of low generation levels to the corresponding normal approximation $t_e \approx \Phi[(\varepsilon - \mu)/\sigma_N]$. For $N = 12$ and $\varepsilon = \mu/10 \approx 3.7\%$, the normal approximation overestimates the duration by a factor of about 30; it also estimates that 19 independent generators would be required to achieve a duration $t_e < 2.7 \times 10^{-5}$ when 12 would suffice.

The nonzero temporal and geographical correlations of real wind speeds may raise concerns about the utility of our model, based as it is on independent random samples. Our model provides an upper bound to variability reductions achievable through geographic diversity because it is based on the assumption that every wind speed is uncorrelated with every other in space and time. However, real wind speeds are correlated with themselves in time and with speeds at sites nearby, reducing diversity. Real duration curves are usually measured over a time interval presumed long enough to characterize the range of typical system behavior. In this case, temporal correlation effects in the real data should average out and the independent nature of the random samples in our model should not affect the realism of the results.

With regard to spatial dependence, the close correspondence between our model results and the data reported by Kempton and colleagues (2010) might be interpreted, however, in light of an *effective sample size*. For their buoy wind-speed data, they found intersite correlation falling as $\rho(r) = e^{-r/\ell}$ with $\ell = 430$ km. From the $L = 2,600$ km end-to-end length of their array, we would expect, by comparison with the diversity factor of Planter and colleagues (1980), that the array would behave as though it comprised $N_{eff} \approx L/(2\ell) = 2,600/860 = 3.0$ independent generators. Comparing the average of the variances we extract from their 11 single-station duration curves, $\langle \sigma_1^2 \rangle = 0.132$, to the average of the variances we extract from their P_{grid} curve, $\sigma^2 = 0.042$ gives another estimate with a similar value: $N_{eff} \approx 0.132/0.042 = 3.1$. We find the agreement between these N_{eff} values and the $N = 4$ of our closest-fitting model curve acceptable given (1) the difficulty in estimating correlation length ℓ from noisy meteorological data and (2) that we generated our Monte Carlo samples from a base distribution adjusted to match only a single "representative" site (S7), ignoring those other sites with lesser (or greater) t_e.

To gain further insight into N_{eff} values, we examined generation data previously available from ERCOT, as analyzed by Katzenstein and colleagues (2010), for 20 individual wind plants lying within a 450 km × 220 km region in Texas. Over the year 2008, the generation-level variances of the individual plants had a capacity-weighted average of 0.084 (range of 0.013 to 0.118), while the total generation had variance 0.055. Thus, comparing the variability of all wind generation in ERCOT to the average of the variabilities of single plants leads to an apparent value for ERCOT of $N_{eff} = 1.5$.

One might expect on the basis of size that a typical U.S. ISO would behave as though it comprised a number of independent generators lying between the $N_{eff} = 1.5$ we found for the 450-km ERCOT subregion and the $N_{eff} = 4$ we found for Kempton's 2,600-km P_{grid}. According to our modeled dependence of firm capacity on the number of independent generators, such an ISO sits just on threshold of having enough diversity to yield the beginning of any firm wind capacity. Utilizing transmission lines to connect several such regions could give nonnegligible firm capacity, although firming with conventional reserve generation would likely be more cost effective.

However, before taking the effective sample size concept too seriously, we need a better understanding of the nature of correlation versus separation for real wind generators. In particular, if the infrequent outages responsible for the outages at low rates are also short, then the appropriate correlation length ℓ may be smaller than it is for longer fluctuations (Ernst, Wan, & Kirby, 1999), resulting in a larger N_{eff} than might be expected from the ratio of variances. Moreover, most work to date on wind-speed correlation versus site separation characterizes correlation only by its single *coefficient*, with no assurance that this number equally represents behaviors at low and average wind speed levels.

12.6 Summary

This chapter presents a simple model for the benefits of geographic diversity of wind generation based on arrays comprising a selectable number N of *statistically independent* wind generators. The N statistically independent generators in the model correspond to a greater number of partially correlated generators in a real array. We emphasize that N is not the number of actual wind turbines or wind plants. For electric power generated by N statistically independent wind plants, the duration of near-mean generation grows at a rate initially proportional to $\sqrt{N}$ that then declines only as the generation level approaches being within the near-mean bounds 100% of the time. The incremental or marginal benefit from each added statistically independent generator declines as $N^{-1/2}$. Duration t_e of near-zero generation decreases approximately as e^{-aN}, where rate a becomes larger as the generation threshold gets closer to zero. Log[t_e] never saturates and has constant marginal improvement from each incremental independent generator.

The concept of effective sample size helps differentiate between the effect of increasing generator *number* and increasing geographic *area*. An array of real generators *within* a region of linear size comparable to correlation length ℓ has high intergenerator correlation and never gives N_{eff} much larger than 1, yielding a variability reduction that saturates rapidly with generator number, as shown systematically by Hasche (2010). On the other hand, to the extent that intergenerator correlation falls off exponentially with separation, there is no limiting distance or region size beyond which further expansion fails to produce growth in achievable N_{eff} and substantial additional variability reduction.

Further work on the nature of wind correlation is needed to determine whether sufficiently large N values are feasible within continental-scale geographic regions to provide substantial firm capacity.

12.7 Appendix: Turbine power curve

Our Monte Carlo model converts each randomly generated wind speed sample v_i to a generation level w_i using the following turbine power function:

$$
w_i = \begin{cases} 0 & v_i < v_1 \text{ or } v_i \geq v_3 \\ \sin^2\left[\dfrac{\pi}{2}\left(\dfrac{v_i - v_1}{v_2 - v_1} + \dfrac{(v_i - v_1)(v_i - v_2)}{k^2}\right)\right] & v_1 \leq v_i < v_2 \\ 1 & v_2 \leq v_i < v_3 \end{cases} ,
$$

with $v_1 = 3$ m/s, $v_2 = 13$ m/s, $v_3 = 25$ m/s, and $k = \sqrt{300}$ m/s, unless otherwise noted. The function provides a continuous, invertible, and differentiable approximation to the manufacturer's data for the GE 1.5 MW turbine.

References

Archer, C.L., & Jacobson, M.Z. (2003). Spatial and temporal distributions of U.S. winds and wind power at 80 m derived from measurements. *Journal of Geophysical Research, 108*, D9.

Archer, C.L., & Jacobson, M.Z. (2007). Supplying baseload power and reducing transmission requirements by interconnecting wind farms. *Journal of Applied Meteorology and Climatology, 46*(11), 1701–1717.

Beyer, H.G., Luther, J., & Steinberger-Willms, R. (1993). Power fluctuations in spatially dispersed wind turbine systems. *Solar Energy, 50*(4), 297–305.

Carlin, J., & Haslett, J. (1982). The probability distribution of wind power from a dispersed array of wind turbine generators. *Journal of Applied Meteorology, 21*(3), 303–313.

Dvorak, M.J., Stoutenburg, E.D., Archer, C.L., Kempton, W., & Jacobson, M.Z. (2012). Where is the ideal location for a U.S. East Coast offshore grid? *Geophysical Research Letters, 39*(6), L06804.

Ernst, B., Wan, Y.-H., & Kirby, B. (1999). *Short-term power fluctuation of wind turbines: Analyzing data from the German 250-MW measurement program from the ancillary services viewpoint.* National Renewable Energy Laboratory report, NREL/CP-500-26722. www.nrel.gov/docs/fy99osti/26722.pdf

Farmer, E. D., Newman, V. G., & Ashmole, P. H. (1980). Economic and operational implications of a complex of wind-driven generators on a power system. *IEE Proceedings A, 127* (5), 289–295.

Fertig, E., Apt, J., Jaramillo, P., & Katzenstein, W. (2012). The effect of long-distance interconnection on wind power variability. *Environmental Research Letters, 7*(3), 034017.

Fisher, S.M., Schoof, J.T., Lant, C.L., & Therrell, M.D. (2013). The effects of geographical distribution on the reliability of wind energy. *Applied Geography, 40,* 83–89.

Hasche, B. (2010). General statistics of geographically dispersed wind power. *Wind Energy, 13*(8), 773–784.

Holttinen, H. (2005). Hourly wind power variations in the Nordic countries. *Wind Energy, 8*(2), 173–195.

Huang, J., Lu, X., & McElroy, M.B. (2014). Meteorologically defined limits to reduction in the variability of outputs from a coupled wind plant system in the Central U.S. *Renewable Energy, 62,* 331–340.

Justus, C.G., & Mikhail, A.S. (1978). *Energy statistics for large wind turbine arrays.* Georgia Institute of Technology report. http://hdl.handle.net/1853/40482

Kahn, E. (1979). The reliability of distributed wind generators. *Electric Power Systems Research, 2*(1), 1–14.

Katzenstein, W., Fertig, E., & Apt, J. (2010). The variability of interconnected wind plants. *Energy Policy, 38*(8), 4400–4410.

Kempton, W., Pimenta, F.M., Veron, D.E., & Colle, B.A. (2010). Electric power from offshore wind via synoptic-scale interconnection. *Proceedings of the National Academy of Sciences, 107*(16), 7240–7245.

Lewis, J.T., & Russell, R. (1997). *An introduction to large deviations for teletraffic engineers.* www.cl.cam.ac.uk/research/srg/netos/measure/tutorial/rev-tutorial.ps.gz

Louie, H. (2014). Correlation and statistical characteristics of aggregate wind power in large transcontinental systems. *Wind Energy, 17*(6), 793–810.

McNerney, G., & Richardson, R. (1992). The statistical smoothing of power delivered to utilities by multiple wind turbines. *IEEE Transactions on Energy Conversion, 7*(4), 644–647.

Molly, J.P. (1977). Balancing power supply from wind energy converting systems. *Wind Engineering, 1*(1), 57–66.

Nanahara, T., Asari, M., Maejima, T., Sato, T., Yamaguchi, K., & Shibata, M. (2004). Smoothing effects of distributed wind turbines. Part 2. Coherence among power output of distant wind turbines. *Wind Energy, 7*(2), 75–85.

Tarroja, B., Mueller, F., Eichman, J.D., Brouwer, J., & Samuelsen, S. (2011). Spatial and temporal analysis of electric wind generation intermittency and dynamics. *Renewable Energy, 36*(12), 3424–3432.

Thomas, P.H. (1945). *Electric power from the wind.* Washington, DC: Federal Power Commission.

13

CYCLING AND RAMPING OF FOSSIL PLANTS, AND REDUCED ENERGY PAYMENTS

13.1 Introduction

Increasingly, we ask the machines that generate our electric energy to change the amount of power they produce; the industry calls this cycling. The growing market penetration of variable renewable generators, such as wind, is causing existing generators to cycle in order to compensate for wind's variability. For large generators, cycling means operating in a less efficient way, leading to higher fuel costs and air emissions. It also means an increased risk of creep and fatigue damage, which leads to higher maintenance costs and more outages (Agan, Besuner, Grimsrud, & Lefton, 2008).

In the peer-reviewed papers this chapter is based on (Katzenstein & Apt, 2009a; Oates & Jaramillo, 2013), we asked three questions about cycling. First, how much do we expect the cycling and ramping of conventional thermal generators (i.e., natural gas and coal plants) to increase as we add wind? Second, how much does cycling reduce the operating cost and emissions reductions we expect to see with high wind penetrations? Finally, in a high-wind future, if power plant operators change the way they bid their startup costs into the electric power market, how would it affect our conclusions about the quantity, costs, and emissions effects of cycling? We have addressed these questions through two case studies: (1) the operations of natural gas power plants used to balance the variability from a renewable energy plant and (2) an entire system, to evaluate the response of all the generators (including coal units) in the system.

When variable energy resources provide a significant fraction of electricity, other generators or rapid demand response must compensate when their output drops (Apt, 2007; Curtright & Apt, 2008; O'Neill et al., 2010). In many locations, natural gas turbines will be used to compensate for variable renewables. When turbines are quickly ramped up and down, their fuel use (and thus CO_2 emissions) may be larger than when they are operated at a steady power level. Systems that mitigate other emissions such as NO_x may not operate optimally when the turbines' power level is rapidly changed. In this case study, we estimated the emissions from a system that combines a variable renewable power plant with fast-ramping natural gas turbines to provide power that meets demand. We used a regression analysis of measured emissions and heat rate data taken at 1-minute resolution from two types of gas turbines to model emissions and heat rate as a function of power and ramp rate. One-minute data resolution is important because hourly data do not capture the minute-to-minute variability that drives the ramping of the natural gas–based generators. The required gas turbine power and ramp rate to fill in the variations in 1-minute data from four wind plants and one large solar photovoltaic (PV) plant were determined; then the emissions were computed from the regression model. The system emissions were compared to the emissions of one or an ensemble of natural gas plants and to the emissions reductions expected from adding a given amount of renewable power, such as wind and solar.

The gas turbine model is subject to physical operating constraints: the upper and lower power limits and how quickly the turbine can change its power output. For the natural gas turbine, we obtained 1-minute resolution emissions data for seven General Electric LM6000 natural gas combustion turbines and two Siemens-Westinghouse 501FD natural gas combined cycle turbines. The LM6000 CTs have a nameplate power limit of 45 MW and utilize steam injection to mitigate NO_x emissions. A total of 145 days of LM6000 emissions data were used in a regression analysis.

The Siemens-Westinghouse 501FD NGCC turbines have a nameplate power limit of 200 MW with GE's Dry Low NO_x (DLN) system (lean premixed burn) and an ammonia selective catalytic reduction (SCR) system for NO_x control. Emissions data for 11 days were obtained for the 501FD NGCC. Each emissions data set contains six variables: date, time, power generated, heat rate, mass of emissions, and a calibration flag. We modeled only NO_x and CO_2 emissions from the turbines.

The renewables data included 1-second, 10-second, and 1-minute resolution from four wind plants and one large solar photovoltaic facility located in the following regions in the United States: Eastern Mid-Atlantic, Southern Great Plains, Central Great Plains, Northern Great Plains, and Southwest. We averaged the wind data to 1-minute resolution to match the time resolution of

---

1. This section is based on Katzenstein and Apt (2009a, 2009b).

the natural gas generator emissions data and scaled each wind or PV data set's maximum observed power generated during the data set to the nameplate capacity of the paired natural gas turbine.

If a given level of penetration of wind or solar energy causes no additional emissions from gas generators, the emissions are expected to be displaced linearly according to the penetration factor of the renewables, an assumption called equivalent displacement.

Consider a system with generators that emit 2 tons of CO_2 per MWh without renewables in the system. Under the equivalent displacement scenario, a 10% penetration of renewable resources would reduce emissions to 1.8 tons of CO_2 per MWh.

The measured emissions of natural gas plants vary as they are ramped up and down to follow variable renewable generators, so that the equivalent displacement assumption does not hold. In addition, one or more of the fossil plants used to compensate for wind or solar variable generation must be running at all times so that it is ready to deliver power when needed; the idle emissions from those generators must be considered.

The analysis finds that a system with renewables that uses LM6000 turbines (45 MW) for fill-in power achieves roughly 80% of the expected CO_2 emissions reductions and 20 to 45% of the expected NO_x emissions reductions.

For the larger 200 MW 501FD turbine, the analysis found similar penalties for CO_2 emissions reductions. However, NO_x emissions increase substantially because the GE Dry Low NOx (DLN) control system of the 501FD turbine is optimized for steady-state high power operations, and emissions increase when the turbine is operated below about half power.

When the analysis was run for a system in which up to 20 200 MW 501FD natural gas turbines provide the reserve capability, with the required number at idle even when the renewables are producing at their maximum power, the CO_2 reductions were found to vary between 76% and 95% of that expected by the displacement assumption (depending on how many gas generators must be dispatched to counter variability).

If system operators recognize the potential for ancillary emissions from gas generators used to fill in variable renewable power, they can take steps to produce a greater displacement of emissions. By limiting generators with GE's DLN system to power levels of 50% or greater, ancillary emissions can be minimized. Operation of DLN controls with existing (but rarely used) firing modes that reduce emissions when ramping may be practical. On a time scale compatible with RPS implementation, design and market introduction of generators that are more appropriate from an emissions viewpoint to pair with variable renewable power plants may be feasible. In fact, recent developments in gas turbine technology suggest that this is the path forward: Siemens Frame H and GE's new FlexEfficiency portfolio of power plants have been designed and built to more efficiently balance variable power.

CASE STUDY *A coal-dominated power system.*[2]

In order to evaluate the effects of variability on the operations of a fleet of power plants, we developed a model of a sample power system. The previous case study evaluated the implications of ramping of natural gas plants to balance wind or solar PV variability. In this case study, we focus on the amount of coal cycling we expect to occur under a high-wind scenario as well as on the production cost and emissions implications. We thus built a cost-minimizing, unit commitment and economic dispatch (UCED) model to simulate the operation of a power system similar to the ones developed by others (EnerNex Corporation, 2011; Valentino et al., 2012). Our UCED models the Western region of the PJM Interconnection (PJM West) and includes 348 generators. The UCED minimizes the cost of meeting the demand for electricity, subject to operating constraints of the units and reserve requirements. We obtained load data for PJM West for 2006 (PJM, 2006a). Spinning reserve requirements were set at 3% of load in each period, approximating NERC requirements (NERC, 2012), plus 5% of wind generation, following a heuristic proposed by NREL (Piwko, Clark, Freeman, & Jordan, 2010).

The analysis used information about average fuel and startup costs for these generators. Startup costs are broken out by the amount of time the generator spends offline, with longer times generally corresponding to higher costs. A range was used, with the low startup costs based on average startup costs bid into PJM (PJM, 2006b), while the elevated startup costs for coal generators are based on calculations from APTECH (Agan, 2008; APTECH, 2009).

In order to model wind power production for high levels of wind penetration, we used simulated wind power output data from NREL's Eastern Wind Interconnection and Transmission Study (EWITS; EnerNex Corporation, 2011). The study used a mesoscale atmospheric model to generate spatially and temporally correlated wind power output based on anemometer data for the years 2005 and 2006. EWITS provided wind power output data for a number of hypothetical wind sites across the Eastern Interconnection. We selected sites from this database, in order of decreasing capacity factor, in order to achieve the desired level of wind for each of our model runs.

At the relatively low ramp rates reachable in our hourly model, the contribution of ramp rate to emissions is small. Cycling emissions are therefore dominated by coal units. For these units, hourly data from the EPA's Continuous Emissions Monitoring System (CEMS) database (EPA, 2006) reveal that ramp rate does not contribute greatly to overall emissions. Due to the relatively slower ramp rates available to coal plants, use of these hourly data is likely justified. We therefore divided generator operations into two states: normal operations, where emissions were modeled as a product of power output and an emissions factor, and shutdown/startup cycling, where emissions were modeled on a per-cycle basis. During normal operations, we modeled emissions as a product of

---

2. This section is based on Oates and Jaramillo (2013).

TABLE 13.1	*Average emissions factors for each unit type during normal operations. These rates correspond to base rates from EPA (2010). Uncontrolled NO_x rates (corresponding to psotcombustion scrubbers turned off) were used during the winter months, and controlled rates were used during summer months.*

	CO_2 Rate (ton/ MMBTU)	SO_2 Permit Rate (lb/ MMBTU)	Uncontrolled NO_x Rate (lb/MMBTU)	Controlled NO_x Rate (lb/MMBTU)
Coal	0.10	2.2	0.50	0.19
NG CC	0.058	0.062	0.10	0.077
NG Steam	0.058	1.3	0.097	0.097
NG CT	0.058	2.0	0.093	0.09
Oil Steam	0.087	1.7	0.26	0.26
Oil CT	0.081	1.6	0.82	0.82
Other	0.069	0.68	1.7	1.6

Source: Reprinted with permission from Oates and Jaramillo (2013).

power output, heat rate, and an input emissions factor. The input emissions factors, shown in Table 13.1, were estimated using data from EPA (2010).

During shutdown/startup cycling, we modeled emissions on a per-cycle basis. We first determined an energy penalty to capture inefficiency during operations at low power outputs. Heat rate penalty curves, of the form shown in Figure 13.1, were constructed for each unit type based on data from the CEMS database (EPA, 2006). Then, using the unit's ramp rate to determine its startup time and integrating the heat rate penalty curve over time, we were able to determine the fuel input requirement of a shutdown/startup cycle. Per-cycle fuel inputs were then multiplied by the emissions factors in Table 13.1 (using the uncontrolled NO_x rate) to determine per-cycle emissions.

Using the model described, we could assess changes in cycling and cycling-related costs and emissions in a high-wind scenario. At the levels of cycling suggested by our results in the high-wind scenario, it would likely be uneconomic to operate a coal plant without fully accounting for these costs. In our analysis, we therefore considered the effect of incorporating these elevated coal startup costs into our UCED. For this purpose, we included start-up costs estimated by APTECH into the costs constraints of the optimization model.

The results of our model indicated that startup/shutdown cycling of coal generators increased by 14% to 640% with high wind under business-as-usual startup costs. However, when elevated startup costs were included at high wind, coal shutdown/startup cycles were substantially reduced. This reduction was balanced by a large increase in shutdown/startup cycles by flexible gas units. Figure 13.2 shows resource use by generator type for each scenario. As previously noted, coal use decreased significantly with wind penetration and

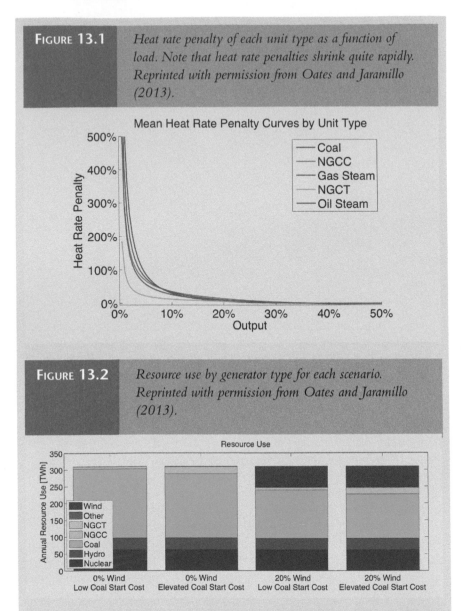

FIGURE 13.1 *Heat rate penalty of each unit type as a function of load. Note that heat rate penalties shrink quite rapidly. Reprinted with permission from Oates and Jaramillo (2013).*

FIGURE 13.2 *Resource use by generator type for each scenario. Reprinted with permission from Oates and Jaramillo (2013).*

decreased marginally more when coal startup costs were increased. NGCC use increased slightly with wind penetration and more dramatically with elevated coal startup costs.

At high wind, elevated coal startup costs increased the value of curtailing wind. In our model, wind could be curtailed economically, and curtailment reached nearly 2% in the elevated coal startup cost scenario. Most of this occurred during the early morning hours, when the UCED found it less costly to curtail wind than to shut down coal generators and restart them soon after.

TABLE 13.2		Emissions of CO_2, NO_x, and SO_2 during normal operations and startups in each scenario.			
		0% Wind, Low Coal Start Cost	0% Wind, Elevated Coal Start Cost	20% Wind, Low Coal Start Cost	20% Wind, Elevated Coal Start Cost
Normal Operations	CO_2 [M ton]	210	200	150	140
	NO_x [M lb]	740	690	470	420
	SO_2 [M lb]	4,500	4,200	3,100	2,800
Startups	CO_2 [M ton]	1.0	0.63	2.0	0.99
	NO_x [M lb]	4.3	2.1	9.5	3.5
	SO_2 [M lb]	18	9.1	42	15

Source: Reprinted with permission from Oates and Jaramillo (2013).

Table 13.2 shows normal operations and startup emissions for each scenario. Startup CO_2, SO_2, and NO_x emissions all increased by approximately a factor of two under the high-wind scenario. However, the contribution of startups to overall emissions was small – less than 2%. Due to this fact, the reductions in normal operations emissions associated with wind more than offset the increase in cycling related emissions. This is in agreement with the results of Apt (2007), Curtright and Apt (2008), and Valentino and colleagues (2012).

Startup and normal operations emissions were further reduced in the elevated startup costs scenario. These reductions were partially due to reduced coal startup/shutdown cycling (and the lower emissions intensity of the NGCC cycling that replaced it) and partially due to reduced utilization of coal generators.

Incorporating cycling costs in the UCED model also had some interesting implications for economic efficiency of operating the system. When low coal startup costs were included in the UCED, total startup costs increased by 80% in the high-wind scenario. Most of these costs were borne by coal units. However, when elevated coal startup costs were included in the dispatch model, the costs borne by coal units decreased. In addition, incorporating the higher coal cycling costs in the optimization had the additional effect of increasing producer surplus at the expense of consumer surplus. The change in consumer surplus associated with adding wind was approximately $1 billion in the low coal start cost scenario but only $100 million in the elevated coal start cost scenario.

13.2 Summary

In a power system using natural gas generators to compensate for wind or solar power's variability, we found that air emissions benefits were somewhat less than those expected based on an average emissions factor calculation. When 20 natural gas plants, with the required number of plants operating as spinning reserve, were used to balance 20% renewables, we found that 87 to 93% of the expected reduction in CO_2 emissions was achieved. Note that emissions penalties may change depending on the reserve requirements associated with wind, the number of plants able to provide balancing services, and the operating constraints of these power plants. It is clear, however, that the ancillary emissions associated with using dedicated natural gas plants to support wind cannot be ignored.

We found that in a large interconnected system, like PJM, coal power plants would be cycled more often under a 20% wind scenario than in the absence of wind. However, incorporating full startup costs into unit commitment prevented coal startups from increasing. We also found that even though there is an efficiency and emission penalty associated with cycling large thermal power plants, these are not very high. The observed reductions in emissions from coal power plants, accounting for cycling, are, on average, 97% of what the emission reductions would be if cycling were not accounted for. This is in good agreement with the 98% found by Valentino and colleagues (2012) for their Illinois system. Further, the emission benefits associated with a 20% penetration of wind power are large: 28% reduction for CO_2, 36% for NO_x, and 31% for SO_2. These reductions are higher than the wind penetration because wind almost always displaced coal. Coal had a much higher emissions factor than the system average, primarily due to the large contributions of nuclear and hydroelectric power to the mix. Note that the size of the emissions reductions is sensitive to the relative prices of coal and natural gas. They are also sensitive to the characteristics of the initial fleet; coal-heavy systems will see larger emissions benefits from wind than gas-heavy systems.

Our hourly model of an entire power grid did not include the emissions associated with load following within the hour. It could thus be argued that the results for the hourly model are biased by our inattention to changes in subhourly ramping of gas turbines. Though it is likely that increased wind penetration will lead to increases in subhourly ramping, the power spectrum of wind power shows that short time scale variability (not captured in our model) will be much smaller than interhour variability (captured in our model). We therefore argue that any such bias would be small.

References

Agan, D.D., Besuner, P.M., Grimsrud, G.P., & Lefton, S.A. (2008). *Cost of cycling analysis for Pawnee Station Unit 1 Phase 1: Top-down analysis.* APTECH Engineering Services report, 08116940-2-1pr. www.dora.state.co.us/pls/efi/efi.show_document?p_dms_document_id=79884&p_session_id=

Apt, J. (2007). The spectrum of power from wind turbines. *Journal of Power Sources, 169*(2), 369–374.

APTECH. (2009). *Integrating wind.* Sunnyvale, CA: Xcel Energy. http://blankslatecommunications.com/Images/Aptech-HarringtonStation.pdf

Curtright, A.E., & Apt, J. (2008). The character of power output from utility-scale photovoltaic systems. *Progress in Photovoltaics: Research and Applications, 16*(3), 241–247.

EnerNex Corporation. (2011). *Eastern wind integration and transmission study*. National Renewable Energy Laboratory report, NREL/SR-5500-47078. www.nrel.gov/docs/fy11osti/47078.pdf

EPA. (2006). *EPA clean air markets data*. http://ampd.epa.gov/ampd/

EPA. (2010). *National electric energy data system (NEEDS) 4th ed*. www.epa.gov/airmarkets/progsregs/epa-ipm/BaseCasev410.html#needs

Katzenstein, W., & Apt, J. (2009a). Air emissions due to wind and solar power. *Environmental Science and Technology*, *43*(2), 253–258.

Katzenstein, W., & Apt, J. (2009b). Response to comment on "Air emissions due to wind and solar power." *Environmental Science and Technology*, *43*(15), 6108–6109.

NERC. (2012). *WECC standard BAL-002-WECC-2*. www.nerc.com/files/BAL-002-WECC-2.pdf

Oates, D.L., & Jaramillo, P. (2013). Production cost and air emissions impacts of coal cycling in power systems with large-scale wind penetration. *Environmental Research Letters*, 8. DOI:10.1088/1748-9326/8/2/024022

O'Neill, R.P., Hedman, K.W., Krall, E.A., Papavasiliou, A., & Oren, S.S. (2010). Economic analysis of the N-1 reliable unit commitment and transmission switching problem using duality concepts. *Energy Systems*, *1*(2), 165–195.

Piwko, R., Clark, K., Freeman, L., & Jordan, G. (2010). *Western wind and solar integration study*. National Renewable Energy Laboratory. www.nrel.gov/docs/fy13osti/55588.pdf

PJM. (2006a). *Hourly day-ahead bid data 2006*. Valley Forge, PA: PJM Interconnection. www.pjm.com/markets-and-operations/energy/day-ahead/hourly-demand-bid-data.aspx

PJM. (2006b). *PJM daily energy market bid data*. Valley Forge, PA: PJM Interconnection. www.pjm.com/markets-and-operations/energy/real-time/historical-bid-data/unit-bid.aspx

Valentino, L., Valenzuela, V., Botterud, A., Zhou, Z., & Conzelmann, G. (2012). System-wide emissions implications of increased wind power penetration. *Environmental Science and Technology*, *46*(7), 4200–4206.

14

STORAGE TO SMOOTH VARIABILITY

14.1 Introduction

Electric power systems today have limited storage capacity. Both in the United States and worldwide, storage makes up less than 3% of generation capacity (EPRI, 2010). This lack of storage forces grid operators to continuously balance generation and load and prevents the electricity sector from operating as a conventional competitive market that relies on inventory. Pumped hydropower storage (PHS) is the predominant storage technology today, making up 99% of all deployed storage capacity. However, PHS requires suitable locations with a height difference between reservoirs. R&D investments have led to rapid improvements in advanced battery technologies. Recent advancements suggest batteries may approach cost parity with PHS at $300/kWh.

Inexpensive electricity storage has the potential to both enable the integration of larger amounts of variable renewable energy and to transform electricity markets. Storage can provide a variety of high-value services, such as frequency regulation (Eyer & Corey, 2010). Although profitable, these relatively small market opportunities are expected to saturate quickly. After those markets are saturated, storage operators and manufacturers will consider larger-volume, lower-value applications. One such application is arbitrage in wholesale energy markets.

In this chapter, we summarize RenewElec research results on both small-scale storage that has the potential to smooth the variable output of wind plants or solar arrays (Section 14.2) and the economics of grid-scale storage (Section 14.3). The grid storage portion includes subsections discussing the economics of four aspects of grid-scale storage: the market prospects of large-scale storage (Section 14.3.1, using the PJM market as an example), pumped hydroelectric storage in Portugal and Norway (Section 14.3.2), compressed air energy storage (Section 14.3.3), and two features of the economics of storage using batteries in vehicles (Sections 14.3.4 and 14.3.5).

14.2 Small-scale storage

We have considered whether adding a small amount of fast-ramping energy storage to wind + natural gas turbine systems can reduce the costs and emissions of smoothing the output from wind generators by providing a small amount of short-duration smoothing (Hittinger et al., 2010). We model a colocated power generation/energy storage block that contains wind generation, a gas turbine, and fast-ramping energy storage. Conceptually, the system is designed to produce near-constant baseload power while still delivering a significant and environmentally meaningful fraction of that power from wind.

Gas generators and storage are used complementarily to smooth wind; energy storage is expensive but is able to ramp extremely quickly and handle high power levels. Gas turbines are able to provide large quantities of fill-in power at a reasonable cost but have important operational limitations. We investigate a hybrid (gas turbine and energy storage) compensation system by modeling both wind power and the wind + natural gas turbine system at a 10-second time resolution.

This research has three main results. First, modeling wind and compensating resources using shorter time scales produces results notably different from modeling them in 1-hour blocks. Studies frequently use 1-hour blocks of time, both because of the availability of such data and because the largest-amplitude wind fluctuations occur over longer time scales (for example, Wan & Bucaneg, 2002). However, nearly all of the time-based operational limitations of natural gas generators occur sub-hourly and, by modeling in 1-hour increments, gas turbines unrealistically appear to be perfect generators capable of fulfilling any power requirements.

Second, a small amount of energy storage colocated with the wind and natural gas turbines can significantly reduce high-frequency power fluctuations. As mentioned, energy storage devices can buffer the power spikes and dips from wind fluctuations. The inclusion of energy storage decreases the quantity and size of power fluctuations externalized to the grid, which then requires less regulation service. Companies are now installing colocated wind and storage at sites such as the Notrees Windpower Project in west Texas. Battery installations of this sort can both mitigate variability and use transmission lines at higher capacity factors than can be achieved with wind alone (Pattanariyankool & Lave, 2010).

Third, a wind/natural gas/energy storage hybrid generation block is capable of delivering a large fraction of wind energy, smoothed to a power variability of less than 0.5%, at a reasonable cost.

14.2.1 Model description

We model the wind + natural gas turbine energy storage system using a time-series operational framework that takes as an input actual wind generation, measured with 10-second time resolution, and a number of operational constraints, including

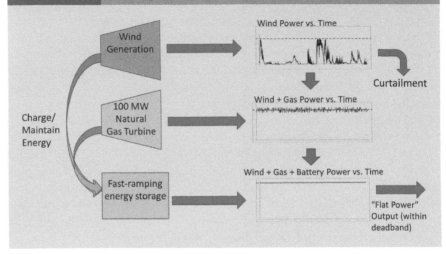

FIGURE **14.1** *Model concept of wind + natural gas + energy storage generation block. The scale of wind generation and the wind generation profile are fixed for each run of the model. The 100 MW natural gas turbine is operated so as to smooth the wind power to the target power output level (red dashed line). Due to the operational constraints of the gas turbine, there may be some residual power transients that are eliminated by a fast-ramping energy storage device, which is sized to the minimum scale required to mitigate the remaining fluctuations. Reprinted with permission from Hittinger and colleagues (2010).*

natural gas ramp rate and system target power output. The model determines the operation of the generation and storage resources required to meet the defined system power requirements. This operational model is used in a scenario analysis, which investigates different system combinations and determines their average cost of electricity and the wind energy content of their power output.

For each combination of wind generation, natural gas generation, and power output, the model determines the quantity of fast-ramping energy storage (grid-scale sodium sulfur [NaS] batteries, flywheels, or supercapacitors) required to produce a fixed power output with constrained variability.

For each system examined, the gas generator is modeled to operate such that it provides maximum fill-in power for the varying wind resource in an effort to bring the combined wind + gas power output to the target power output (Figure 14.1). If the gas turbine is unable to provide all of the required fill-in power due to insufficient ramping capability or cold-start limitations, the residual power is provided by an energy storage device. This residual power includes both positive and negative

FIGURE **14.2** *A sample of the operational output from the wind +
natural gas + NaS battery model. This shows a 24-hour
period of operation of a system with 100 MW of natural
gas capacity, 66 MW of wind capacity, and a target power
output of 100 MW. Positive values for the battery power
indicate discharge, while negative values are charging events.
The battery is required infrequently and generally for short,
sharp charges/discharges. As the wind power increases in
hours 4 through 7, the natural gas turbine ramps down to
its low operating limit of 40 MW and the excess energy is
used to charge the battery. Reprinted with permission from
Hittinger and colleagues (2010).*

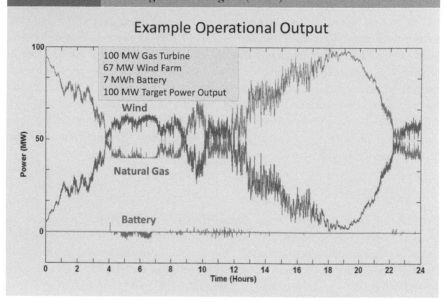

power requirements from the energy storage, which represent both the discharge
energy from the device as well as the required charge energy. Actual 10-second time
resolution wind data are used to model the wind generation (Southern Great Plains
United States wind plant, sum of seven turbines, 15 days, 10-second resolution, 46%
capacity factor during this period).

Once the gas turbine has provided all of the smoothing allowable by its operational constraints, the minimum size of the required energy storage device can be
directly determined. Given the wind + gas generation, the residual power that must
be handled by an energy storage device is calculated, including both charge and
discharge energy (Figure 14.2).

Once the operation of the wind generation, natural gas turbine, and energy storage device has been determined, the emissions and costs of the system over the studied time frame can be calculated. The emissions calculation uses results from Katzenstein and Apt (2009) showing the effect of partial load conditions on efficiency and CO_2 and NO_x emissions of a Siemens-Westinghouse 501FD gas turbine. Capital, variable, and average costs of electricity are also calculated for each potential composite system, including amortized capital costs, other fixed costs, and variable costs of the wind generation, the gas turbine, and the energy storage device. NO_x and CO_2 prices are included in the cost calculation.

14.2.2 Results

Figure 14.3 shows the average cost of electricity from wind + gas + battery systems for different wind/gas capacity ratios, demonstrating that the contribution to electricity cost due to the required NaS batteries is negligible over a wide range of wind penetrations. This result also shows that the average price of electricity stays fairly constant as wind penetration increases, up to 30%, with a noticeable transition around 30% wind energy. This change is due to a change in the operation of the gas turbine: while the turbine is ramped up and down in all scenarios, it is occasionally forced to shut down entirely with systems that have greater than 30% wind energy. The need to start up and shut down the turbine produces notably lower efficiencies and requires more energy storage. Table 14.1 summarizes the results and includes the CO_2 and NO_x emissions from each system.

14.2.3 Summary

In most electric markets, policies encourage the deployment of wind generation. As part of this encouragement, coupled with the limited deployment of wind resources, there are currently few restrictions on the variability of the power produced by wind plants. But, as the penetration of wind power increases, the variability of wind power will become an increasingly important issue. Already, electrical systems that utilize a relatively large fraction of wind energy, such as ERCOT and Nord Pool, are considering enacting or have enacted limitations on the ramp rate of wind power (ERCOT, 2010; Morthorst, 2003). Determining who bears the responsibility for dealing with the variability of wind will become an important policy decision in the coming decades. But regardless of who is responsible, compensating for large-scale penetrations of wind energy requires careful planning.

In the near term, there are other applications for the described systems, such as small electrical grids that are unable to rely on a large base of traditional generators to provide compensation. A higher price of electricity, a desire for increased renewable penetration, and a smaller generator base in these electric grids make them candidates for systems similar to those described herein. The costs for integrating wind are shown to be reasonable and the required technology can be colocated

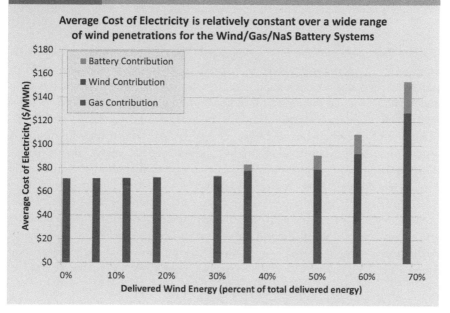

FIGURE 14.3 *Average cost of power in the wind + gas + NaS battery system as a function of delivered wind energy and divided according to system component. Each bar represents a system with 100MW of gas generation and corresponds to the lowest cost of power for a particular wind/gas ratio. NaS batteries are not required until after 12% delivered energy from wind. There is a slight discontinuity after 30% wind energy, corresponding to the point at which wind power fluctuations are so great that they force the gas turbine to shut down at times to prevent the generation of excess power. Reprinted with permission from Hittinger and colleagues (2010).*

with the wind generation, avoiding the need to rely on a large base of traditional generation resources that is nonexistent in small electrical grids such as Hawaii.

Our results suggest a different policy guideline for large electrical systems attempting to integrate wind generation, especially those with flexible traditional generators such as ERCOT. While the hybrid wind + gas + storage systems are shown to be a financially viable option, the scenario analysis results also show that a small energy tolerance (0.5%) and the availability of compensating generation permit wind energy fractions of up to 12% before any storage is required. This

TABLE 14.1	Cost, emissions, and NaS battery capacity for wind + gas + NaS battery systems.			
Wind Nameplate Capacity (MW)[a]	0	25	43	67
Delivered Wind Energy	0%	12%	19%	30%
Average cost of electricity ($/MWh)[b]	62	64	65	67
Contribution of NaS battery to average cost of electricity (percentage)	0%	0%	0.5%	1%
Average CO_2 emissions (tonnes/MWh)	0.34	0.31	0.29	0.26
Average NO_x emissions (g/MWh)	50	44	40	164
NaS battery capacity (MWh)	0	0	10	21

[a] All systems include a 100MW gas turbine.
[b] The average cost of electricity includes emissions prices of $25/tonne for CO_2 and $750/tonne for NO_x.
Source: Reprinted with permission from Hittinger and colleagues (2010).

suggests that energy storage may not be needed, on a system level, until approximately 10% of energy is produced by wind. Despite this, large electricity markets may still find a use for fast-ramping energy storage as a substitute for the close coordination required to provide fill-in power through the market. We have shown that the use of energy storage to smooth the sharpest fluctuations, allowing a gas turbine to provide the remaining fill-in energy, is a cost-effective application. As a result, complex electricity markets might consider enacting lightly binding limitations on the bus-bar ramp rate of wind generators, which could then motivate the deployment of small energy storage systems colocated with wind generation.

This model of wind + gas + energy storage generation systems demonstrates a potential method for integrating significant quantities of wind energy while reducing power fluctuations to a small deadband and maintaining a reasonable cost of electricity. Furthermore, over a wide range of wind penetrations, relatively little energy storage is needed, and this energy storage acts to mitigate potentially harmful transient pulses.

14.3 Economics of grid storage

14.3.1 Storage in the largest electricity market, PJM

14.3.1.1 Introduction

For at least the past 35 years, it has been recognized that storage can significantly help the integration of variable renewable power (see, for example, Sørensen, 1978). Inexpensive electricity storage also has the potential to transform the electricity sector by enabling arbitrage in wholesale electricity markets. For small amounts of storage, this arbitrage will not affect prices or generator dispatch order. How large amounts of storage will change wholesale markets is less well understood. Although storage can provide benefits on ancillary service markets, these markets are small compared to wholesale energy markets and will saturate quickly (PJM, 2012). We therefore focus our analysis on the effects of storage on wholesale energy markets.

TABLE 14.2	*Modeled storage technologies. Costs in 2010 dollars.*		
Technology	Duration (hours)	% Round Trip Efficiency (total cycles)	Cost [$/kWh]
Aqueous hybrid ion battery	4–20	80–90 (10,000)	300
Sodium sulfur battery	6–8	75–86 (4,500)	535–550
Pumped hydropower	4–12	70–85 (>13,000)	250–430

Source: Connolly, 2009; EPRI, 2010; Wiley, 2012.

To investigate this question, we have developed a reduced-form unit commitment model to simulate the PJM Interconnection day-ahead energy market (Lueken & Apt, 2014). We use the model to estimate the value of large storage deployments. We investigate how storage will affect consumers, generators, storage operators, and emissions of CO_2 and other pollutants.

14.3.1.2 Methods

We have developed a unit commitment and economic dispatch (UCED) model to simulate the PJM day-ahead wholesale market. This software, titled PHORUM, is open source and is designed to be easily used by other researchers.[1] PHORUM simulates all generators in PJM to find the least-cost combination to meet load at every hour, thereby minimizing the total overall cost of providing electricity. PJM and all major restructured markets use UCEDs to dispatch power.

We simulate two promising electrochemical storage technologies: aqueous intercalation batteries and sodium sulfur batteries. In our model, we use the aqueous hybrid ion (AHI) battery, a type of aqueous intercalation battery. We simulate deployments of up to 80 GW of capacity (60% of PJM peak demand). To incorporate uncertainties about the cost and performance, we model both a lower bound (low duration, efficiency) and upper bound (high duration, high efficiency) case for each technology.

14.3.1.3 Results

14.3.1.3.1 Effect on consumers

Storage benefits consumers in two ways: by reducing costs in the wholesale energy market and by reducing reliance on expensive oil and gas peaking generators. Storage reduces wholesale energy costs by lowering prices at high-load hours. As large amounts of storage are deployed, fewer peaking plants are needed and could, in

---

1. PHORUM can be downloaded at https://github.com/rlueken/PHORUM.

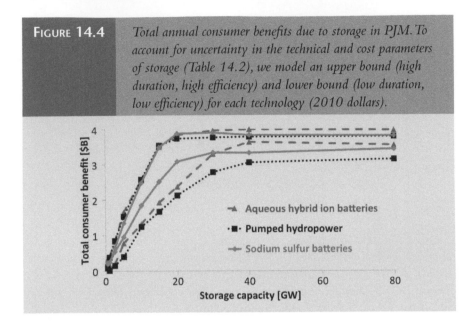

| FIGURE 14.4 | *Total annual consumer benefits due to storage in PJM. To account for uncertainty in the technical and cost parameters of storage (Table 14.2), we model an upper bound (high duration, high efficiency) and lower bound (low duration, low efficiency) for each technology (2010 dollars).* |

theory, be decommissioned. We quantify the savings to consumers if these unused plants were decommissioned with the PJM 2010/2011 capacity auction price of $175/MW-day (PJM, 2012). Figure 14.4 shows how total consumer benefits increase as more storage is deployed. Total consumer benefits approach $4 billion annually, equivalent to 10% of sales on the PJM day-ahead wholesale energy market.

Net consumer benefit is the total consumer benefit minus the annualized cost of storage. A positive net consumer benefit means that the total benefits storage provides on energy and capacity markets exceed the capital costs of storage. Net consumer benefit varies depending on the amount of storage deployed and assumptions of storage parameters and costs. Figure 14.5 shows that AHI batteries can provide positive net consumer benefits depending on parameter assumptions, while the net consumer benefits of sodium sulfur batteries are always negative. The net benefits of AHI are similar to that of traditional pumped hydro storage. We assume an 8% cost of capital.

14.3.1.3.2 Effect on generators

By reducing prices on the wholesale energy market and reducing reliance on peaking generators, storage negatively affects the revenues of all generators. As shown in Figure 14.6, generation from peaking plants (combustion turbine, oil/gas steam, and combined cycle) falls dramatically as they are displaced by storage. Output from coal steam plants increases as they charge storage at off-peak hours. Generation from baseload nuclear and hydro plants is unchanged. Revenues to all generators fall as storage capacity increases; total revenues fall by more than 10% in high storage cases. Peaking plants see the largest revenue drop, as they are largely replaced by storage.

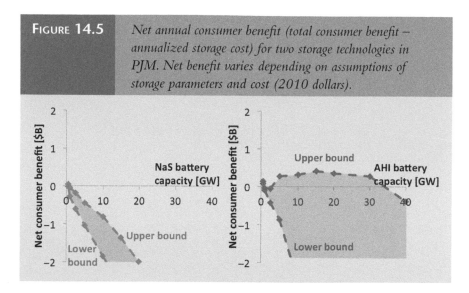

FIGURE **14.5** *Net annual consumer benefit (total consumer benefit – annualized storage cost) for two storage technologies in PJM. Net benefit varies depending on assumptions of storage parameters and cost (2010 dollars).*

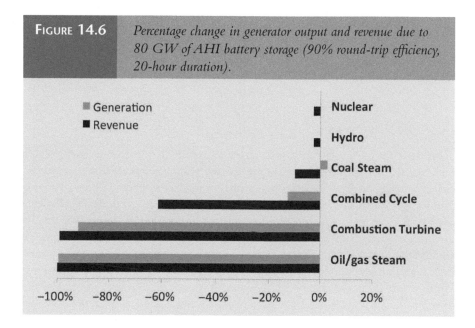

FIGURE **14.6** *Percentage change in generator output and revenue due to 80 GW of AHI battery storage (90% round-trip efficiency, 20-hour duration).*

14.3.1.3.3 Effect on storage operators

Storage owners make profits in this application by arbitraging between high and low prices. As more storage is deployed, fewer arbitrage opportunities are available, and revenues drop. Net annual profits are negative, regardless of storage technology or capacity deployed, if used only for arbitrage. Capital costs greatly exceed the revenues operators receive.

TABLE 14.3	Annual net profits of storage technologies to storage operator. Bracketed variation accounts for uncertainty in technology parameters. 2010 dollars.	
Capacity [GW]	AHI Battery Annual Profits [$B]	SS Battery Annual Profits [$B]
1	[−0.5, 0.0]	[−0.3, −0.5]
10	[−4.9, −0.9]	[−4.2, −2.6]
80	[−40.3, −8.0]	[−28.2, −20.0]

TABLE 14.4	Annual emission increases due to storage in the 2010 PJM wholesale energy market. Storage technology is aqueous hybrid ion batteries, 90% round-trip efficiency, 20-hour duration.		
AHI Battery Capacity [GW]	Change in Emissions [MT] (%)		
	CO_2	NO_x	SO_2
1	2,400,000 (0.5%)	2,500 (0.6%)	16,600 (0.9%)
10	5,400,000 (1.2%)	4,700 (1.2%)	58,000 (3.0%)
80	6,600,000 (1.5%)	5,600 (1.4%)	71,800 (3.7%)

14.3.1.3.4 Effect on emissions

Table 14.4 shows the effect of storage on emissions in the base scenario (2010 fuel prices). Storage increases emissions for two reasons. First, storage is primarily charged during off-peak hours by coal plants, which have higher emissions than the peaking gas plants they replace. Second, because storage is not perfectly efficient, additional electricity must be generated to compensate for the losses inherent in storing electricity. Although storage increases emissions in the 2010 base scenario, we find that in other scenarios, storage decreases total emissions. In a scenario with 20% wind, storage enables more wind but slightly increases total grid emissions. We therefore do not conclude that increases in storage are directly correlated with increases in emissions. Instead, storage's effect on emissions depends on the underlying market dynamics.

14.3.1.4 Summary

Storage provides substantial benefits to consumers in two ways: by reducing prices in the wholesale energy market and by reducing the need for peaking generators. The net consumer benefit, or total benefit minus annualized storage cost, varies greatly depending on the technology parameters assumed. Net benefits of storage on wholesale markets are likely negative due to high capital costs. However, benefits may be positive for aqueous hybrid ion batteries under optimistic performance and cost assumptions (Figure 14.5). Storage reduces the profitability of all generators. Large deployments of storage cause generation from peaking gas and oil plants to decrease and generation from coal plants to increase. Storage also reduces the number of generator startups by up to 90%. Storage modestly increases system emissions of CO_2 and other pollutants.

We find that no current storage technologies are profitable if solely used for arbitrage on the PJM day-ahead (DAH) market. Although storage provides substantial benefits to consumers (Figure 14.4), net consumer benefits are positive only under optimistic cost and technical assumptions. PJM could incentivize the deployment of socially beneficial storage in one of four ways: (1) establish rules that allow profit-maximizing behavior for storage; (2) directly subsidize storage for the consumer benefits it provides; (3) have public institutions such as municipalities operate the storage devices for the benefit of consumers; or (4) remove all price caps and allow for high price spikes.

14.3.2 Pumped hydropower storage

Renewables such as wind and solar are variable across a broad range of time scales (see Chapter 1 and Chapter 8). While technologies such as batteries and flywheels can balance the highest-frequency variability, their cost and technical aspects render them impractical for balancing renewable power output over the course of hours or days. Both compressed air energy storage (Section 14.3.3) and pumped hydropower storage (PHS) are well suited to balancing both the higher-frequency and lower-frequency fluctuations of renewable power. PHS, the focus of this section, is a mature technology that ramps quickly enough to balance wind, has low startup costs in both pumping and generation modes, and has efficiencies comparable to those of other storage technologies (Hadjipaschalis, Poullikkas, & Efthimiou, 2009).

This chapter presents research on the economics and operation strategies of PHS in Portugal and Norway, both in response to increased wind capacity. Portugal plans to install new PHS to better integrate new renewable generation within Portugal. "Case study: The case of Portugal" uses nonlinear optimization to show the differences in how a social planner would operate storage as compared to a thermal generator fleet owner in Portugal. Norway, on the other hand, would use increased PHS capacity to help balance large increases in installed wind capacity in the rest of Europe. "Case study: The case of Norway" uses real options to analyze an investment opportunity in PHS located in Norway but operating in the German market and finds that such a system would be profitable but that waiting to invest, to resolve uncertainty on the effect of wind on daily price volatility, would be economically optimal.

The case of Portugal: Who benefits from investing in storage?

Portugal plans to increase the nation's pumped hydro storage (PHS) capacity to better integrate wind energy into the electricity system and insulate against price volatility in imported fuel. Most thermal generation and hydropower in Portugal and Spain participate in the MIBEL electricity market, which traded on average 24.6 GWh of electricity per hour in 2011 (MIBEL, 2012a). Renewables, small hydro,[2] and some cogeneration operate in a "special regime," receiving feed-in tariffs for power production (EDP, 2012c). To fulfill stringent renewable energy consumption targets, Portugal plans to more than double its current hydropower capacity to 9.5 GW (European Union, 2012; Silva, 2011). Of the new hydro capacity, five new dams comprising 636 MW of generation capacity will be equipped with storage capability, an increase of 64% (Silva, 2011). Portugal intends for the new storage to balance variable electricity generation from renewable energy and distributed generation (EDP, 2012a). In 2010, Portugal had 3.8 GW of wind capacity (EIA, 2012).

This case study focuses on how the motivations of the storage owner can affect storage operations. We compare the effects of new pumped hydro capacity in Portugal if it were operated by a hypothetical social planner and alternatively by a thermal generation fleet owner. The effects of the two owner-ship cases on the market price of electricity, consumer expenditures, thermal generator profits, and CO_2 emissions vary significantly.

In fact, the companies building Portugal's new PHS capacity own thermal generation fleets. It is important that Portugal be aware that thermal generation companies that own storage may increase market price volatility and therefore increase emissions, which is not the outcome from PHS that Portugal intends.

Method

Data

We use 2011 price and market volume data from MIBEL, the Iberian Electricity Market (MIBEL, 2012b). To model the pumped hydro storage units, we use technical specifications from Portuguese electricity generation companies and PNBEPH (Programa Nacional de Barragens de Elevado Potencial Hidroeléctrico), the Portuguese organization for hydroelectric power development (EDP, 2012b; PNBEPH, 2012).

Model

Using a nonlinear optimization program to model the storage system, we calculate the changes to consumer expenditure on electricity, thermal generator profits, and CO_2 emissions that result from storage-influenced changes to

---

2. Small hydro generally refers to run-of-river hydropower systems and capacities below 10 MW (Paish, 2002).

market prices. We solve the optimization problem using the Tomlab NPSOL solver in Matlab. The amount of energy discharged from or added to the dams in each hour is related to the flow rate of the pump-turbine at each dam, its efficiency, and the head height.

We consider two cases: a social planner operating storage and a thermal generation fleet owner operating storage. In the case where one entity owns thermal generation and storage, the owner maximizes arbitrage profits, $P_{thermal}$, from the storage plus thermal profits:

$$P_{thermal} = \max \sum_{i=1}^{N} \left[\sum_{k=1}^{5} \left(\frac{-p_i g h_{i,k} q_{i,k}^{up}}{\eta_k^{up}} + \eta_k^{dn} p_i g h_{i,k} q_{i,k}^{dn} \right) + \sum_{j=1}^{J} \left(p_i - r_j \right) w_{i,j} \right] \quad (14.1)$$

The cost of operating storage is the term in the first parentheses, where p_i represents the price of electricity in hour i, $h_{i,k}$ is the head height of reservoir k in hour i, η_k^{up} and η_k^{dn} are efficiencies, g is the gravitational constant, and $q_{i,k}^{up}$ and $q_{i,k}^{dn}$ represent the flow rate of water stored or released in hour i at reservoir k, respectively.

Thermal profits are the difference between the hourly market price of electricity, p_i, minus the marginal cost of operating generator j, r_j, multiplied by the energy supplied by the generator j in hour i, $w_{i,j}$. When the market price is less than the marginal cost of operating the generator, the generator does not run. Storage profits are added to the thermal generator profits.

$$C_{consumer} = \min \sum_{i=1}^{N} \left[p_i d_i + \sum_{k=1}^{5} \left(\frac{p_i g h_{i,k} q_{i,k}^{up}}{\eta_k^{up}} - \eta_k^{dn} p_i g h_{i,k} q_{i,k}^{dn} \right) \right] \quad (14.2)$$

For social-planner storage operations, the objective function minimizes the consumer expenditure on electricity plus the cost of operating storage, with the assumption that consumers pay the market price of electricity, p_i, multiplied by the traded volume of electricity in each hour, d_i.

Assumptions and limitations

It is necessary to simulate the relationship between quantity of electricity and price to determine the effect of storage on market prices. We use two linear best-fit lines per day to approximate the relationship between traded volume and price in the Iberian electricity market. We split the day into two segments: 1 a.m. to 9 a.m. and 10 a.m. to midnight. Spees (2008) used fitted curves to relate price to load in PJM. We found that using two daily linear equations has a lower absolute mean y-error compared to using only one best-fit line or using a quadratic curve. We chose this break because it had the lowest absolute mean

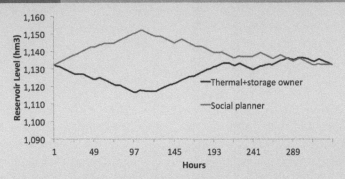

Figure 14.7 *Storage levels of one Portugal PHS facility under operation by an owner of thermal generation and operation by a social planner during the Winter case. The different objectives lead to vastly different operation patterns.*

y-error (2.04 €/hr) compared to all other hours of the day, using a single best-fit line (2.56 €/hr) or using a curve (2.29 €/hr).

We simulate storage operations in 2-week horizon optimization segments throughout the year. In our engineering model of the PHS facilities, we use a linear approximation of the relationship between head height and reservoir volume. This assumption is reasonable because the dams operate near full capacity where the change in volume due to generation and pumping has a very small effect on head height.

Results
Storage operations vary based on the motivations of the owner

We examine two cases of storage ownership to illustrate how storage operations change depending on the incentives of the operator. Figure 14.7 illustrates how the different ownership scenarios create different decisions about how to operate the PHS. Under certain price conditions, the thermal storage owner may decide to store electricity when demand is high, thus driving up the price and the profit margin of thermal generators. A different objective would seek to minimize consumer electricity costs and therefore would not operate in a way that drives up electricity prices at peak times.

Storage ownership affects market prices

While a single PHS unit is effectively a price taker in MIBEL, adding 636 MW of new storage capacity to the grid has the potential to change market prices in MIBEL.

FIGURE 14.8 *A daily price-spread duration curve for the no-storage, social-planner, and thermal-storage owner scenarios (where price spread equals the maximum price minus the minimum price in a day). Data were available only for four 2-week periods during the year. The thermal-storage owner scenario creates bigger price spreads since this would help thermal plants increase profits. Lower price spreads (right side of graph) exhibit less variation between scenarios because storage is not used as much when the price spread is small in either of the storage ownership scenarios.*

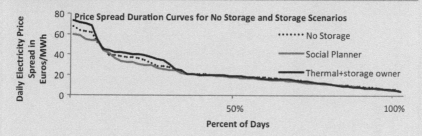

Figure 14.8 contains a price spread duration curve for each of the ownership scenarios. The price spread is defined as the difference between the maximum and minimum daily price of electricity. The consumer-focused operator (called a social planner in these figures) reduces price spreads under optimal operations, but thermal storage owners sometimes increase them.

Effects of storage ownership on thermal generator profits and CO_2 emissions

We assume thermal generators operate when the market price of electricity exceeds their marginal cost of operation. This assumption disregards startup cost considerations and is meant to be a first-order approximation of how large-scale grid storage affects CO_2 emissions. We calculate changes in profit based on a combination of EDP's, Endesa's, and Iberdrola's thermal generator fleets in Portugal and Spain, which consists of 42% coal and 58% natural gas.

Figure 14.9 shows that thermal generator profits are highest under the scenario when storage is owned by a company that also owns thermal generation units. Profits consist of arbitrage profits plus thermal generator profits. In the winter, most of the profit consists of arbitrage, which reduces CO_2 emissions, as fossil-based generators run less often. Our analysis accounts only for CO_2 emissions changes due to thermal generators and not changes in emissions based on avoided wind curtailment.

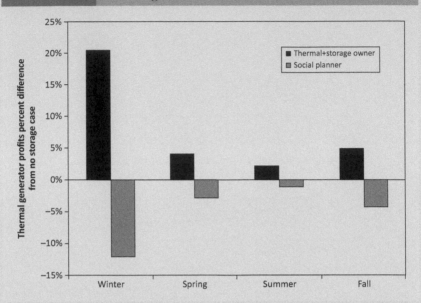

FIGURE 14.9 *Percentage change in thermal generator profits under different storage ownership conditions compared to the no-storage case.*

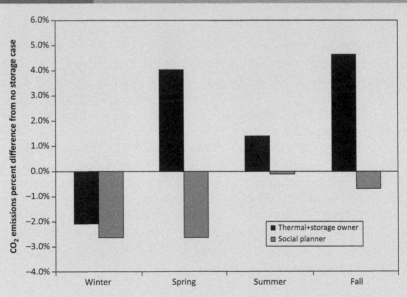

FIGURE 14.10 *Percentage change in CO_2 emissions from thermal generators under different storage ownership conditions compared to the no-storage case.*

Effects of storage on consumers' electricity expenditure

Changes to market prices will affect the total cost of providing electricity to consumers. We assume that the consumer expenditure on electricity related to energy traded in the MIBEL market for each hour is equal to the product of traded volume and price. Under the thermal ownership scenario, the consumer expenditure on electricity increases (Figure 14.11). When a social planner owns the PHS, consumer expenditure decreases.

Summary and policy implications

If operated in a manner that reduces price spreads, new storage may eliminate the need for some of the most inefficient and rarely used capacity. A social planner interested in reducing consumer expenditure would operate this way. However, our results show that a storage owner with thermal generation assets may be encouraged by the market to operate storage to create the opposite effect. We have shown that thermal fleet owners who also operate storage will do so in a way that does not meet the stated goals for investing in storage. To influence thermal storage owners to operate in a way that satisfies Portugal's requirements, Portugal must provide incentives to the storage operators that exceed the benefits of operating in a profit-maximizing way.

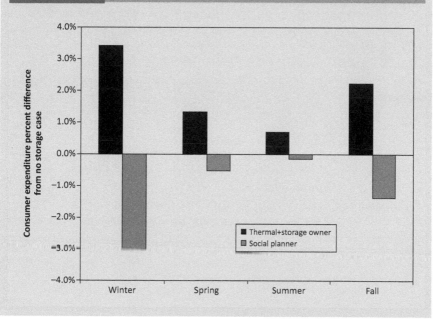

FIGURE **14.11** *Percentage change in consumer expenditures on electricity under different storage ownership conditions compared to the no-storage case.*

The case of Norway:[3] Optimal investment timing and capacity choice for pumped hydropower storage.

Wind power capacity in Germany is expected to grow in the coming decades far beyond its current value of 27 GW. While PHS in Germany could currently profit from a certain degree of daily arbitrage, increased wind capacity could cause a shift in the generator mix of the rest of the system, which would increase price volatility and arbitrage revenue, motivating investment in storage.

Germany currently has 7 GW of PHS generation capacity, 40 GWh of reservoir capacity, and an additional 4.5 GW of hydropower capacity without pumping that utilizes 300 GWh of storage capacity. There is little possibility for further expansion of domestic hydropower resources because most suitable sites are already in use. Across the North Sea, Norway is a much smaller system but has 27 GW of hydropower production capacity with 84 TWh of storage. With increased HVDC transmission between the two countries, the Norwegian system could substantially help balance increased German wind capacity.

Taking the perspective of a Norwegian investor, we analyze an investment opportunity in a PHS facility in southern Norway and accompanying HVDC transmission line. The effect of increased wind capacity on prices in Germany is highly uncertain but expected to change over time as wind capacity grows. A simple now-or-never net present value (NPV) investment model would fail to capture the effect of both uncertainty and intertemporal choice on the optimal investment decision. Therefore, we show here how real options theory can be used to analyze the value of the investment opportunity and the optimal time to build.

Model

The analysis consists of three steps: (1) sample price paths are generated for the German electricity market using volatility-adjusted, bootstrapped historical data, (2) the price paths are used as inputs to a hydropower scheduling model that calculates revenue, and (3) the revenue streams are used in a real options analysis to find the profit-maximizing investment time and capacity choice for the project.

Generating future price paths

Since short-term price volatility rather than mean price determines the arbitrage revenue of a pumped hydropower storage system, the aim of the price model is to capture the change in price volatility over time. Price paths are constructed first by randomly sampling data, in yearly increments with replacement, using 2003 to 2010 German historical prices. The standard deviation of the hourly prices is then increased (around a monthly mean) by a factor that is stochastic

---

3. This section is based on Fertig and colleagues (2014).

but tends to increase with time. The central tendency of this factor is derived from prices in the futures market, which show increasing spread between peak and off-peak prices for contracts with maturity between 2011 and 2016. For the base scenario derived from the futures market, the growth rate in volatility was found to be 8 percentage points per year.

While a bottom-up model of the German power system could be a better source of simulated future prices, such models are constructed to capture other features so that they underestimate intraday price volatility. As result, the price model used in this analysis accounts for the change in price volatility but not for the change in price structure due to an increased share of renewables. The assumed linear growth rate of price volatility also neglects long-lasting shocks due to policy interventions or increased transmission capacity. Nevertheless, the magnitude of intraday price volatility is more important than the functional form of its increase, supporting the use of the linear model. Under the base scenario, the expected real price volatility increase of 50% by 2020 is near that predicted by a detailed power systems model of Europe in 2020 (60% above 2010 values) under no substantial increase in transmission capacity (Schaber et al., 2012).

Pumped storage scheduling model

Using a price path as input, the PHS scheduling model generates the optimal production schedule and expected revenue for a hydropower plant of a given capacity. Operation of the pumped storage system is optimized over a planning horizon of 168 hours (1 week). For the first 24 hours, the spot price is known, and for the remaining hours, the spot price is approximated as the mean price for the corresponding hour exactly 1 week earlier and 2 weeks earlier. Optimizing production over a week instead of a day gives the system the flexibility to store energy on weekends, for example, when prices are lower and produce energy at times of week when prices are higher. Operating the system for a single day before reoptimizing allows new price information to be incorporated on a daily basis without assuming future knowledge of prices beyond a day. Since generators make price-dependent bids, scheduling based on a day of known prices is approximately equivalent to scheduling based on knowledge of the price distribution, a weaker assumption than perfect price knowledge and producing nearly the same results.

Real option valuation

Classical methods of project planning dictate that investment should be made if expected NPV is positive. This investment strategy ignores the value of postponing investment in a positive-NPV project to gather more information or take advantage of a favorable anticipated change. For projects in which these factors are important, real options capture the value of flexible decision-making and intertemporal choice.

The value of the investment opportunity expiring T years in the future is calculated as

$$ROV = \max_{t \leq T}(\mathbb{E}[e^{-r \cdot t} \cdot (PV_t - F)], 0) \tag{14.3}$$

where ROV signifies real option value, PV_t is the present value of investing in the maximum-value project at time $t \leq T$, F is the investment cost, r is the discount rate, and E is the initial expected value operator. If the NPV (equal to $PV_t - F$) at the optimal investment time is positive, then the option value equals the NPV and the investment should be made; if the NPV is negative, the option is worthless.

The option is valued using the least-squares Monte Carlo method (Longstaff, 2001), based on comparing the current NPV of the maximum-value project with the expected value of postponing investment (the continuation value) and investing only if the NPV is the higher of the two. The continuation value at the end of the 10-year option lifetime is defined as 0. For prior times, the continuation value is calculated numerically for each price path, after which the continuation values are regressed against a basis function of the present values of investment. The resulting function defines the decision rule for investing versus waiting. The algorithm steps backward in time, continuation values are calculated by following the decision rule previously generated, and the process continues.

Norway and Germany

The investment opportunity considered is a unique, previously secured right to construct a pumped storage facility, modeled after the Tonstad power plant in southern Norway, which operates solely on the German power market through an HVDC connection. Upriver of Tonstad, there are three reservoirs, which are small compared to others in the system and serve mainly as short-term storage for the existing power plant. These form the upper reservoir capacity in this analysis (Figure 14.12). We assume the pumped storage plant has no inflow and operates in isolation from the rest of the reservoir system. Since the reservoirs already exist and do not require adaptation, their cost is set to zero. The cost of securing the investment option is negligible since there is no permitting fee, the current Tonstad plant owner Sira Kvina already possesses the majority of the necessary land rights, and the cost of assembling an application is very low, approximately 130,000 euro ($175,000).

The investor can choose among pumped storage plants of capacities between 480 and 2,400 MW, in 480 MW increments (the proposed upgrade to Tonstad is 960 MW, consisting of 2 × 480 MW pump/generator units). The projects are mutually exclusive, with no option to upgrade a smaller project to a

FIGURE **14.12** *(a) Diagram of the investment opportunity, with the Tonstad power plant at (1) and an HVDC cable extending from (2) to (3). (b) The PHS facility has upper reservoirs (1) through (3) and a lower reservoir (4), labeled with operating water levels (m).*

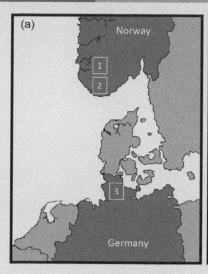
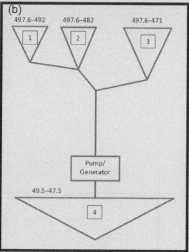

larger project. Although the pumped storage system is considered an upgrade to the existing Tonstad plant, it would involve the construction of a new water tunnel and the installation of new machinery such that its cost is commensurate with building an entirely new plant. The investment opportunity includes construction of a high-voltage direct current (HVDC) transmission line from the power plant to the coast and a subsea HVDC cable between Norway and Germany. We assume that the investor can purchase HVDC transmission of equal capacity to the pumped storage plant and that the cost is proportional to that of the proposed 1,400 MW NorGer cable between Norway and Germany (NorGer, 2009). Since licenses issued by the Norwegian Water Resources and Energy Directorate last 5 years with the possibility of a 5-year extension (NVE, 2010), we assume the investment opportunity lasts 10 years beginning in 2011. Revenue begins 3 years after construction costs are incurred, and the lifetime of the power plant and the HVDC cable is set to 40 years.

Capital cost amounts to 0.66 million euro ($0.9 M) per MW of installed capacity for the pumped storage plant and transmission to the Norwegian coast, and with the subsea HVDC cable, the cost becomes 1.71 million euro ($2.3 M) per MW. We neglect economies of scale with respect to capacity,

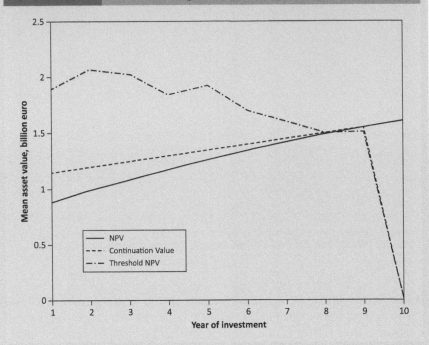

FIGURE **14.13**   *NPV, continuation value (expected value of postponing investment), and threshold NPV for the base scenario. For values above the threshold NPV, investment would be optimal if NPV could be observed. In the eighth year, the expected optimal investment time, the threshold is equal to the mean NPV.*

which could shift decisions toward larger projects. We assume a before-tax, real discount rate of 6%, in accordance with the Norwegian government's discount rate for high-risk projects (NDR, 2011).

Results

A profit-maximizing investor would wait to invest in Norwegian PHS

Figure 14.13 shows the NPV of the maximum-value project and the continuation value at each year of the 10-year option lifetime. Under the base case assumptions, the optimal project in which to invest is always the largest one. Since the NPV at the start of the investment period is positive, classical project planning would dictate immediate investment. However, using real options to value flexible decision making shows that postponing investment by approximately 8 years will add an expected 25% to the value of the investment opportunity.

Figure 14.13 shows that the ability to delay investment adds less value as the option nears expiration and that the NPV and continuation value become equivalent at the optimal investment time. The threshold NPV curve shows the NPV above which investment would be optimal at each time step; since only expected NPV can be observed, this curve cannot directly form the basis of a decision rule and the optimal strategy must be updated as more information becomes available.

Option value, investment time, and project size are sensitive to assumed growth in price volatility

The price effect of wind is highly uncertain, so the growth rate of price volatility was treated with sensitivity analysis. Since profitability of the PHS system is driven by price arbitrage, option value and NPV increase with growth in short-term price volatility. Table 14.5 gives the NPV, option value, mean investment time, and optimal project size for a range of growth rates in price volatility.

Higher growth in price volatility promotes early investment in large projects since revenue grows quickly in early years and is large enough to overcome a high capital cost. With earlier optimal investment times, the option to postpone adds less value relative to the initial NPV. For nominal growth rates in price volatility of 8 percentage points and higher, almost all simulated NPVs are positive and NPV grows linearly with the growth rate; at lower values of the growth rate, the linear relationship does not hold.

While the optimal project size for expected volatility growth rates of 8 percentage points and higher is always 2,400 MW, the largest available, reducing the growth rate to 6 or 4 percentage points introduces a low-profit region in which investment in the smallest (480 MW) project is optimal. If high revenue is expected, immediate investment in the 2,400 MW project is optimal, and at intermediate expected profit, waiting to determine the maximum-value project is optimal. For lower growth rates of 0 or 2 percentage points, the arbitrage revenue is insufficient to overcome the capital cost of a plant of any size and the investment option is worthless.

TABLE 14.5	*Sensitivity to percentage-point growth rates in short-term price volatility (base scenario in bold).*							
Growth in price volatility	0.00	0.02	0.04	0.06	**0.08**	0.10	0.12	0.14
NPV (€M)	−530	−380	−190	65	**880**	1,900	3,000	4,100
Option value (€M)	0	0	0.77	240	**1,100**	2,100	3,200	4,200
Mean investment time (y)	–	–	8.5	9.9	**8.4**	7.1	6.0	5.2
Project size (MW)	–	–	480	2400	**2,400**	2,400	2,400	2,400

Source: Adapted from Fertig et al., 2014.

Further sensitivity analyses show that option value and capacity choice are both robust to reservoir size and that the option value sharply declines at higher discount rates, approaching zero at discount rates of 11% and above.

Summary and policy implications

Under a base scenario of growth in price volatility derived from the European Energy Exchange (EEX) futures market, a prospective investor in a pumped hydropower storage facility in Norway seeking to profit from arbitrage in the German market should hold the option for approximately 8 years and then invest, updating the strategy as more information becomes available. The option to postpone investment provides substantial added value, due to both deterministic growth in price volatility as well as greater uncertainty in revenue combined with the flexibility to avoid investment if a project proves to be unprofitable. Using the options approach rather than an NPV analysis quantifies the benefit of waiting to invest, and robustness of the result to doubling the option lifetime shows that the waiting period is not simply an artifact of the options problem framing. Progress toward resolving uncertainty on the price effect of wind will enable investors to refine optimal investment strategies.

The effect of increased wind power penetration on short-term price volatility will depend on factors such as the viability of base load generators in the changing market, the amount of new transmission capacity integrating European electricity markets, the price of natural gas, and the greenhouse gas regulatory regime. The published literature appears to contain neither explicit modeling of these effects nor price scenarios from bottom-up models that sufficiently capture uncertainty, have fine enough time resolution, and have long enough time spans for use in this study. Updating the price model used in this work with better information on the listed effects will strengthen the conclusions and is left for future work. This work serves best as a decision support tool to provide guidance on the optimal investment strategy in a pumped storage system given investor expectations on the mean increase in short-term price volatility.

14.3.3 Compressed air energy storage (CAES)

While technologies such as batteries and flywheels are best suited to balancing short-term fluctuations, they lack sufficient energy capacity to balance the longer-term fluctuations that occur when renewable generators over- or underproduce for hours or days on end. The cost-effective and technically feasible energy storage technologies capable of balancing renewable power output over longer time scales are pumped hydropower storage, advanced batteries, and compressed air energy storage (CAES). This section focuses on CAES.

The principle behind CAES is that during the storage phase, electricity is used to compress ambient air into a pressure-sealed vessel or underground cavern. During the generation phase, the air is mixed with natural gas and combusted to power a gas turbine generator.[4] CAES plants generally have power capacities on the order of hundreds of MW and energy capacities of a GWh or more, allowing them to effectively balance the power output of a large wind plant over the course of several days.

A CAES plant combined with a wind or solar plant could act as a baseload generator in place of fossil fuel and nuclear plants or could be dispatched as a peak-shaving or shoulder-load plant. The operating flexibility of CAES also enables the system to provide ancillary services such as frequency regulation, spinning reserve, capacity, voltage support, and black-start capability (Gyuk, 2004).

The remainder of this section describes the mechanics of CAES, both proven and developmental designs; the geology required for underground CAES; existing and proposed CAES plants; cost estimates for CAES units; and research results on the economics of CAES in Texas.

14.3.3.1 Mechanics of CAES

Figure 14.14 shows a schematic diagram of a CAES plant, separated into compression and expansion stages. The compression stage of CAES begins with the

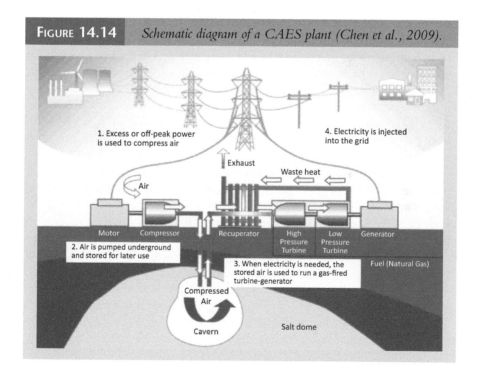

FIGURE 14.14   *Schematic diagram of a CAES plant (Chen et al., 2009).*

---

4. Adiabatic CAES plants that do not use natural gas have been proposed, and the ADELE project in Germany may reach the deployment stage.

intake of air at ambient pressure and temperature. A motor, drawing electricity from the grid, renewable generator, or other source, runs a series of progressively higher-pressure compressors and intercoolers to bring the air to its storage pressure and temperature. By cooling the air after each compression stage, the intercoolers reduce both the power necessary for compression and the storage volume required for a given mass of air (Gyuk & Eckroad, 2003).

In the commercial plants operating as of this writing, the compressed air is stored in an underground cavern. Above-ground CAES designs have also been explored but have been found to be cost effective only for systems storing less than about 100 MWh (Gyuk & Eckroad, 2003). This is too little to balance wind and solar plants for long periods of time, so this section focuses on CAES with underground air storage.

The generation phase begins with release of air from the cavern. The air pressure is reduced to the inlet pressure of the first, high-pressure expansion turbine and is heated in a recuperator (a heat exchanger that captures exhaust heat). The air is mixed with natural gas and combusted in the high-pressure and then the low-pressure turbines, driving the generator to produce electricity and completing the CAES cycle.

In a conventional gas plant, 55 to 70% of the electricity produced is used to compress air in preparation for combustion and expansion (Nease & Adams, 2013). Since CAES uses excess electricity to power the compression stage, the heat rate is about 4,300 BTU/kWh compared with 6,700 BTU/kWh for widely deployed high-efficiency natural gas combined-cycle turbines (Klara & Wimer, 2007). In a new, potentially less costly and more efficient design proposed by the Electric Power Research Institute (EPRI), only the low-pressure turbine is combustion based; the high-pressure turbine is similar to a steam turbine. The heat rate of the EPRI design is estimated to be 3,800 BTU/kWh (Schainker, 2008).

CAES has a high ramp rate and quick startup time, rendering it well suited for smoothing power output from renewable generators. A CAES plant with one or more 135 MW generators starts up in 7 to 10 minutes and, once online, ramps at about 4.5 MW per second (or 200% per minute; EPRI, 2004). In the compression phase, a CAES plant starts up in 10 to 12 minutes and ramps at 20% per minute. For comparison, the ramp rate of a natural gas turbine of roughly the same size is about 7% per minute (Western Governors' Association, 2002).

Although not yet demonstrated, the concept of adiabatic CAES would eliminate the use of fossil fuel in CAES. Rather than dissipating the heat of compression, as in current CAES designs, adiabatic CAES would store the heat and use it to reheat the air before the expansion stage. The efficiency of the system would be approximately 80%. Adiabatic CAES, however, is not likely to be cost effective in the United States at current low natural gas prices and in the absence of a carbon price, but industry experts affirm that the technology required to build a viable adiabatic CAES demonstration plant is within reach (Bullough et al., 2004). Pacific Northwest National Laboratory (PNNL) recently proposed a hybrid CAES plant, which would use a geothermal reservoir as well as molten salt thermal storage to absorb the heat of

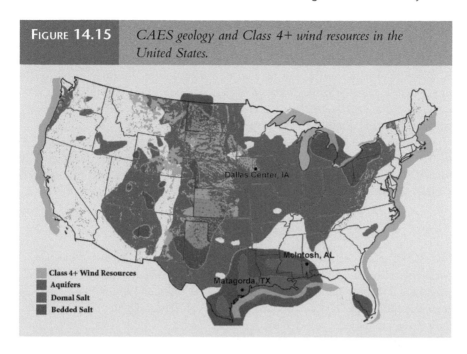

FIGURE 14.15 *CAES geology and Class 4+ wind resources in the United States.*

compression and reheat the air during the expansion phase, avoiding the use of natural gas (McGrail et al., 2013).

14.3.3.2 Geology of CAES

CAES is feasible in three broad types of geology: solution-mined salt caverns, aquifers of sufficient porosity and permeability, and mined hard-rock caverns. Figure 14.15 shows a map of geographical range of each type of CAES geology in the United States. While this map provides a broad indication of possible locations for CAES development, it is not definitive because siting a CAES plant depends largely on local geological characteristics and preexisting land use patterns.

Solution-mined salt caverns are the least costly and most proven geology for CAES. The two currently operational CAES plants, in McIntosh, Alabama, and Huntorf, Germany, both use these types of caverns for air storage. These structures are advantageous for CAES due to the low permeability of salt, which enables an effective pressure seal, and the speed and low cost of cavern development. The caverns are formed by dissolution of underground halite (NaCl) in water and subsequent removal of the brine solution. The resulting caverns are usually vertical, with the vertical dimension roughly four times that of the horizontal. The CAES plant injects and removes air through a single well connecting the salt cavern and turbomachinery. In the United States, salt domes, which are the best type of geology for CAES cavern development, occur primarily in the Gulf Coast and East Texas Basin (Hovorka, 2009).

In addition to solution-mined salt caverns, underground storage for CAES is also feasible in an aquifer-bearing sedimentary rock. The rock must be sufficiently permeable and porous to allow water displacement and air cycling and lie beneath an anticline of impermeable cap-rock to stop the buoyant rise of air and impede fingering (Succar & Williams, 2008). A bubble in the aquifer, developed by pumping air down multiple wells, serves as the air storage cavern and remains at a nearly constant size throughout the CAES cycle.

An early study on the use of aquifers for CAES was based on the success of storing natural gas in porous formations and on the assumption that the techniques of storing air and natural gas are identical. This study found that most of the central United States contained possible aquifer CAES sites (Allen, 1985).

14.3.3.3 Existing, planned, and terminated CAES plants

Two CAES plants are currently operational: one in Huntorf, Germany, and one in McIntosh, Alabama. The first was completed in 1978 and is used primarily for peak shaving, as a supplement to hydroelectric storage facilities, and as a means to supplement the ramp rate of coal plants. The plant was originally designed to provide black-start services to nuclear plants and as a source of inexpensive peak power. The original 2 hours of storage were sufficient for those purposes, but the plant has since been modified for 4 storage hours (Gyuk & Eckroad, 2003). In addition, it now helps mitigate power fluctuations from wind plants in North Germany (Bullough et al., 2004).

The Alabama Electric Cooperative completed the McIntosh CAES plant in 1991 after 30 months of construction (Gyuk & Eckroad, 2003). After initial problems with the underground storage were solved, the McIntosh plant reached 91.2% and 92.1% starting reliability and 96.8% and 99.5% running reliability over 10 years for the generation and compression cycles, respectively (Bullough et al., 2004).

Plans for two CAES plants in Texas were announced in 2012. Apex Energy is building the 317 MW Bethel Energy Center, and Chamisa Energy is building a 270 MW facility in the Texas panhandle. Both will use salt caverns for air storage and earn revenue through intraday price arbitrage.

The Iowa Stored Energy Park, a CAES plant with aquifer storage whose purpose was to balance wind, was in planning stages for much of the past decade but was eventually abandoned in 2011 due to unfavorable geology. A CAES plant in Norton, OH, with air storage in a former limestone mine has been in planning stages for more than a decade but has not moved forward due to low electricity market prices and financing issues.

14.3.3.4 Cost estimates

The capital cost of CAES includes the plant's turbo-machinery (high- and low-pressure expanders, compressor, and recuperator), underground storage facility, and the balance-of-plant (including site preparation, building construction, and electrical and controls).

| FIGURE 14.16 | *Total capital cost and 95% prediction interval (the bounds between which 95% of data are expected to fall) for the capital cost of salt cavern and aquifer CAES. The latter shows greater variability due to greater differences among sites. Reprinted with permission from Fertig and Apt (2011).* |

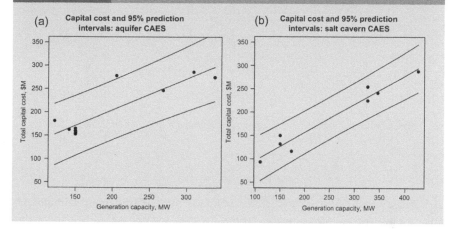

If the underlying geology is suited to a solution-mined salt cavern, storage cavern capital costs include the cost of drilling the wells, the leaching plant, cavern development and dewatering, the brine pipe (to transport the solution away from the site), and water. Development costs associated with aquifer CAES include the cost of drilling multiple wells, the gathering system, the water separator facility, and the electricity used to run an air compressor to initially create the air-storage bubble in the aquifer (Swensen & Potashnik, 1994).

CAES cost data from planning studies and extant plants was regressed against expander capacity (Figure 14.16; Haug, 2004; Hydrodynamics Group, 2009; Schainker, 2008; Swensen & Potashnik, 1994). The capital cost of a CAES plant with salt-cavern storage is close to linear with expander capacity. The capital cost of a CAES plant with aquifer storage is more variable due to the high site specificity of the underground storage cost.

The marginal cost per kWh of energy storage in an aquifer is \$0.10 to \$0.20, which reflects the cost of electricity required to expand the bubble such that the generation phase produces an additional kWh. The marginal cost to expand a solution-mined salt cavern to produce an additional kWh is \$1 to \$2 (Schainker, 2008).

14.3.3.5 Economics of CAES in Texas

In a 2011 paper, we examined the economics of CAES to integrate wind power in Texas (Fertig & Apt, 2011). In the first decade of this century, Texas rapidly increased

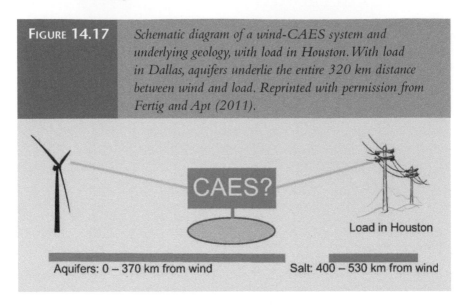

FIGURE 14.17 | *Schematic diagram of a wind-CAES system and underlying geology, with load in Houston. With load in Dallas, aquifers underlie the entire 320 km distance between wind and load. Reprinted with permission from Fertig and Apt (2011).*

CAES?

Load in Houston

Aquifers: 0 – 370 km from wind   Salt: 400 – 530 km from wind

its installed wind power capacity, leading to local price volatility and curtailments due to transmission congestion. We analyzed whether a wind–CAES system, consisting of two large wind plants in western Texas and a CAES plant with flexible location, could profitably provide firm power to load in Houston and Dallas. We found the profitability of the wind-CAES system to be highly dependent on infrequent price spikes, such that a risk-averse investor would be unlikely to adopt the project.

Figure 14.17 shows a diagram of the wind-CAES system we modeled. Using cost as an objective function, the analysis optimized the location of the CAES plant (given that air storage in an aquifer is more expensive than in salt), the capacities of the CAES compressor and expander, the storage capacity, and the capacities of both transmission lines. While CAES near wind would enable a smaller, more efficiently used transmission line spanning the length from wind to load, CAES near load would minimize losses if the operators of the system choose to buy grid power for intraday arbitrage.

Three price scenarios were examined. The first used unadjusted 2007 through 2009 historical prices from the hourly Balancing Energy Services (BES) market, which is analogous to a spot market with lower trading volume (most transactions in ERCOT are bilateral). The second scenario capped these prices at \$300/MWh to remove anomalous spikes and instituted a capacity payment of \$100/MWd. The third scenario assumed a contract price of \$63/MWh, equivalent to the 2008 mean BES price.

For the first two price scenarios, the system was dispatched according to a heuristic strategy that found two profit-maximizing price thresholds per month. For any given hour, if the price is below the lower threshold, the CAES system stores as much wind power as possible; if the price is between the two thresholds, power is sold directly from the wind plant and the CAES system is idle; if the price is above the upper threshold, the CAES plant produces power to be sold in addition to the wind plant output. If the hourly price is below the marginal cost of producing wind power (set to 0), the wind plant is curtailed. The hourly production of energy from the wind–CAES system, e_i, is expressed as follows:

$$
e_i \begin{cases}
0 & \text{if } p_i < MC_W \\
\min(T_W, T_C, w_i - x_i) \text{ if } w_i > x_i, \text{ else } 0 & \text{if } MC_W < p_i < p_s \\
\min(T_W, T_C, w_i) & \text{if } p_s < p_i < p_d \\
\min(T_W + y_i, T_C, w_i + y_i) & \text{if } p_d < p_i
\end{cases} \quad (14.4)
$$

Decision variables are the upper and lower price thresholds p_d and p_s, the capacities of the wind–CAES and CAES-load transmission lines T_W and T_C, and the CAES compressor, expander, and storage capacities. The parameter x_i represents the amount of energy the CAES system is capable of storing in hour i (constrained by the compressor capacity, T_W, and the cavern pressure), w_i is the amount of wind power produced, y_i is the amount of power the CAES system can produce (constrained by the expander capacity, T_C, and the cavern pressure), and MC_W is the marginal cost of wind power.

For the contract price scenario, the price-based heuristic strategy is not feasible, so the optimization was performed with the constraint that the capacity factor of the second transmission line be at least 80%.

For all pricing scenarios, the objective function of the optimization is annual profit, calculated as

$$
\prod = \sum (p_i \cdot e_i - g_i \cdot r_i \cdot HR) - A \cdot (\text{CAES cost} + \text{transmission cost})
$$

(14.5)

in which p_i is the hourly electricity price, g_i is the cost of gas, r_i is the energy produced by the CAES system, HR is the heat rate, and A is an annualization factor (determined with a 10% discount rate).[5] A simulated annealing algorithm was used to perform the optimization.

Results showed that the only two electricity price regimes under which the CAES system added value to the wind plant were unadjusted 2008 historical prices and 2008 capped prices. The year 2008 saw an exceptional number of price spikes on the order of $1,000/MWh, motivating the construction of quite large CAES systems to sell as much energy as possible while prices were high. We note that the two announced CAES systems in Texas are each approximately 300 MW. Figure 14.18 shows a sample week of operation of the wind–CAES system under

---

5. Details on the cost models for CAES and transmission can be found in Fertig and Apt (2011).

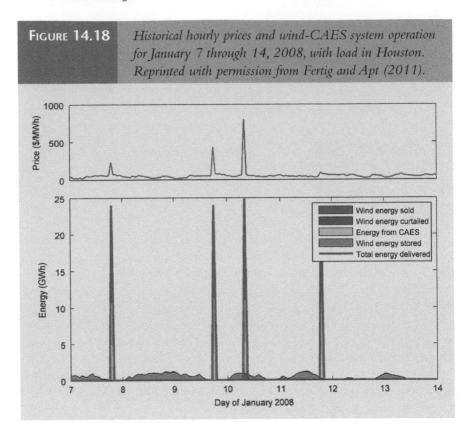

FIGURE 14.18 | *Historical hourly prices and wind-CAES system operation for January 7 through 14, 2008, with load in Houston. Reprinted with permission from Fertig and Apt (2011).*

unadjusted 2008 prices. The system almost always stores wind power and sells it only when prices spike. The optimal CAES expander size is an unrealistic 24 GW. The optimal CAES location in this scenario is near load in Houston, enabling a shorter high-capacity CAES-load transmission line and making use of domal salt for air storage.

For all other price scenarios considered, the wind–CAES system generated less profit than the standalone wind plant. Allowing the system to participate in regulation markets did not change this result. After 2008, ERCOT began overprocuring generation capacity, which has significantly dampened the price spikes necessary to make CAES profitable. We conclude that a profit-maximizing investor would not pair a CAES system with a wind plant under current economic conditions.

We performed an approximate analysis of the social benefits of CAES and found that the combined effects of improved air quality, reduced greenhouse gas emissions, and economic surplus were unlikely to outweigh the private cost of CAES.

In 2012, two plans for two CAES plants were announced in Texas. Since this study was conducted, ERCOT has switched from a zonal to a nodal pricing scheme, increasing the price volatility in wind-rich areas and increasing the incentive for storage. The price cap in the ERCOT market has also been raised substantially. In 2011, it was \$2,500/MWh; in 2013 it was \$5,000; in 2015 it will be \$9,000. Gas prices were substantially lower in 2012 than they were from 2007 to 2009 and are expected to remain low; however, since operating cost of the wind–CAES system was found to be less than 10% of total cost, low gas prices would not change the conclusions of the study. Due to uncertainty in the cost and technical feasibility of CAES, and the track record of failed projects, CAES remains a risky investment.

14.3.4 Plug-in hybrid electric vehicle (PHEV) storage

It has long been suggested that electric vehicles might be used for grid energy services (Kempton & Letendre, 1997; Letendre & Kempton, 2002); these might include storage to support variable energy generators. United States' interest in reducing petroleum consumption and emissions in the transportation sector has resulted in an effort to promote electric vehicles. The 2007 Energy Independence and Security Act provided loan guarantees for advanced battery research. The act also provided grant programs to manufacturers of plug-in hybrid electric vehicles (PHEVs, which use a mix of electricity and gasoline) and battery electric vehicles (BEVs, which use electricity only and have no gasoline backup) to help make them economically feasible as mass-market vehicles (U.S. Government, 2007).

To help overcome obstacles to electric vehicle adoption, policy makers have provided purchase incentives based on battery size. The Energy Improvement and Extension Act of 2008 and the American Recovery and Reinvestment Act of 2009 provide a tax credit of \$2,500 to \$7,500 for vehicles (depending on battery size) with a gross vehicle weight less than 14,000 pounds (U.S. Government, 2009). This subsidy for a specific manufacturer's vehicles declines by 50% then 75% in a phase-out period, which begins in the second calendar quarter after that manufacturer has sold 200,000 vehicles. The subsidy then expires after four calendar quarters (U.S. Government, 2009). Some states provide additional tax credits.

Other incentives that indirectly encourage electric vehicles have been enacted in the United States. In August 2012, the National Highway Traffic Safety Administration (NHTSA) finalized rules raising the Corporate Average Fuel Efficiency (CAFE) standards to 40.3 miles per gallon (mpg) by 2021, and 48.7 mpg by 2025 (NHTSA, 2012). In addition to CAFE standards, the Environmental Protection Agency (EPA) released in August 2012 an average industry-fleet-wide combined cars and trucks greenhouse gas (GHG) standard of 163 gr CO_2 per mile phased in between 2017 and 2025. This standard allows for manufacturers to make improvements not related to fuel efficiency, such as a reduction in coolant leakage. The EPA standard further includes a number of incentives to encourage the use of "game changing" technologies. These incentives begin with credits for model year 2012 through 2016 vehicles that implement specific technologies. For example, alternative fuel vehicles such as PHEVs are permitted a multiplier for model years 2017 through 2021 in the GHG calculations. The rule also treats emissions from electric travel as 0 g/mile in model year 2022 through 2025. Thereafter, upstream fuel emissions will be taken into account.

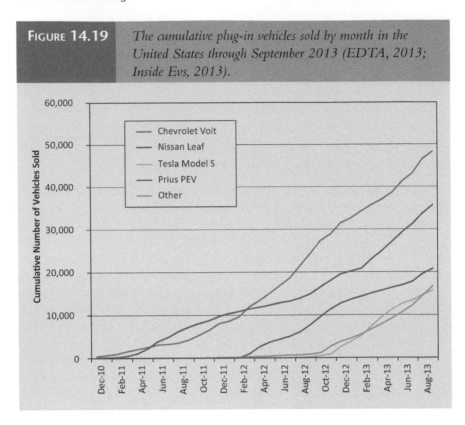

| **FIGURE 14.19** | *The cumulative plug-in vehicles sold by month in the United States through September 2013 (EDTA, 2013; Inside Evs, 2013).* |

At the end of 2010, General Motors introduced the Volt (a PHEV), while Nissan introduced the Leaf (BEV). Other automakers have since begun to offer PHEV and BEV models. The Chevrolet Volt, Toyota Prius Plug-in Hybrid Vehicle, and Ford C-Max Energi are PHEVs, while the Nissan Leaf and Tesla Model S are BEVs. PHEVs and BEVs are collectively referred to as plug-in electric vehicles (PEVs) in the rest of this chapter. Figure 14.19 shows how the sales of PEVs are growing under a supportive policy environment. According to the Electric Drive Transportation Association, an industry association dedicated to promoting the use of electric vehicles in the U.S., the total number of PEVs sold between January 2010 and September 2013 was approximately 140,000 (EDTA, 2013).

The Volt has a 16 kWh pack and the Leaf a 24 kWh pack. Figure 14.20 shows that the cumulative storage capacity of the Chevy Volt and Nissan Leaf batteries has reached more than 1,600 MWh. This figure does not include information about the capacity of the Tesla Model S batteries, which come in different sizes, nor the batteries from other vehicle brands, which are much smaller.

The batteries in PEVs could be used for energy or power support to the grid. The energy available from PEV batteries at any given time depends on when vehicles are plugged in, their current state of charge, what sort of infrastructure is available, and whether the vehicle owners and manufacturers are willing to allow

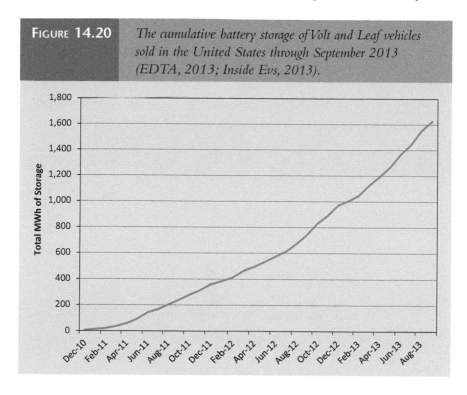

| FIGURE 14.20 | *The cumulative battery storage of Volt and Leaf vehicles sold in the United States through September 2013 (EDTA, 2013; Inside Evs, 2013).* |

vehicle batteries to be used for grid support. Many vehicles are not driven at all on a given day, and at any given hour the majority of vehicles are likely parked at home (Peterson & Michalek, 2013; Weverstad, 2009). If vehicles were plugged in when at home and at work, the proportion of time vehicles could be available to provide grid support would be more than 50%.

The most lucrative market for grid services at present is the ancillary services market. Services such as frequency regulation and spinning reserve have the potential to be quite profitable: some work has suggested that annual revenue may be $1,000 to $2,500 per vehicle (Tomić & Kempton, 2007). However, only a small number of vehicles will saturate those markets (for California, fewer than 200,000 vehicles for regulation and a comparable number for spinning reserve; Letendre & Kempton, 2002). While first movers in these markets may receive higher revenues, the number of vehicles that can benefit is typically less than 1% of the total vehicles owned.

Another energy service possible with PEVs is energy arbitrage. Sourcing energy from vehicles during periods of high demand can help avoid starting electric generation plants, building more peaking plants, and potentially decrease the need for spinning reserves. This market would allow many more vehicles to participate than the ancillary services market. However, RenewElec research (Sioshansi, Denholm, Jenkin, & Weiss, 2009) has shown that the likely profits from participating in the energy market are quite

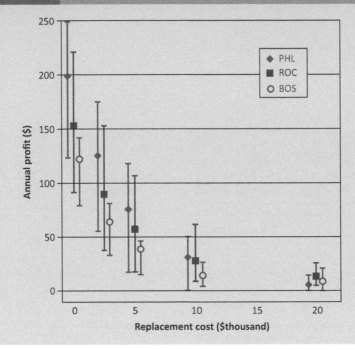

FIGURE 14.21 — *Energy arbitrage profit sensitivity to battery pack replacement cost with perfect forecast of market information in Philadelphia, Rochester, NY; and Boston. The symbol indicates the median annual profit from 2003 through 2008, and the range indicates the most and least profitable years. The profit in each city is calculated given replacement of 16 kWh battery at the cost of $0, $2,500, $5,000, $10,000, and $20,000 (Peterson, Whitacre, & Apt, 2010).*

small (Figure 14.21). It appears to be about an order of magnitude less profitable than the ancillary service market if battery degradation is not included ($0 replacement cost) and less otherwise.

Figure 14.22 shows the percentage of days in the year that energy arbitrage would be a potentially profitable service. Given the costs necessary to allow PEV adoption, it seems likely that in approximately half of the days in a year, energy arbitrage would be profitable but that the profits summed over a year for any vehicle owner would be quite small.

Recent estimates of net social welfare provided by energy arbitrage in one electricity market total $8 per vehicle (U.S. Government, 2007). Avoiding the cost of

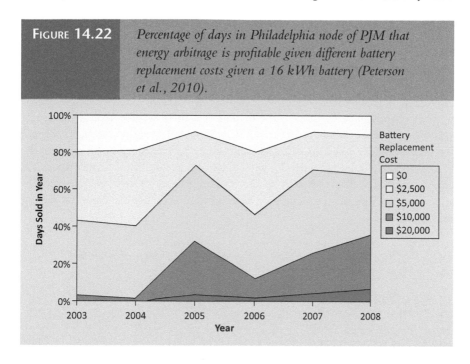

FIGURE **14.22** *Percentage of days in Philadelphia node of PJM that energy arbitrage is profitable given different battery replacement costs given a 16 kWh battery (Peterson et al., 2010).*

new generation plants could result in approximately $200 for participating vehicles (Sioshansi et al., 2009). Some of the costs of integrating higher levels of variable renewable power could also be offset using vehicles. Overall, the total net social welfare benefits that could be transferred to owners of PEVs are quite modest, approximately $300 to 400 annually (Peterson et al., 2010).

PEV storage differs from the previous types of storage discussed in this chapter in a number of important ways. Availability of natural resources does not constrain the spatial availability of PEV storage; PEV storage could be available wherever electric vehicles are driven. Batteries will be purchased with the vehicle regardless of whether they are used for grid support, so compensation would be required only for the additional degradation, not the entire installation cost associated with more traditional storage. Vehicle batteries are typically high-performance batteries and could respond with quick power demand, allowing almost instantaneous ramping. Vehicles are also distributed, so it is possible that investment in transmission lines could be avoided.

The most significant obstacle to using PEVs for grid support is the loss of warranty from vehicle and battery manufacturers. Vehicle manufacturers are testing repurposing vehicle batteries for grid support after they have been deemed unusable for driving purposes. It is possible that this will be the path to using vehicle batteries to provide grid support.

14.3.4.1 Summary

Electric vehicles sold through mid-2013 in the United States contain more than a gigawatt-hour of electric energy storage capacity. If the penetration of BEVs or PHEVs grows, this will increase the potential for grid energy services. Frequency regulation, spinning reserves, and other ancillary services are profitable for a modest number of storage participants in the market. It is these services that are most relevant to countering wind power variability. At large scale, energy arbitrage, charging the vehicle battery when prices are low and selling back to the grid when prices are high, is unlikely to be profitable for vehicle owners. A significant barrier to vehicle battery grid use is that it would void the owner's warranty. Use of vehicle batteries after they have degraded to a point where they are no longer suitable for driving may be practical.

14.3.5 Mitigating renewable energy variability with controlled electric vehicle charging

14.3.5.1 Introduction

The previous section examined the possibility of using electric vehicle batteries in grid energy services such as storage support for variable energy generators. Issues with profitability and vehicle warranty most likely make these types of services unappealing to owners. This section examines whether controlling the charging of BEV or PHEV batteries can be used in integrating variable energy resources into the grid.

Electric vehicles at large scale would add electricity demand that could be shifted to better match the production of different types of renewable energy. This demand would otherwise add to peak electricity demand, increasing the use of the most expensive power plants. Therefore, in addition to pairing well with renewable energy resources, shifting the added demand from electric vehicles may also significantly reduce the cost of adding these vehicles to the electricity system. We use the term *controlled charging* to mean that the grid operator controls when the vehicle battery is charged, within certain constraints set by the owner. *Uncontrolled charging* is used to mean that the owner charges the vehicle at a time entirely of their choosing.

To investigate the benefits of controlled charging, we use two cases studies, one of the New York Independent System Operator (NYISO) and one of the PJM power system (consisting of Pennsylvania, New Jersey, Maryland, West Virginia, Virginia, Ohio, and parts of other states), to determine the costs of generating electricity both with and without significant wind resources, with electric vehicles operated with uncontrolled and controlled charging. In the New York case, we also examine the value of controlled charging in situations in which new generating capacity needs to be built.

CASE STUDY *NYISO.*

To model the New York power system, we use a cost-minimizing unit commitment and economic dispatch (UCED) hourly model that runs over a sample period of 20 days of the year and includes the behavior of the plug-in hybrid electric vehicles and the ability to build new power plants. The 20 days are taken from the four different seasons, capturing the different patterns of wind and load, and include the peak demand of the year. The model requires that the generators meet the load in every hour, a wind generation target is met across the year, and system-level reserve margins are available, while making sure that each generator does not violate its operating constraints. These constraints include minimum and maximum levels of generation, ramp rate constraints, and minimum on and off times.

Our capacity expansion scenario uses the existing power plants in the New York Independent System Operator (NYISO) region but excludes the hydro plants, which represent a significant portion of the generators but could not be modeled in this case. The model can choose to build new gas combustion turbine, gas combined-cycle, and coal power plants for the annualized capital cost.

For the fixed capacity scenario, we scale up the existing fossil fuel generators to be able to cover the entire load without additional power plants. All fuel and capital costs are based on EIA forecasts for 2025.

In both scenarios, vehicles can be included in the controlled charging program for a fixed payment to the vehicle owners per year, and their charging is optimized along with the generator operations. The objective function for the optimization model is shown as follows:

$$\min Vehicle\ Paymets + \sum_{new\ plants} Capital\ Costs + \sum_{time} \sum_{plants} startup\ costs$$
$$+ shutdown\ costs + fuel\ costs \tag{14.6}$$

The vehicles are based on a Chevrolet Volt plug-in hybrid. They are driven according to representative profiles for passenger vehicles from the National Household Travel Survey (NHTS). Uncontrolled charging was calculated based on the entire dataset of 900,000 passenger vehicles; constraints for controlled charging come from 20 representative profiles chosen to best match the aggregate properties of all 900,000 passenger vehicles in the dataset. These aggregate characteristics included the average number of miles driven in each hour of the day, the average cumulative number of miles driven by each hour of the day, the percentage of vehicles parked at home at each hour, and the percentage of vehicles parked anywhere in each hour, as shown in Figure 14.23.

In Table 14.6, we compare the cost savings from controlled electric vehicle charging in NYISO in the fixed capacity scenario and capacity expansion scenario

FIGURE 14.23 *Aggregate characteristics for all passenger vehicles in the NHTS dataset and best-match 20 optimally weighted vehicle profiles drawn from the NHTS dataset over 1 million random draws. The percentage of vehicles at home dips during the day, and only a small percentage of the fleet is driving at any time.*

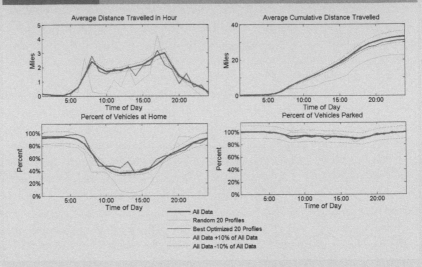

for a 0% and 20% RPS, given different charging scenarios: uncontrolled charging, which uses the entire set of vehicles from the NHTS and begins as soon as the vehicle arrives home for the day, and controlled charging, which uses the weighted set of 20 representative vehicles and optimally charges each vehicle as part of the dispatch optimization, given a $0 payment to vehicle owners for participation, to calculate the maximum possible savings. The maximum savings are calculated as the difference between the uncontrolled and controlled charging system costs. The system costs for each system without electric vehicles are given as a reference, and reduction in vehicle integration costs is found by dividing the difference in costs between uncontrolled charging versus controlled charging with difference in costs between uncontrolled charging versus no vehicles.

As shown in Table 14.6, the cost savings from controlled charging make up a very significant percentage of the total charging costs of the vehicles. There is also a significant increase in the cost savings in the capacity expansion scenario. Having a 20% renewables portfolio standard (RPS) only slightly increases the cost savings from controlled charging.

TABLE 14.6 *NYISO cost savings comparison.*

	Fixed Capacity Scenario (Starting Capacity: 34,700 MW)		Capacity Expansion Scenario (Starting Capacity: 27,500 MW)	
	0% RPS	20% RPS	0% RPS	20% RPS
A. System costs with no electric vehicles (reference)	3.56 $billion/year	4.42 $billion/year	4.05 $billion/year	4.89 $billion/year
B. System costs with uncontrolled charging	3.69 $billion/year	4.53 $billion/year	4.20 $billion/year	5.04 $billion/year
C. System costs with 100% controlled charging and $0 payment to vehicle owners	3.62 $billion/year	4.46 $billion/year	4.10 $billion/year	4.93 $billion/year
Maximum cost savings with controlled charging [B – C]	65 $million/year	69 $million/year	97 $million/year	110 $million/year
Operational cost savings %, capital cost savings %	100%, 0%	100%, 0%	–27%, 127%	30%, 70%
Reduction in vehicle integration costs with controlled charging [(B – C)/(B – A)]	54%	63%	66%	73%

In the second case study, we examine another power system, PJM, using an open-source hourly UCED model of the system called PHORUM, similar to the model used in the NYISO case but focusing on the operational time scale. This model optimizes each day at a time, stepping through the entire year, with no option for capacity expansion. Each day is optimized with a 48-hour window before stepping ahead 24 hours to the next day. The objective function is shown below:

$$\min \sum_{time} \sum_{plants} startup\ costs + shutdown\ costs + fuel\ costs \tag{14.7}$$

We use the same driving profiles and PHEV characteristics as in the NYISO case study, along with PJM power plants and fuel prices by state from 2010. This model also includes transmissions constraints between 5 major areas within PJM, as shown in Figure 14.24.

As before, the results show that controlled charging can reduce the cost of vehicle integration by about 50%, but the cost savings do not increase significantly with a high penetration of wind (see Table 14.7). The reductions in the costs of operating the power system represent a slightly smaller portion of the vehicle integration costs in this case because the modeled PJM system already has more flexibility built in, in the form of pumped hydro storage. The addition of transmission constraints did not significantly increase the value of controlled charging, even with high wind penetration added to the western portion of the PJM region.

FIGURE 14.24	*PHORUM reduced-form model of the PJM region, divided into five transmission regions, connected by the transmission lines shown in white.*

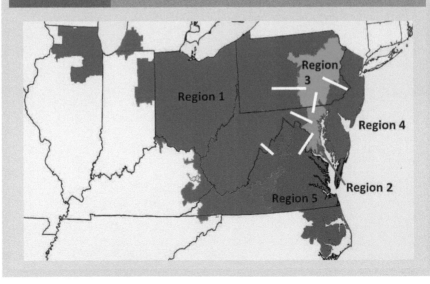

TABLE 14.7	Comparison of cost savings from controlled electric vehicle charging in PJM for a 0% and 20% RPS for controlled and uncontrolled charging. The costs savings are calculated in the same manner as in Table 14.6.		
		0% RPS	**20% RPS**
A. System costs with no electric vehicles (reference)		17.3 $billion/year	13.5 $billion/year
B. System costs with uncontrolled charging		17.6 $billion/year	13.8 $billion/year
C. System costs with 100% controlled charging and $0 payment to vehicle owners		17.5 $billion/year	13.6 $billion/year
Maximum cost savings with controlled charging [B − C]		127 $million/year	144 $million/year
Reduction in vehicle integration costs with controlled charging [(B − C)/(B − A)]		41%	52%

14.3.5.2 Summary

Controlled charging significantly reduces the cost of vehicle charging, but the savings in high-wind-penetration scenarios are not much larger than with low wind penetrations. Controlled charging does not provide much additional value in mitigating the variability of renewable energy in these cases. Future work is needed on the potential cost reductions that could be achieved from using the demand from electric vehicles to compensate for high-frequency variability as well as for errors in wind forecasting. These applications likely add additional value to the service of controlled vehicle charging.

References

Allen, K. (1985). CAES: The underground portion. *IEEE Transactions on Power Apparatus and Systems, PAS-104*, 809–813.

Bullough, C., Gatzen, C., Jakiel, C., Koller, M., Nowi, A., & Zunft, S. (2004). Advanced adiabatic compressed air energy storage for the integration of wind energy. *Proceedings of the European Wind Energy Conference.* www.theeestory.com/files/EWEC_Paper_Final_2004.pdf

Chen, H., Cong, T.N., Yang, W., Tan, C., Li, Y., & Ding, Y. (2009). Progress in electrical energy storage system: A critical review. *Progress in Natural Science, 19*(3), 291–312.

Connolly, D. (2010). The Integration of Fluctuating Renewable Energy Using Energy Storage. PhD dissertation, Department of Physics and Energy, University of Limerick, Limerick, Chapter 5. http://vbn.aau.dk/files/66664679/David_Connolly_PhD_2010_Updated_Journal_Appendices_2012_.pdf

EDP. (2012a). *Complementaridade Hídrica-Eólica.* www.a-nossa-energia.edp.pt/centros_produtores/complementariedade_hidroeolica_he.php

EDP. (2012b). *Novas Barragens.* www.a-nossa-energia.edp.pt/centros_produtores/empreendi mento_type.php?e_type=nb

EDP. (2012c). *Special Regime.* www.edp.pt/en/aedp/unidadesdenegocio/producaodeelectri cidade/Pages/RegimeEspecial.aspx

EDTA. (2013). *Electric drive vehicle sales figures in the U.S. market.* Washington, DC: Electric Drive Transportation Association. http://electricdrive.org/index.php?ht=d/sp/i/20952/pid/20952

EIA. (2012). *Independent statistics and analysis: Portugal.* www.eia.gov/countries/country-data. cfm?fips=PO

EPRI. (2004). *Wind power integration technology assessment and case studies.* EPRI report, 1004806.

EPRI. (2010). *Electric energy storage technology options: A white paper primer on applications, costs, and benefits.* EPRI report, 1020676. Palo Alto, CA: Electric Power Research Institute.

ERCOT. (2010). *ERCOT protocols.* www.ercot.com/mktrules/nprotocols/current

European Union. (2012). *The EU climate and energy package.* http://ec.europa.eu/clima/policies/package/index_en.htm

Eyer, J., & Corey, G. (2010). *Energy storage for the electricity grid: Benefits and market potential assessment guide*, SAND2010-0815. Sandia National Laboratory. www.sandia.gov/ess/publications/SAND2010-0815.pdf

Fertig, E., & Apt, J. (2011). Economics of compressed air energy storage to integrate wind power: A case study in ERCOT. *Energy Policy, 39,* 2330–2342.

Fertig, E., Heggedal, A.M., Doorman, G., & Apt, J. (2014). Optimal investment timing and capacity choice for pumped hydropower storage. *Energy Systems, 5*(2), 285–306.

Gyuk, I. (2004). *EPRI-DOE handbook supplement of energy storage for grid connected wind generation applications.* EPRI report, 1008703.

Gyuk, I., & Eckroad, S. (2003). *EPRI-DOE handbook of energy storage for transmission and distribution applications.* EPRI report, 1001834.

Hadjipaschalis, I., Poullikkas, A., & Efthimiou, V. (2009). Overview of current and future energy storage technologies for electric power applications. *Renewable and Sustainable Energy Reviews, 13*(6–7), 1513–1522.

Haug, R. (2004). *The Iowa stored energy plant.* www.sandia.gov/ess/docs/pr_conferences/2006/haug.pdf

Hittinger, E., Whitacre, J.F., & Apt, J. (2010). Compensating for wind variability using co-located natural gas generation and energy storage. *Energy Systems, 1*(4), 417–439.

Hovorka, S.D. (2009). *Characterization of bedded salt for storage caverns: Case study from the Midland Basin.* Austin, TX: Texas Bureau of Economic Geology.

Hydrodynamics Group. (2009). *Norton compressed air energy storage.* http://hydrodynamics-group.net/norton.html

Inside EVs. (2013). *Monthly plug-in sales scorecard.* http://insideevs.com/monthly-plug-in-sales-scorecard

Katzenstein, W., & Apt, J. (2009). Air emissions due to wind and solar power. *Environmental Science and Technology, 43*(2), 253–258.

Kempton, W., & Letendre, S.E. (1997). Electric vehicles as a new source of power for electric utilities. *Transportation Research, 2*(3), 157–175.

Klara, J.M., & Wimer, J.G. (2007). *Cost and performance baseline for fossil energy plants, vol. 1.* DOE/NETL-2007/1281. http://www.netl.doe.gov/File%20Library/Research/Energy%20Analysis/OE/BitBase_FinRep_Rev2a-3_20130919_1.pdf

Letendre, S.E., & Kempton, W. (2002). The V2G concept: a new model for power? *Public Utilities Fortnightly, 140*(4), 16–26.

Longstaff, F.A. (2001). Valuing American options by simulation: A simple least-squares approach. *Review of Financial Studies, 14*(1), 113–147.

Lueken, R. & Apt, J. (2014). The effects of bulk electricity storage on the PJM market. *Energy Systems*, in press.

McGrail, B.P., Davidson, C.L., Bacon, D.H., Chamness, M.A., Reidel, S.P., Spane, F.A., Cabe, J.E., Knudsen, F.S., Bearden, M.D., Horner, J.A., Schaef, H.F., & Thorne, P.D. (2013). *Techno-economic performance evaluation of compressed air energy storage in the Pacific Northwest*. Pacific Northwest National Laboratory, PNNL-22235. http://caes.pnnl.gov/pdf/PNNL-22235.pdf

MIBEL. (2012a). *The Iberian energy derivatives exchange*. www.omip.pt/Downloads/tabid/104/language/en-GB/Default.aspx

MIBEL. (2012b). *Market price and traded volume data*. www.omip.pt/Downloads/Derivados-deElectricidade/tabid/104/language/en-US/Default.aspx

Morthorst, P.E. (2003). Wind power and the conditions at a liberalized power market. *Wind Energy, 6*(3), 297–308.

NDR. (2011). *Veileder i samfunnsøkonomiske analyser* [Guide to economic analysis]. Olso, Norway: Department of Finance.

Nease, J., & Adams II, T.A. (2013). Systems for peaking power with 100% CO_2 capture by integration of solid oxide fuel cells with compressed air energy storage. *Journal of Power Sources, 228*, 281–293.

NHTSA. (2012). *2017 and later model year light-duty vehicle greenhouse gas emissions and corporate average fuel economy standards*. www.nhtsa.gov/staticfiles/rulemaking/pdf/cafe/2017-25_CAFE_Final_Rule.pdf

NorGer. (2009). *Konsesjonssøknad, likestrømsforbindelse (1400 MW) mellom Norge og Tyskland, [Application for license for constructing a direct current transmission line (1400 MW) between Norway and Germany]*, 200704983. Olso, Norway: Norwegian Water Resources and Energy Directorate.

NVE. (2010, March). *Konsesjonshandsaming for vasskraftsaker, [Processing of license applications for hydropower plants]*. Olso, Norway: Norwegian Water Resources and Energy Directorate.

Paish, O. (2002). Small hydro power: Technology and current status. *Renewable and Sustainable Energy Reviews, 6*(6), 537–556.

Pattanariyankool, S., & Lave, L.B. (2010). Optimizing transmission from distant wind farms. *Energy Policy, 38*(6), 2806–2815.

Peterson, S.B., & Michalek, J.J. (2013). Cost-effectiveness of plug-in hybrid electric vehicle battery capacity and charging infrastructure investment for reducing US gasoline consumption. *Energy Policy, 52*, 429–438.

Peterson, S.B., Whitacre, J.F., & Apt, J. (2010). The economics of using PHEV battery packs for grid storage. *Journal of Power Sources, 195*(8), 2377–2384.

PNBEPH. (2012). Homepage. http://pnbeph.inag.pt/np4/home.html

PJM. (2012). *PJM 2015/2016 RPM base residual auction results*. www.pjm.com/markets-and-operations/rpm/~/media/markets-ops/rpm/rpm-auction-info/20120518-2015-16-base-residual-auction-report.ashx

Schaber, K., Steinke, F., & Hamacher, T. (2012). Transmission grid extensions for the integration of variable renewable energies in Europe: Who benefits where? *Energy Policy, 43*, 123–135.

Schainker, R. (2008). *Compressed air energy storage system updated cost assessment*. EPRI report, 1016004.

Silva, R. (2011). *Merits of hydroelectric pumped-storage schemes in the management of intermittent natural energy resources in Portugal*. Norwegian Power Production Conference, Oslo. www.energinorge.no/getfile.php/FILER/KALENDER/Foredrag%202011/Torsomi nar_PTK/07_Silvia.pdf

Sioshansi, R., Denholm, P., Jenkin, T., & Weiss, J. (2009). Estimating the value of electricity storage in PJM: Arbitrage and some welfare effect. *Energy Economics, 31*, 269–277.

Sørensen, B. (1978). On the fluctuating power generation of large wind energy converters, with and without storage facilities. *Solar Energy, 20*(4), 321–331.

Spees, K. (2008). *Meeting electric peak on the demand side: Wholesale and retail market impacts of real-time pricing and peak load management policy.* PhD thesis, Carnegie Mellon University, Pittsburgh, PA. http://wpweb2.tepper.cmu.edu/electricity/theses/Kathleen_Spees_PhD_Thesis_2008.pdf

Succar, S., & Williams, R.H. (2008). *Compressed air energy storage: theory, resources, and applications for wind power.* Princeton, NJ: Princeton Environmental Institute.

Swensen, E., & Potashnik, B. (1994). *Evaluation of benefits and identification of sites for a CAES plant in New York State.* EPRI report, TR-104268. www.epri.com/abstracts/Pages/ProductAbstract.aspx?ProductId=TR-104268.

Tomić, J., & Kempton, W. (2007). Using fleets of electric-drive vehicles for grid support. *Journal of Power Sources, 168*(2), 459–468.

U.S. Government. (2007). *Energy independence and security act of 2007.* www.gpo.gov/fdsys/pkg/PLAW-110publ140/html/PLAW-110publ140.htm

U.S. Government. (2009). *The American recovery and reinvestment act of 2009.* http://frwebgate.access.gpo.gov/cgi-bin/getdoc.cgi?dbname=111_cong_bills&docid=f:h1enr.pdf

Wan, Y., & Bucaneg, D. (2002). Short-term power fluctuations of large wind power plants. *Journal of Solar Energy Engineering, 124*(4), 427–431.

Western Governors' Association, Northwest Power Planning Council. (2002). *New resource characterization for the fifth power plan: Natural gas combined-cycle gas plants.* www.westgov.org/wieb/electric/Transmission%20Protocol/SSG-WI/pnw_5pp_02.pdf

Weverstad, A. (2009). *Fuel displacement & CO_2 benefits of vehicle electrification.* www.epa.gov/air/caaac/mstrs/may2009/gm.pdf

Wiley, T. (2012). Aquion Energy, personal communication.

15

THE COST EFFECTIVENESS OF DYNAMICALLY LIMITING WIND TURBINES FOR SECONDARY FREQUENCY REGULATION CAPACITY

15.1 Introduction

In the modern alternating-current electrical grid, electrical current oscillates with a nominal frequency of 60 Hz in North America and parts of South America, or 50 Hz in most of the rest of the world. The grid frequency is constant when the supply of electric power (generation) matches the demand (load), but the frequency falls when generation is lower than load and rises when generation is greater than load. There are a few other technical reasons that cause the frequency to fluctuate, but the imbalance between generation and load is the main one. If the frequency deviates too far from its nominal value, the grid can become unstable. Grid operators prevent that from happening by adjusting generation to match the changing load – they continuously start up or shut down power plants or adjust their power output.

This continuous adjustment of generation to control the grid frequency is typically divided into two time scales: primary frequency control (sometimes called frequency response), in which generators automatically adjust their power output to frequency changes within seconds, and secondary frequency control (another term is frequency regulation), in which generators respond to commanded changes within 5 to 10 minutes. We focus here on secondary frequency control because some grid operators in regions with high wind penetration, such as Denmark, Ireland, and Great Britain, require that new wind plants be able to increase or decrease their power output over several minutes to help control the grid frequency (Eir-Grid, 2009; Ramtharan, Jenkins, & Ekanayake, 2007; Singh & Singh, 2009; Tsili & Papathanassiou, 2009). Here we call this dynamic limiting.

It is straightforward for a generator to decrease its power output to decrease the grid frequency (for example, a wind turbine can change the angle at which its blades meet the wind to reduce aerodynamic efficiency), but a generator must be operating below its maximum in order to be able to increase power output. The reserve of power created when a generator operates below its maximum is known as capacity; we will refer to it as up-regulation capacity to denote that it is used to increase the grid frequency. That reserve comes at a price – the generator sacrifices revenue it could have earned by generating more power. For that reason, generators that provide secondary frequency regulation are generally paid separately for the up-regulation capacity they provide and the energy they generate. Most restructured electricity markets (e.g., ERCOT in Texas or PJM in the Mid-Atlantic) have separate markets for energy and regulation. At the time of this writing, wind plants in the United States are not allowed to participate in the markets for frequency regulation because their power output is not considered sufficiently controllable.

Prior research has demonstrated that it is technically feasible to control the power output of a wind plant for *primary* frequency control (de Almeida & Lopes, 2007; Holdsworth, Ekanayake, & Jenkins, 2004; Ramtharan et al., 2007; Teninge et al., 2009) or dynamically limit a wind plant to reduce the variation in power output or limit power ramp rates (Lubosny & Bialek, 2007; Rawn, Lehn, & Maggiore, 2007; Viguera-Rodriguez et al., 2009). Secondary frequency control, which we analyze here, uses the same equipment and methods. Many modern wind plants already have most or all of the necessary equipment, and many others can be retrofitted inexpensively. However, the revenue a wind plant foregoes due to curtailment may be greater than the cost of procuring the same secondary frequency control from traditional sources such as gas turbines or hydroelectric power plants or using an energy storage technology such as a battery to store a reserve of energy (Holdsworth et al., 2004; Teninge et al., 2009). Kirby and colleagues (2010) use a simple wind plant simulation to demonstrate the technical feasibility of dynamically limiting a wind plant for secondary frequency control, but they analyze historical energy and regulation prices in Texas to show that wind would infrequently be able to supply regulation at competitive prices.

In this chapter, which is based on work by Rose and Apt (2013), we calculate how much a wind plant must dynamically limit to provide a given quantity of up-regulation capacity. We then compare the wind plant's opportunity cost to historical market prices to estimate how often wind plants can be cost-effectively dynamically limited for up-regulation capacity for secondary frequency control.

15.2 Method

15.2.1 Simulating a wind plant

We simulate the operation of a dynamically limited 100-MW wind plant with realistic wind speed data. The wind plant consists of 20 5-MW pitch-regulated turbines (Grunnet et al., 2010; Jonkman, Butterfield, Musial, & Scott, 2009) arranged in a line perpendicular to the prevailing wind direction. Each one is driven by wind speed data that are a hybrid of measured low-frequency wind speed data and simulated high-frequency turbulence (Rose & Apt 2012). A central wind plant controller regulates the aggregate power output of the wind plant by continuously adjusting the blade pitch of each turbine to reduce its power output. The Danish grid code refers to this control scheme as

"Delta production constraint," where Δ is the fixed difference between the possible and actual power outputs (Elkraft & Eltra, 2004). We simulate the power output of the wind plant for levels of curtailment $0 \leq \Delta \leq 30$ MW with identical wind inputs.

Each turbine in these simulations dynamically adjusts several times per second to changes in the wind speed and commands from the central controller. This method differs from previous studies of wind power variability (De Tommasi, Gibescu, & Brand, 2010; Sørensen et al., 2007, 2008) that simulated turbines using a simple power curve: the power output is a static function of the wind speed. A dynamic simulation better represents real turbines that cannot respond instantly to rapid changes in wind speed and cannot be dynamically limited below their minimum power limits. In our simulations, we do not allow the wind plant controller to command any wind turbine to dynamically limit below its lower operating limit (LOL) of 20% of its rated power, but a turbine's power output may go below that limit when the wind speed is low.

The wind plant is simulated with 60 days of wind speed data sampled randomly from a given year at a given location. We use historical wind speeds from 2008 at a wind plant site in west Texas as our baseline. We test the sensitivity of our results to those specific wind speed data by running the same simulations with 2007 and 2009 data from the same site in west Texas and 2008 data from a site in the northern Great Plains and a site in Ontario, Canada. More details are given in Rose and Apt (2013).

15.2.2 Analysis of the wind plant simulations

We analyze the wind plant simulations to calculate the amount of up–regulation capacity that a dynamically limited wind plant can provide and at what cost. The up–regulation capacity that a wind plant can supply for given curtailment Δ in a given period is the smallest difference between the dynamically limited power and the power possible if the wind plant were not dynamically limited. For example, Figure 15.1 shows a 300-second (5-minute) period when the wind plant power is dynamically limited to 5 MW (green line) below the possible power (solid blue line). At point A, the central wind plant controller is dynamically limiting the wind plant by the desired 5 MW. At point B, the central wind plant controller can dynamically limit the power by only 3.9 MW, either because the wind speed changed faster than the turbines could adjust or because some turbines reached their lower operating limit and could not be dynamically limited further. The up–regulation capacity available in this period (grey band) is determined by point B, where the minimum difference between the dynamically limited and possible power occurs.

The average cost (AC) of up–regulation capacity from a dynamically limited wind plant in a given period is the amount of energy generation sacrificed divided by the amount of up–regulation capacity provided, as shown in equation (15.1).

$$Average\ Cost = \frac{Energy\ production\ sacrificed}{Upregulation\ capacity} \tag{15.1}$$

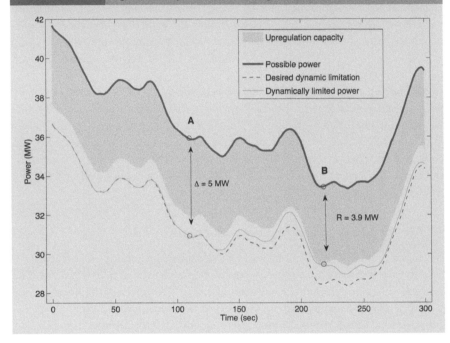

FIGURE **15.1** *Method for calculating the available regulation capacity from dynamic limiting. In this example, the 100 MW-capacity wind plant is dynamically limited by Δ = 5 MW (point A), but the available regulation capacity is 3.9 MW (grey band) because point B is the smallest difference between the possible power and the dynamically limited power output of the wind plant (point B). Reprinted with permission from Rose and Apt (2013).*

We measure sacrificed energy generation in MWh and up-regulation capacity in MW-h, so AC is measured in MWh/MW-h. We also calculate the marginal cost (MC) as the additional energy generation that must be sacrificed to increase up-regulation capacity by 1 MW-h.

15.3 Results

15.3.1 Cost of dynamically limiting for up-regulation capacity

We use the wind plant simulations described above to calculate the cost of up-regulation capacity in terms of sacrificed energy production. A wind plant must sacrifice at least 1.0 to 1.1 MWh of energy production for every 1 MW-h of up-regulation capacity in optimal conditions and often must sacrifice significantly more. A wind plant must sacrifice more energy production than it gets in regulation

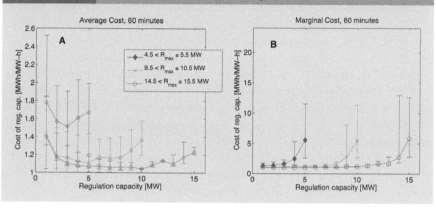

FIGURE 15.2 *The average cost (A) and marginal cost (B) of regulation capacity for 60-minute dispatch intervals. Each line represents cost data for dispatch intervals with maximum available regulation capacity in a certain interval described in the figure key. Circles denote the median and error bars show the 5th and 95th percentile costs. These results are calculated for a 100-MW wind plant simulated with wind speed data from west Texas in 2008. Reprinted with permission from Rose and Apt (2013).*

capacity because the turbines cannot instantly respond to wind speed changes; turbines must be dynamically limited more deeply to ensure they can reliably deliver the desired regulation capacity. Figure 15.2 plots the average and marginal costs of up-regulation for a range of conditions and shows several consistent trends:

1 Cost is lower in periods with higher available up-regulation capacity R_{max};
2 Cost is lowest when a wind plant provides approximately half of the available up-regulation capacity (R_{max});
3 Cost is lower when a few turbines are dynamically limited deeply than when all turbines are dynamically limited equally.

These trends are consistent for wind data from the three different sites (west Texas, northern Great Plains, and Ontario) and several different years.

The cost of up-regulation capacity is lower in periods with higher available capacity. The maximum available up-regulation capacity R_{max} is proportional to the power output of the wind plant, so higher wind speeds mean higher R_{max}, which in turn means lower costs.

Figure 15.2A plots three average cost curves for periods with ranges of R_{max} of approximately 5, 10, and 15 MW-h. The cost curve for the period with $R_{max} \approx 5$ MW shows higher costs for all levels of regulation capacity than the curves for periods with larger R_{max} (10 MW-h or 15 MW-h). For example, the lowest point of the AC curves for a 60-minute period is 1.5 MWh/MW-h for $R_{max} \approx 5$ MW, 1.1 MWh/MW-h for $R_{max} \approx 10$ MW, and 1.1 MWh/MW-h for $R_{max} \approx 15$ MW. The cost of up-regulation capacity increases as R_{max} decreases because the dynamically limited turbines are more likely to reach their lower operating limit (LOL). When turbines reach their LOL, even for a short time, it limits the up-regulation capacity for the entire interval, as described in equation (15.1). However, sacrificed energy production is cumulative over the entire interval, so it is not affected much when turbines briefly reach their LOL. Thus the cost of regulation capacity, calculated with equation (15.1), increases sharply when the denominator decreases (because turbines reach their LOL), but the numerator changes very little.

The cost of up-regulation capacity is lowest when a wind plant provides approximately half of the available up-regulation capacity R_{max}, as shown in Figure 15.2A. For example, in 60-minute periods where the maximum available up-regulation capacity is approximately 10 MW-h, the average cost is 1.1 MWh/MW-h for 5 MW-h of capacity. Notice in Figure 15.2A that the average cost increases very slowly from its minimum for a wide range of up-regulation capacities. This means that the cost of up-regulation capacity is not very sensitive to the estimate of R_{max}, which is calculated assuming perfect foresight of future wind conditions. However, the AC increases sharply for capacities near R_{max}. For example, the AC is 2.8 MWh/MW-h for 10 MW-h of capacity because turbines are more likely to reach their LOL as described. The AC also increases for capacities near zero: the AC is 1.4 MWh/MW-h for 1 MW-h of capacity because it is difficult for turbines to consistently dynamically limit by a small amount.

The cost of a small amount of up-regulation capacity can be significantly reduced by dynamically limiting a few turbines deeply instead of all turbines equally. For example, the average cost of 1 MW-h of capacity in a 15-minute period is 1.5 MWh/MW-h if all turbines are dynamically limited equally, but only 1.1 MWh/MW-h if half the turbines are dynamically limited and 1.0 MWh/MW-h if a quarter of them are dynamically limited (we did not perform enough simulations to confidently estimate these costs for 60-minute periods). This is particularly important for grid codes such as Ireland's, which require wind plants to dynamically limit their power output by a small amount when the grid frequency is less than some threshold (EirGrid, 2009). However, concentrating the curtailment on a few turbines reduces the maximum available regulation capacity; for example, concentrating the curtailment on 25% of the turbines reduces R_{max} by a factor of four.

The results plotted in Figure 15.2 are calculated for wind speed data from west Texas in 2008, but results calculated for the other years (2007 and 2009) and other locations (northern Great Plains and Ontario) show the same trends. We exclude all 1-hour periods when the mean wind plant power is less than 20 MW or greater than 50 MW. The average power was greater than 50 MW in approximately 20% of the 1-hour periods we examined.

15.3.2 Cost effectiveness of dynamically limiting for up-regulation capacity

We calculate the monetary cost of dynamically limiting a wind plant for up-regulation capacity by multiplying the energy production sacrificed per unit of capacity (calculated in the previous section) by the opportunity cost for each MWh of energy. We compare the monetary cost in each 1-hour period to market prices for up-regulation capacity in Texas. Figure 15.3, which plots the cumulative distribution function (CDF) for the cost premium for up-regulation from a dynamically limited wind plant, shows that a wind plant is rarely cost competitive with the conventional generators that set the market price.

FIGURE 15.3	*A cumulative distribution function of the cost premium for frequency regulation capacity from a dynamically limited wind plant, as compared to ERCOT market prices for up-regulation. The dynamically limited wind plant can provide up-regulation at less than the market price in fewer than 1% of the 1,440 1-hour periods studied. We assume an opportunity cost for curtailment of $62/MWh. Reprinted with permission from Rose and Apt (2013).*

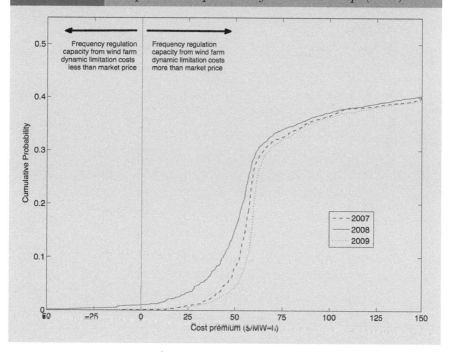

The opportunity cost of energy production varies from wind plant to wind plant depending on local prices for electricity. We assume a fixed opportunity cost of $62/MWh, an average value estimated by Wiser and Bollinger (2011). It consists of a wholesale energy price of $40/MWh, a federal production tax credit (PTC) of $21/MWh in 2010 ($23 as of 2013), and a renewable energy certificate (REC) price of $1/ MWh (Holt et al., 2011). The cost effectiveness is very sensitive to this opportunity cost, so we plot the percentage of hours when dynamically limiting a wind plant is cost effective against the opportunity cost in Figure 15.4.

Figure 15.3 plots the CDF of cost premium for up-regulation capacity from a dynamically limited wind plant. For each hour, we assume the wind plant operator curtails the wind plant optimally to get the minimum average cost possible in the given wind conditions. The energy production sacrificed in that hour is multiplied by the fixed opportunity cost above to calculate the monetary cost of up-regulation capacity. To find the cost premium, we subtract the market price of up-regulation capacity from ERCOT (the Texas market) for the same hour from the wind plant monetary cost (ERCOT, 2012). We use time-synchronized wind speed and market price data to capture any correlations between wind conditions and market activity. A negative premium means up-regulation capacity from the wind plant is less

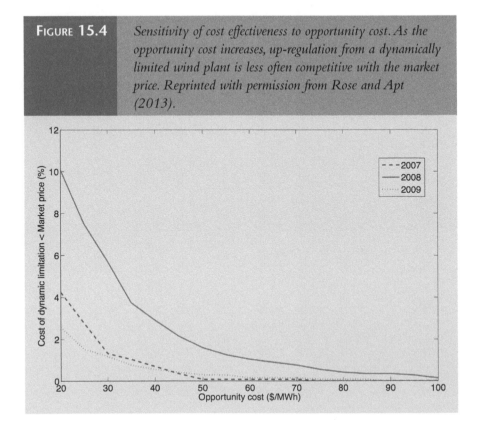

FIGURE **15.4**   *Sensitivity of cost effectiveness to opportunity cost. As the opportunity cost increases, up-regulation from a dynamically limited wind plant is less often competitive with the market price. Reprinted with permission from Rose and Apt (2013).*

expensive and a positive premium means the market price is less expensive; the vertical line in Figure 15.3 indicates a premium of zero. The cumulative probability does not reach 1 since we exclude approximately 60% of the hours because the available up-regulation capacity was nearly zero or because the average wind plant power was outside the 20 to 50 MW power range we studied.

The cost of up-regulation capacity from a dynamically limited wind plant would have been less than the market price in approximately 1% of the 1,440 1-hour periods studied in 2008 and approximately 0% of the dispatch intervals in 2007 and 2009. For a premium up to $50/MW-h, the wind plant can provide regulation capacity in 8% of the 1-hour periods in 2007, 15% in 2008, and 5% in 2009. To examine the sensitivity of the results to the opportunity cost, we plot in Figure 15.4 the percentage of 1-hour periods when up-regulation capacity from a dynamically limited wind plant costs less than the market price against the opportunity cost of curtailment. These results show that a wind plant with a lower opportunity cost will be competitive in the up-regulation capacity market more often than a wind plant with higher opportunity costs.

The results in Figure 15.3 and Figure 15.4 are calculated for wind speed data from west Texas, but results calculated for the other locations show similar trends. We compare regulation capacity costs calculated with wind speed data from the other sites to regulation capacity market prices from Texas, so the results do not account for any correlation between regulation market prices and wind conditions. Regulation capacity costs calculated with northern Great Plains wind data are approximately half of the costs calculated with wind data from the other two sites, though they are still only competitive with the market price in 3.5% of the 1-hour periods studied. We believe the costs are lower because a wind plant at the northern Great Plains site has a significantly higher capacity factor than one at the other two sites – 41%, as compared to 33% for the west Texas site in 2008 and 28% for the Ontario site.

15.4 Summary

Dynamically limiting a wind plant for up-regulation capacity is rarely cost competitive with the market price for up-regulation; we estimate that a dynamically limited wind plant is competitive with market-procured up-regulation less than 1% of the time. There are two reasons wind plants are so rarely competitive in the up-regulation market. First, a wind plant must sacrifice more energy production than a conventional generator to supply the same amount of capacity. Second, the opportunity cost of sacrificed energy is higher for a wind plant than for the conventional generators that set the market price.

A wind plant must sacrifice at least 1.0 to 1.1 MWh of energy production for 1 MW-h of up-regulation capacity, whereas a conventional generator (e.g., a gas turbine) would need to sacrifice only 1 MWh of energy production. Both wind plants and gas turbines can easily dynamically limit their power output, and the up-regulation capacity they can offer is usually limited by their lower operating limit

(LOL). However, dynamically limited wind plants reach their LOL much more frequently because they dynamically limit from a power level determined by the wind conditions, which is typically less than their maximum rated power output.

The opportunity costs of unproduced energy are also different for wind plants and conventional generators. In the results section above, we compare the opportunity cost of dynamically limiting a wind plant to the market price for up-regulation capacity. We assume the up-regulation capacity market in Texas is competitive, which means the generators that participate in that market bid their marginal costs. The marginal costs of up-regulation capacity for both wind plants and conventional generators are approximately their opportunity costs. However, the opportunity cost of dynamically limiting a wind plant is higher because it loses the value of energy plus the value of production subsidies, whereas a conventional generator such as a gas turbine loses the value of energy *minus* the cost of fuel it would have used to generate that energy.

However, there are times and places when it may be necessary for wind plants to provide up-regulation. For example, the Hawaiian Islands use expensive fast-ramping conventional generators (often diesel) to provide frequency regulation, and the penetration of wind power is increasing on those islands. If it is necessary to dynamically limit a wind plant to provide up-regulation capacity, there are several ways to minimize the cost. First, wind plants with low opportunity costs should be curtailed first – plants that sell power on low-price long-term contracts or that no longer receive production subsidies. Second, wind plants should be curtailed to provide approximately half of their maximum available up-regulation capacity. Third, if a wind plant is required to supply a small amount of up-regulation capacity like in the case of the Irish grid code (EirGrid, 2009), a few turbines should be curtailed deeply instead of dynamically limiting all turbines equally.

Finally, it is not unreasonable for grid operators to require that wind plants have the *capability* to be dynamically limited for up-regulation capacity. Adding that capability to a wind plant is simple and inexpensive, and it gives grid operators additional flexibility as the penetration of wind power increases. However, that capability should rarely be used because it is quite expensive to *operate* a wind plant that way.

References

de Almeida, R., & Lopes, J. (2007). Participation of doubly fed induction wind generators in system frequency regulation. *IEEE Transactions on Power Systems*, 22(3), 944–950.

De Tommasi, L., Gibescu, M., & Brand, A.J. (2010). A dynamic aggregate model for the simulation of short term power fluctuations. *Procedia Computer Science*, 1(1), 269–278.

Elkraft and Eltra System. (2004). *Wind turbines connected to grids with voltage above 100kV*. Denmark. www.folkecenter.net/mediafiles/folkecenter/pdf/GRID_Windturbines_con nected_to_grids_with_voltages_below_100kV_2004.pdf

EirGrid. (2009). *EirGrid grid code: WFPS1, controllable wind farm power station grid code provisions, version 3.3.* www.eirgrid.com/media/2009%2009%2014%20EirGrid%20Gird%20 Code%20v3.3%20Wind%20Grid%20Code%20Only.pdf

Electric Reliability Council of Texas (ERCOT). (2012). *Day-ahead ancillary services market clearing prices for capacity archives.* www.ercot.com/mktinfo/prices/mcpc.

Grunnet, J.D., Soltani, M., Knudsen, T., Kragelund, M., & Bak, T. (2010). *Aeolus toolbox for dynamics wind farm model, simulation and control.* Presented at 2010 European Wind Energy Conference & Exhibition, Warsaw.

Holdsworth, L., Ekanayake, J., & Jenkins, N. (2004). Power system frequency response from fixed speed and doubly fed induction generator-based wind turbines. *Wind Energy, 7*(1), 21–35.

Holt, E., Sumner, J., & Bird, L. (2011). *The role of renewable energy certificates in developing new renewable energy projects.* National Renewable Energy Laboratory technical report, NREL/TP-6A20-51904. www.nrel.gov/docs/fy11osti/51904.pdf

Jonkman, J., Butterfield, S., Musial, W., & Scott, G. (2009). *Definition of a 5-MW reference wind turbine for offshore system development.* National Renewable Energy Laboratory technical paper, NREL/TP-500-38060.

Kirby, B., Milligan, M., & Ela, E. (2010). *Providing minute-to-minute regulation from wind plants.* NREL conference paper NREL/CP-5500-48971. Presented at 9th Annual International Workshop on Large-Scale Integration of Wind Power into Power Systems, Quebec.

Lubosny, Z., & Bialek, J. (2007). Supervisory control of a wind farm. *IEEE Transactions on Power Systems, 22*(3), 985–994.

Ramtharan, G., Jenkins, N., & Ekanayake, J.B. (2007). Frequency support from doubly fed induction generator wind turbines. *IET Renewable Power Generation, 1*(1), 3–9.

Rawn, B., Lehn, P., & Maggiore, M. (2007). Control methodology to mitigate the grid impact of wind turbines. *IEEE Transactions on Energy Conversion, 22*(2), 431–438.

Rose, S., & Apt, J. (2012). Generating wind time series as a hybrid of measured and simulated data. *Wind Energy, 15*(5), 699–715.

Rose, S., & Apt, J. (2013). The cost of curtailing wind turbines for secondary frequency regulation capacity. *Energy Systems,* in press. DOI: 10.1007/s12667-013-0093-1

Singh, B., & Singh, S. (2009). Wind power interconnection into the power system: A review of grid code requirements. *Electricity Journal, 22*(5), 54–63.

Sørensen, P., Cutululis, N.A., Vigueras-Rodríguez, A., Jensen, L.E., Hjerrild, J., Donovan, M.H., & Madsen, H. (2007). Power fluctuations from large wind farms. *IEEE Transactions on Power Systems, 22*(3), 958–965.

Sørensen, P., Cutululis, N.A., Vigueras-Rodríguez, A., Madsen, H., Pinson, P., Jensen, L.E., Hjerrild, J., & Donovan, M.H. (2008). Modelling of power fluctuations from large offshore wind farms. *Wind Energy, 11*(1), 29–43.

Teninge, A., Jecu, C., Roye, D., Bacha, S., Duval, J., & Belhomme, R. (2009). Contribution to frequency control through wind turbine inertial energy storage. *IET Renew. Power Gener., 3*(3), 358–370.

Tsili, M., & Papathanassiou, S. (2009). A review of grid code technical requirements for wind farms. *IET Renew. Power Gener., 3*(3), 308–332.

Vigueras-Rodríguez, A., Sørensen, P., Cutululis, N.A., Viedma, A., Gómez-Lázaro, E., & Martin, S. (2009). *Application of ramp limitation regulations for smoothing the power fluctuations from offshore wind farms.* Presented at 2009 European Wind Energy Conference and Exhibition, Marseille.

Wiser, R., & Bolinger, M. (2011). *2010 wind technologies market report.* Oak Ridge, TN: U.S. Department of Energy.

16

QUANTIFYING THE HURRICANE RISK TO OFFSHORE WIND POWER

16.1 Introduction

Wind is the renewable resource with the largest installed capacity growth in the last 5 years, with U.S. wind power capacity increasing from 8.7 GW in 2005 to 60 GW in 2012 (Wiser & Bolinger, 2013). All of this development has occurred onshore, although nearly 1.3 GW of onshore wind turbines are operating within 10 miles of the south Texas coast (U.S. National Renewable Energy Laboratory, 2013; Wiser & Bolinger, 2013).

U.S. offshore wind resources may also prove to be a significant contribution to increasing the supply of renewable, low-carbon electricity. The National Renewable Energy Laboratory (NREL) estimates that offshore wind resources can be as high as four times the 2010 U.S. electricity generating capacity (Schwartz, Heimiller, Haymes, & Musial, 2010). Although this estimate does not take into account siting, stakeholder, and regulatory constraints, it indicates that U.S. offshore wind resources are significant. No offshore wind projects have been developed in the United States, but there are 10 offshore wind projects in the planning process, with an estimated capacity of 2.8 GW (Wiser & Bolinger, 2013) and more proposed (OffshoreWind. net, 2012). The U.S. Department of Energy's 2008 report, *20% Wind by 2030*, envisions 54 GW of shallow offshore wind capacity to optimize delivered generation and transmission costs (U.S. DOE, 2008).

The United States has good wind resources along the Atlantic, Pacific, and Great Lakes coasts. Many areas along the Atlantic coast and Gulf coast are particularly attractive because they have high average wind speeds, are in relatively shallow water, and are close to major population centers. Wind resources at depths shallower than 30 m along the entire the Atlantic coast, from Georgia to Maine, are estimated to be 538 GW; the estimate for these resources in Texas is 157 GW (Schwartz et al., 2010). These estimates ignore restrictions such as marine sanctuaries, shipping lanes, and military practice areas. For comparison, the 2012 net summer generation capacity for the entire U.S. was 1,067 GW (EIA, 2013).

Offshore wind turbines in these areas will be at risk from Atlantic hurricanes. Wind turbines are vulnerable to hurricanes because the maximum wind speeds in those storms can exceed the design limits of wind turbines. Modern offshore wind turbines are designed to survive sustained wind speeds of up to 50 m/s (97 knots or 112 mph) at the top of their towers (International Electrotechnical Commission, 2009). That wind speed, equivalent to a weak hurricane, is the maximum wind speed expected in a storm in the North Sea, where most offshore wind turbines have so far been built. However, intense hurricanes with wind speeds above the design limit sometimes strike parts of the southern U.S., especially along the Gulf Coast. Parts of southern China, Taiwan, southern Japan, and the Philippines face a similar risk from intense typhoons. In 2003, a wind plant of seven utility-scale turbines in Okinawa, Japan, was destroyed by typhoon Maemi, which had an estimated maximum sustained wind speed of 60 m/s (Takahara et al., 2004; equivalent to a Category 4 hurricane), and several turbines in China were damaged by typhoon Dujuan (Clausen et al., 2007).

If the offshore wind resources along the U.S. coast are to be successfully developed, the hurricane risk to wind turbines should be analyzed and understood. Hurricanes strike the U.S. coast only infrequently, so statistical analysis based on historical data is problematic. Here we model the risk by simulating thousands of years of hurricane activity using a model based on numerical weather techniques (Emanuel, Ravela, Vivant, & Risi, 2006) and analyzing their effects on simulated offshore wind turbines. We use this model to estimate the risk to a wind plant over its 20-year life in four representative locations in the coastal United States. We also use the model to estimate the fraction of wind power in a region simultaneously offline due to hurricane damage. This model was first presented by Rose and colleagues (2013).

16.2 Method

16.2.1 Risk of turbine tower buckling as a function of wind speed

Wind turbines can be damaged or destroyed by winds stronger than their design limit, and the probability of damage increases sharply with wind speed. In Figure 16.1, we plot the probability that a turbine tower buckles as a function of the maximum sustained wind speed at the top of its tower (called the hub height). That figure shows the cumulative distribution function (CDF) for two load cases: wind hitting the nacelle of the turbine head on, assuming the turbine can actively yaw to track the wind direction, and wind hitting the nacelle broadside, assuming the turbine is unable to track the wind direction. Figure 16.1 also plots a vertical line at 97 knots to show the design limit for a typical offshore wind turbine.

Each point plotted in Figure 16.1 represents the average probability of buckling for 10 simulations at the same average wind speed. We calculate the probability of the turbine tower buckling by comparing simulated stresses from

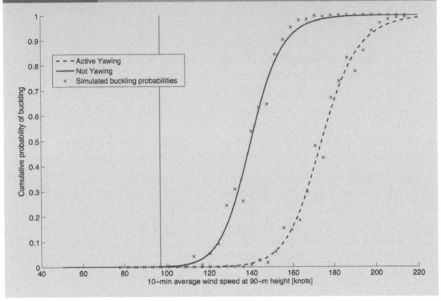

FIGURE 16.1 *Log-logistic functions fitted to probability of tower buckling as a function of wind speed. The vertical red line at 97 knots is the 10-minute sustained wind speed design limit for offshore wind turbines. Reprinted with permission from Rose and colleagues (2013).*

aerodynamic forces on a 5-MW offshore wind turbine (Jonkman et al., 2009) to the stochastic resistance to buckling proposed by Sørensen and colleagues (Sørensen & Tarp-Johansen, 2005). The turbine is shut down with its blades feathered as it would be at wind speeds higher than approximately 25 m/s (48 knots). These simulations do not account for wave loads or damage mechanisms other than tower buckling. Rose and colleagues give additional details about the simulations (Rose et al., 2012).

Figure 16.1 shows that a turbine that can actively yaw has a significantly lower probability of buckling at a given wind speed than a turbine that cannot yaw. Nearly all modern turbines use large electric motors to track the wind direction. These motors are powered from the electrical grid, so turbines will not be able to yaw if a hurricane causes a blackout on the local grid. The primary design standard for offshore wind turbines requires that they be designed to withstand extreme winds hitting the nacelle broadside unless they have 6 hours of backup power for the yaw motors (International Electrotechnical Commission, 2009). Rose and colleagues estimated that adding backup power would add $30,000 to $40,000 (2010 prices) to the price of a turbine and 1,400 to 2,400 kg to its weight (Rose et al., 2012). However, the turbine must be able to yaw fast enough to track wind direction changes in a hurricane, and it must have a sensor that can accurately measure the direction of hurricane-force winds.

16.2.2 Simulated hurricanes

The historical record of hurricanes in the United States is insufficient to confidently estimate the risk of intense hurricanes. A hurricane more intense than category 2 strikes the United States on average only once every 2 years. For this reason, we estimate the risk using hurricanes simulated with the method of Emanuel and colleagues that generates hurricanes with statistical properties that agree with the historical record (Emanuel et al., 2006). Emanuel's method simulates hurricanes by first seeding weak vortices randomly in space and time (Emanuel et al., 2008a). These vortices follow a track stochastically determined by ambient winds, and their intensity evolves along that track as a deterministic function of wind and ocean conditions known as the Coupled Hurricane Intensity Prediction System (CHIPS); some vortices grow into hurricanes, but most dissipate (Emanuel et al., 2008b).

For each hurricane, we calculate the maximum 1-minute sustained wind speed at each offshore wind turbine location using the wind profile proposed by Holland and colleagues (Holland et al., 2010). Our model for the probability of tower buckling takes the 10-minute average wind speed as its input, so we convert the 1-minute average to 10-minute average by dividing by a factor of 1.11, as suggested by Harper and colleagues (Harper, Kepert, & Ginger, 2010).

We modify two aspects of the hurricanes simulated with Emanuel's method. In order to quantify the uncertainty in the simulations, we vary the annual rate of hurricane occurrence randomly in a narrow range determined by the number of simulated hurricanes in our dataset. We also scale the radius of maximum winds of each hurricane by a lognormal-distributed random variable to make the radii more similar to historically observed hurricanes. Details about both of these modifications are given by Rose and colleagues (Rose et al., 2013).

16.2.3 Risk to a single wind plant

We estimate the hurricane risk to a single 50-turbine wind plant over its 20-year life by simulating 5×10^4 years of hurricane activity and calculating the number of turbines buckled in four representative locations. Each hurricane is simulated according to the procedure described in the section *Simulated hurricanes*. For each hurricane, we calculate the maximum 1-minute sustained wind speed at each turbine location; the wind speed allows us to estimate the probability of each turbine buckling using the log-logistic function described in Rose and colleagues (2012). Turbines that buckle are not replaced until the end of a 20-year period.

16.2.4 Risk to regional wind power

We calculate the hurricane risk to all the offshore wind power in a region by simulating 50 5,000-year periods of hurricane activity along the U.S. coast (a total of 2.5×10^5 years). Each hurricane is simulated according to the procedure described in *Simulated hurricanes*. For each hurricane, we calculate the maximum 1-minute sustained wind speed at each turbine location; the wind speed allows us to estimate the probability of each turbine buckling using the log-logistic function described in Rose and colleagues (2012). After towers buckle, we assume it takes 2 years to rebuild them in case several hurricanes strike the same area in a short period (for example, seven hurricanes made landfall in Florida in 2004–2005, with two each year striking the same area). We use simulated hurricanes in order to base our risk calculations on much longer periods of hurricane activity than are available in historical records, which reduces the uncertainty of our estimates.

We place offshore wind turbines in all feasible locations along the U.S. East Coast and Texas. A feasible location must have sufficient wind resource, shallow water, and not conflict with other uses such as marine sanctuaries, military areas, shipping lanes, or oil/gas leases. Maps and further details about turbine placement are given in Rose and colleagues (2013).

16.3 Results

16.3.1 Lifetime risk to a single wind plant

We estimate the lifetime hurricane risk to a single wind plant in four locations: Galveston County, Texas; Dare County, North Carolina; Atlantic County, New Jersey; and Dukes County, Massachusetts. For each location, we calculate the lifetime risk as the distribution of the simulated number of turbine towers buckled by hurricanes in 20 years (the typical design life of wind turbines) if buckled turbines are not replaced. The results for Galveston and Dare counties are shown in Figure 16.2, where the lines plot the median risk for all periods and all wind plant sites near a particular county. The error bars, which plot the 5th and 95th percentile risks, represent the random variation in the number and intensity of hurricanes occurring in any given 20-year period, as well as slight differences in risk to wind plants at different locations in a given county. Solid lines plot the risk to nonyawing turbines, and dashed lines plot the risk to yawing turbines. We do not show plots of the results for Atlantic County, NJ, and Dukes County, MA, because the risks are much smaller.

The risks are greatest for wind plants in areas prone to intense hurricanes, especially the Gulf of Mexico. In Galveston County, TX, there is a 30 to 40% probability that at least one tower will buckle in 20 years and a 2 to 6% probability that more than half the towers will buckle if the turbines cannot yaw. In Dare County, NC, there is a 4 to 11% probability that at least one tower will buckle and less than 2% probability that more than half the towers will buckle if the turbines cannot yaw. If the turbines can yaw to track the wind direction, the risks are much lower. The probability that at least one turbine is buckled is 4 to 9% in Galveston and less than 2% in Dare County, NC. The risks to nonyawing turbines are significantly lower in the Mid-Atlantic and New England, where intense hurricanes are extremely rare. In Atlantic County, NJ, there is a 0.4 to 3% probability and in Dukes County, MA, a 0.2 to 2% probability that at least one tower will buckle. The probability that more than half the turbines buckle is too small to estimate with the number of hurricanes

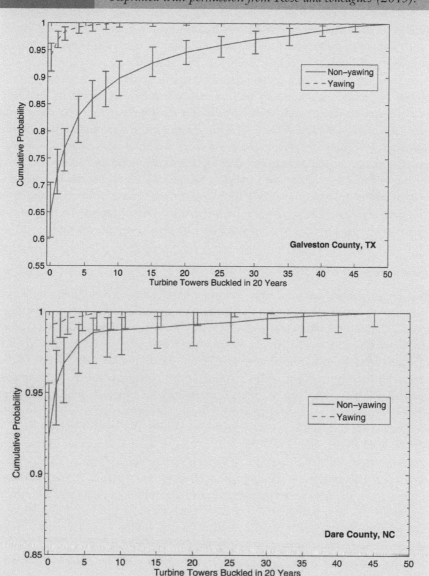

FIGURE 16.2 *Cumulative distribution of number of turbine towers buckled in (a) Galveston County, TX, and (b) Dare County, NC, by hurricanes in 20 years if buckled towers are not replaced. Dashed lines plot the distribution for the case that the turbines can yaw to track the wind direction, and solid lines plot the distribution for the case that turbines cannot yaw. The error bars represent the 90% uncertainty range. Reprinted with permission from Rose and colleagues (2013).*

we simulated. The probabilities that in those counties any turbines that have the capability to yaw into the wind even when grid power is unavailable are buckled are too small to estimate with the number of hurricanes we simulated.

The risks discussed, which are taken from Rose and colleagues (2013), are significantly lower than risks estimated in a previous paper by Rose and colleagues (2012). The results in the older paper likely overestimate the risk to a wind plant because that paper incorrectly assumed that hurricanes maintain their peak intensity as they approach the shore and that any hurricane making landfall in a county affects the entire coastline of that county with its maximum winds (Powell & Cocke, 2012). The results we present here also include a necessary conversion from 1-minute average to 10-minute average wind speeds that the older paper did not include.

To show the geographical distribution of risk, we plot the expected (mean) probability of a turbine buckling in any 20-year period for locations along the U.S. coast in Figure 16.3. This map shows the relative risk for different locations, but the plotted expected risk values may be misleading because they represent the mean risk of many 20-year periods with very low buckling probabilities and a few with very high buckling probabilities. This is illustrated for the case of Galveston by the histogram in Figure 16.3. Galveston has a mean buckling probability of approximately 8%, but the buckling probability for a turbine in that location is near zero for the vast majority of 20-year periods we simulated. Note that the histogram is not directly comparable to the CDF for Galveston in Figure 16.2 because, although they both represent risks in a 20-year period at the same location, the CDF accounts for the correlation of risks to nearby turbines but the histogram does not.

16.3.2 Wind power simultaneously offline

We estimate the amount of wind power in a region that could be offline simultaneously due to hurricane damage, assuming that destroyed turbines require 2 years to rebuild. Figure 16.4 and Figure 16.5 plot results for Texas and the Southeast (Georgia, South Carolina, and North Carolina) as the return period of different fractions of wind power being offline simultaneously. The return period is the inverse of the annual probability: an event with a 50-year return period has a 1 in 50 chance of occurring in any given year. In Texas (Figure 16.4), the median amount of wind power offline with a 100-year return period is 10% with a range of 7.7 to 14% for nonyawing turbines. For a 50-year return period, the median fraction offline is 5.7% with a range of 4.3 to 7.3%. If the turbines were able to yaw to track the wind direction, the median amount of wind power offline in Texas is 0.33% with a 100-year return period and 0.091% with a 50-year return period. In the Southeast (Figure 16.5), the median amount of wind power offline with a 100-year return period is 1.7%, with a range of 1.0 to 2.6% for nonyawing turbines. For a 50-year return period in the Southeast, the median fraction offline is 0.59% with a range of 0.35 to 0.82%. We do not show results for the Southeast if turbines were able to yaw; the median wind power offline is 0.01% with a 100-year return period and 0% with a 50-year return period.

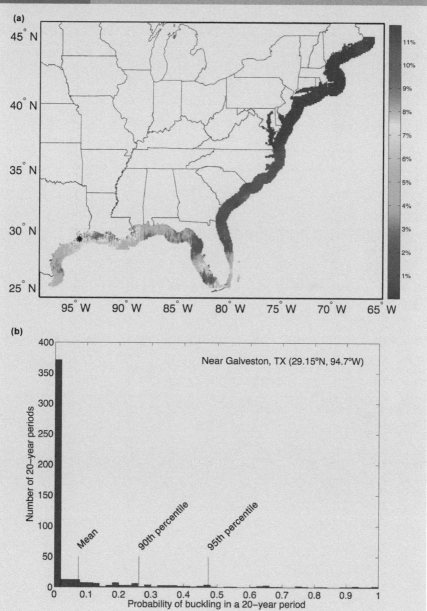

FIGURE 16.3 (a) Map of mean probabilities of a wind turbine buckling in any given 20-year period at various locations along the U.S. coast and (b) a histogram of buckling probability in 20 years for one of those locations: Galveston, TX.

FIGURE 16.4

Return period for fraction of wind power offline due to hurricane damage in Texas, assuming turbines are placed in all locations described previously (total capacity of 87 GW). Figure 16.4(a) gives risks for nonyawing turbines; Figure 16.4(b) gives risks for yawing turbines. Each of the "Simulated hurricanes" lines represents one of the 50 5,000-year periods of simulated hurricanes. The "Historical hurricane" line represents risk calculated from the historical hurricane record (1900–2011), and "H_{Lo}" and "H_{Hi}" represent the lower and upper confidence bounds for the historically based estimates. Reprinted with permission from Rose and colleagues (2013).

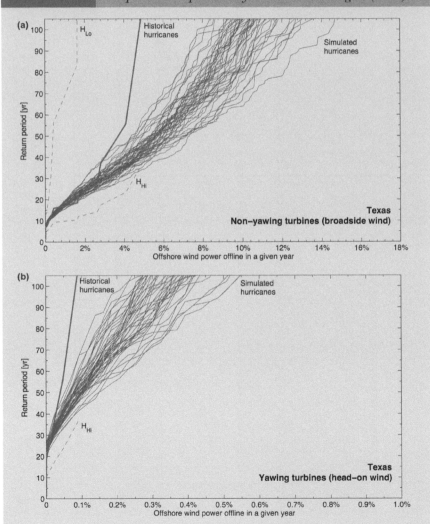

FIGURE 16.5 *Return period for fraction of wind power offline due to hurricane damage in the Southeast (GA, SC, NC), assuming turbines are placed in all locations described (total capacity of 104 GW) and the turbines cannot yaw. We do not show the results for yawing turbines because the risks are negligible. Each of the "Simulated hurricanes" lines represents one of the 50 5,000-year periods of simulated hurricanes. The "Historical hurricane" line represents risk calculated from the historical hurricane record (1900–2011), and "H_{Lo}" and "H_{Hi}" represent the lower and upper confidence bounds for the historically based estimates. Reprinted with permission from Rose and colleagues (2013).*

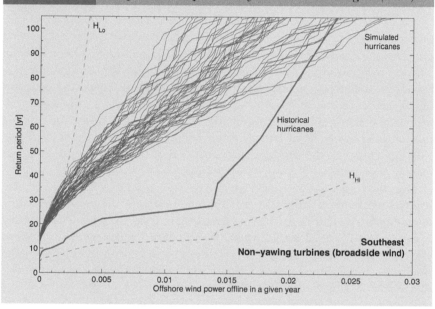

The expected fraction of wind power offline due to hurricane damage is smaller than the expected fraction of wind power offline due to normal component failures (~3%; Harman, Walker, Wilkinson, 2008) for most of these cases, though turbines destroyed by hurricanes take much longer to restore to service. We also do not show results for the Mid-Atlantic and New England because the risks are too small to estimate with the 3,285 simulated landfalling hurricanes we used in our simulations. These results are most sensitive to the rebuilding time and uncertainties in the size of the hurricanes.

Our damage model predicts that historical hurricanes would have been similarly destructive if offshore wind turbines had existed in the locations we describe. The most destructive would have been Hurricane Carla, which struck Texas in 1961.

We calculate it would have buckled 7.9% of the turbines (6.8 GW) in Texas if they were unable to yaw and 0.4% (0.38 GW) if they could yaw fast enough to always point into the wind. Similarly, Hurricane Helene in 1958 would have destroyed 1.8% of the nonyawing turbines (1.8 GW) in the Southeast, and Hurricane Gloria in 1985 would have destroyed 0.2% of the nonyawing turbines (0.1 GW) in the Mid-Atlantic. The one exception is Hurricane Gerda in 1969, which we calculate would have destroyed 8.8% of the nonyawing turbines (3.4 GW) in New England, significantly more than our simulations predict for even a 1,000-year return period (we predict a median of 1.5% and maximum of 4.2%). For comparison, the only other historical hurricane that would have caused measureable simulated damage in New England was Hurricane Esther in 1961, which would have destroyed 0.2% (0.08 GW) of the nonyawing turbines.

16.4 Summary

Our results suggest that hurricanes pose a significant but manageable risk to wind plants built to current designs in the Gulf of Mexico and the Southeast but pose little risk to plants in the Mid-Atlantic and New England. We also find that hurricanes are unlikely to simultaneously damage enough offshore wind power to significantly affect the adequacy of the electricity supply, even in the higher-risk regions. The risks can be mitigated by strengthening turbine designs or ensuring that turbines can yaw to track the wind direction even if grid power is lost. These risks may change as the climate changes, but it is unclear whether the risks will increase or decrease.

From the perspective of an electrical grid operator, hurricane damage to offshore wind power may affect system adequacy, which is a measure of the generation capacity to meet future load. In Figure 16.4 we show that, in Texas, there is a 2% probability of having 4.3 to 7.3% of installed wind capacity offline simultaneously in any given year due to hurricane damage and a 1% probability of 7.7 to 14% offline. These losses would have a small effect on system adequacy in Texas because wind power is a variable resource and can therefore contribute only a fraction (known as "capacity value") of its rated power output to system adequacy. There has been significant work on estimating the capacity value of wind power (Ensslin et al., 2008; Hasche, Keane, & O'Malley, 2011; Keane et al., 2011) based on wind conditions and turbine reliability, but none has considered the risk of losses in wind plant installations resulting from natural hazards. We thus suggest that calculations of the capacity value of offshore wind resources in the Gulf Cost should incorporate the risk of losses due to hurricanes.

To mitigate the risks of hurricanes, offshore wind turbines can be designed for higher maximum wind speeds, designed to track the wind direction (yaw) quickly enough to match wind changes in a hurricane even if grid power is cut off, or placed in areas with lower hurricane risk, as discussed by Rose and colleagues (2012). Efforts are underway to determine design standards for offshore wind turbines in hurricane prone areas (Yu, Samuelsson, & Tan, 2011). Battery backup is a low-cost way to maintain yawing capability when grid power is interrupted (Rose et al., 2012).

Approximately 90% of planned offshore wind development in the United States in the next 10 to 20 years will occur in low-risk areas such as New England and the Mid-Atlantic states. A U.S. Department of Energy report envisions a scenario with 54 GW of offshore wind from North Carolina to Maine by 2030 (U.S. DOE, 2008) and the U.S. Bureau of Ocean Energy Management (BOEM) is planning to auction offshore wind leases from Massachusetts to Virginia (Brannen, 2012). However, the state of Texas has strongly encouraged onshore wind development and has signed a lease for a wind power development near Galveston (U.S. DOE, 2005).

The model developed for this chapter uses simulated hurricane tracks and intensities based on climatological conditions from 1979 to 2011; it does not assess the effects of climate change.

References

Brannen, P. (2012). Offshore wind farms will be encouraged in tracts along the east coast. *Washington Post*, July 23. www.washingtonpost.com/national/health-science/offshore-wind-farms-will-be-encouraged-in-tracts-along-the-east-coast/2012/07/23/gJQAD-2Pu4W_story.html

Clausen, N.-E., Candelaria, A., Gjerding, S., Hernando, S., Nørgård, P., Ott, S., & Tarp-Johansen, N.-J. (2007). Wind farms in regions exposed to tropical cyclones. In *Conference proceedings (online)*, EWEA, Brussels.

EIA. (2013). *Electric power monthly February 2013 with data for December 2012*. R. Hankey (ed.). U.S. Energy Information Administration. www.eia.gov/electricity/monthly/current_year/february2013.pdf

Emanuel, K.A., Ravela, S., Vivant, E., & Risi, C. (2006). A statistical deterministic approach to hurricane risk assessment. *Bulletin of the American Meteorological Society, 87*(3), 299–314.

Emanuel, K.A., Sundararajan, R., & Williams, J. (2008a). Hurricanes and global warming: Results from downscaling IPCC AR4 simulations. *Bulletin of the American Meteorological Society, 89*(3), 347–367.

Emanuel, K.A., Sundararajan, R., & Williams, J. (2008b). *Downscaling hurricane climatologies from global models and re-analyses*. Presented at 28th Conference Hurricanes and Tropical Meteorology, Orlando, FL.

Ensslin, C., Milligan, M., Holttinen, H., O'Malley, M., & Keane, A. (2008). Current methods to calculate capacity credit of wind power, IEA collaboration. *IEEE Power and Energy Society General Meeting*, DOI: 10.1109/PES.2008.4596006

Harman, K., Walker, R., & Wilkinson, M. (2008). *Availability trends observed at operational wind farms*. Presented at European Wind Energy Conference, Brussels.

Harper, B.A., Kepert, J.D., & Ginger, J.D. (2010). *Guidelines for converting between various wind averaging periods in tropical cyclone conditions*. WMO/TD-1555, World Meteorological Organization, Geneva.

Hasche, B., Keane, A., & O'Malley, M. (2011). Capacity value of wind power, calculation, and data requirements: The Irish Power System case. *IEEE Transactions on Power Systems, 26*(1), 420–430.

Holland, G.J., Belanger, J.I., & Fritz, A. (2010). A revised model for radial profiles of hurricane winds. *Monthly Weather Review, 138*(12), 4393–4401.

International Electrotechnical Commission. (2009). *Wind turbines – part 3: design requirements for offshore wind turbines*, 61400–3, 1st ed. Geneva, Switzerland: International Electrotechnical Commission. ISBN 978-2-88910-514-4.

Jonkman, J.M., Butterfield, S., Musial, W., & Scott, G. (2009). *Definition of a 5-MW reference wind turbine for offshore system development.* National Renewable Energy Laboratory technical paper, NREL/TP-500-38060.

Keane, A., Milligan, M., Dent, C.J., Hasche, B., D'Annunzio, C., Dragoon, K., Holttinen, H., Samaan, N., Soder, L., & O'Malley, M. (2011). Capacity value of wind power. *IEEE Transactions on Power Systems, 26*(2), 564–572.

OffshoreWind.Net. (2012). *North American Offshore Wind Project information.* www.offshorewind.net.

Powell, M.D., & Cocke, S. (2012). Hurricane wind fields needed to assess risk to offshore wind farms. *Proceedings of the National Academy of Sciences, 109*(33), E2192–E2192.

Rose, S., Jaramillo, P., Small, M.J., & Apt, J. (2013). Quantifying the hurricane catastrophe risk to offshore wind power. *Risk Analysis, 33*(12), 2126–2141. DOI:10.1111/risa.12085

Rose, S., Jaramillo, P., Small, M.J., Grossmann, I., & Apt, J. (2012). Quantifying the hurricane risk to offshore wind turbines. *Proceedings of the National Academy of Sciences, 109*(9), 3247–3252.

Schwartz, M., Heimiller, D., Haymes, S., & Musial, W. (2010). *Assessment of offshore wind energy resources for the United States.* National Renewable Energy Laboratory technical paper, NREL/TP-500-45889. www.nrel.gov/docs/fy10osti/45889.pdf

Sørensen, J.D., & Tarp-Johansen, N.-J. (2005). Reliability-based optimization and optimal reliability level of offshore wind turbines. *International Journal of Offshore and Polar Engineering, 15*(2), 141–146.

Takahara, K., Mekaru, T., Shinjo, F., Ishihara, T., Yamaguchi, A., & Matsuura, S. (2004). *Damages of wind turbine on Miyakojima Island by typhoon Maemi in 2003.* Paper presented at the European Wind Energy Conference and Exhibition, London, November 22–25.

U.S. DOE. (2005). Company plans large wind plant offshore of Galveston, Texas. *Energy Efficiency & Renewable Energy Network News,* November 2. http://apps1.eere.energy.gov/news/news_detail.cfm/news_id=9502

U.S. DOE. (2008). *20% wind energy by 2030.* DOE/GO-102008-2567. S. Lindenberg, B. Smith, K. O'Dell, E. DeMeo, & B. Ram (eds.). Washington, DC: U.S. Department of Energy.

U.S. National Renewable Energy Laboratory. (2013). *Map of wind farms.* http://en.openei.org/w/index.php?title=Map_of_Wind_Farms&oldid=593614

Wiser, R., & Bolinger, M. (2013). *2012 wind technologies market report.* DOE/GO-102013-3948. Oak Ridge, TN: U.S. Department of Energy.

Yu, Q., Samuelsson, L., & Tan, P.-L. (2011). *Design considerations for offshore wind turbines in US waters – the American way.* Offshore Technology Conference, Houston, TX.

PART III

Review of large-scale wind integration studies

17

A CRITICAL REVIEW OF LARGE-SCALE WIND INTEGRATION STUDIES IN THE UNITED STATES

17.1 Introduction

This chapter reviews 11 major quantitative regional and national wind integration studies from the United States. Each of these is summarized and compared with regard to data sources, methods, and conclusions. We place particular emphasis on the operational impacts of wind integration and changes in reserve requirements necessary to maintain system reliability, as well as the statistical challenges associated with developing and analyzing time-series wind power profiles. Based on comparisons among these studies, we suggest areas in which improvements in methods are warranted in future studies and areas in which additional research is needed to facilitate future improvements in wind integration studies.

To our knowledge, this is only the second broad review of industrial wind integration studies. The first, by Holttinen and colleagues (2011), focused primarily on European studies completed before 2007. Since that time, a number of large and important studies have been published, which have a number of improvements (as well as some areas that still need improvement); we review these. This review also takes a more detailed look at the statistical challenges associated with wind integration studies, suggesting several directions for methodological improvements and future research.

We review 11 U.S. integration studies published since 2005. These include the U.S. Department of Energy's (DOE's) national *20% Wind by 2030* report (U.S. DOE, 2008), two DOE-sponsored follow-up studies covering the eastern (EnerNex Corporation, 2011) and western United States (GE Energy, 2010), and a recent national study by the U.S. National Renewable Energy Laboratory (NREL; NREL, 2012). We also review seven state and regional studies covering New York (GE Energy, 2005a; NYISO, 2010), Texas (GE Energy, 2008), Minnesota (EnerNex Corporation, 2006), California (KEMA, 2010), Nebraska (EnerNex, Ventyx, and Nebraska Power Association, 2010), and the South-central United States (Charles River Associates, 2010).

While the specific research questions addressed by these studies differ, they generally fall into three broad categories. The first category is the assessment of potential wind resources. These assessments frequently included the determination of when, where, and how much wind could be harvested within a given region as well as the creation of net load (load power minus wind power) profiles for the study period. Some studies also calculated other metrics of wind potential such as regional variations in capacity factor or effective load-carrying capacity (ELCC), a measure of wind's ability to contribute to meeting peak demand (Keane et al., 2011).

The second broad category focuses on the effects of wind (and in some cases solar) integration on operational procedures and resources. As part of the assessment of the effect of wind on operational procedures, almost all of the reviewed studies estimated the amount of power balancing resources (reserves) required to maintain adequate reliability.

Third, some studies also assessed infrastructure adequacy, looking at the need for and the effect of new investments in transmission, generation, and control/storage technology. Most of these studies also estimated the financial costs of operational changes and infrastructure investments.

Unfortunately, integration studies do not use uniformly consistent terminology when discussing reserve requirements and integration costs. The terminology describing short-term balancing reserves is particularly inconsistent, reflecting both new challenges associated with increased net load variability and current regional differences in reserve requirements and practices. In most balancing areas, subhourly balancing typically involves a combination of subhourly dispatch adjustments (often referred to as "load following") and automatic adjustments via automatic generation control (AGC) systems. However, the language used to describe this short-term balancing process is not consistent across studies. For example, several of the studies used the term "regulating reserves" to describe any balancing that occurs on subhourly time scales, while others used this term exclusively in reference to fast adjustments made in response to AGC signals.

To avoid this potential confusion, we use the terms "regulation" and "regulating reserves" exclusively to refer to adjustments made in response to the AGC system and use "balancing reserves" to refer to all subhourly load/wind following reserves inclusive of regulating reserves. A third category of reserves, contingency reserves, must be available to compensate for unexpected plant failures; most studies reported that wind integration did not increase the need for contingency reserves. The system costs resulting from large-scale wind integration come primarily from increased balancing costs (the costs of procuring larger reserve margins, increased plant ramping, etc.) and the costs of additional investments in grid infrastructure. As observed in Holttinen and colleagues (2011), only in a few cases did the studies also describe the financial benefits, in terms of fuel costs or emissions reductions, of wind integration.

Conducting large-scale wind integration studies involves massive data collection and modeling efforts. While each study selected data and modeling methods based on its specific objectives, there were common patterns in the data and methods that illustrate the challenges of large-scale wind integration studies as well as the opportunities for scientific advances going forward. Section 17.2.1 of this chapter discusses the various approaches to wind and net load data collection. Once data were obtained from the data collection and modeling processes, most of the studies performed detailed statistical analysis on the resulting data. Section 17.2.2 reviews the various statistical approaches to characterizing net load variability in some detail. Section 17.2.3 discusses the power system modeling methods used to assess the operational and grid stability aspects of wind integration.

Because different studies approached their questions in different ways, their conclusions were also diverse. Section 17.3 of this chapter highlights the specific research questions raised in each study and the particular methods used to address them. In Section 17.4, we quantitatively compare the key conclusions from the studies. Because operational issues are a common theme across a majority of the studies, this comparison focuses particularly on the suggested operating policy changes and the estimates of additional reserves required to support wind integration at varying penetration levels. Section 17.5 provides our conclusions from this review and suggests several topic areas in which methodological improvements and additional research are needed to facilitate greater insight from future integration studies.

17.2 Data and methods for wind integration studies

Wind integration studies generally follow a similar format. After defining a set of research questions, wind data are collected for a set of potentially viable plant locations using a combination of meteorological models, anemometer measurements, and historical wind plant output. Next, since both wind and load can have correlated seasonal and weather-related components, historical load data are collected for the same time period. From these synchronized datasets (see Section 17.2.1), time-series net load data are typically calculated. Wind and load data then are used in a variety of statistical analyses and power system modeling tools, which are used to assess the costs (and, in a few cases, benefits) of wind integration.

When used well, this process can provide valuable insight into the operational effects of large-scale wind integration, which can facilitate effective investment and policy planning. However, some statistical and modeling methods may lead to misleading conclusions and, eventually, suboptimal planning outcomes. For example, assuming that wind and load are uncorrelated or that wind forecast errors are distributed according to Gaussian distributions could lead to an underestimation of balancing resource requirements. Additionally, modeling the transmission system such that power flows can be directed could lead to an underestimation in transmission needs. This section provides an overview and analysis of the data sources, statistical methods, and modeling tools used in large-scale integration studies. In several places, we compare methods to empirical data to understand the strength and limitations of particular approaches.

17.2.1 Sources of wind and load data

Gathering representative data for potential future wind plants is one of the most significant challenges to a successful wind integration study. Three types of wind data were used in the reviewed studies: historical wind plant output data, wind speed measurements from anemometers or LIDAR systems, and mesoscale numerical weather prediction (NWP) model data. Data from each of these sources offer trade-offs in terms of quality, time resolution, generalizability, and availability.

Data availability is particularly challenging because historical wind plant performance data are almost always proprietary. Efforts to release data more openly substantially advance wind integration research. For example, NREL's efforts to publish much of the data behind their EWITS (EnerNex Corporation, 2011) and WWSIS (GE Energy, 2010) studies, as well as BPA's publication of historical 5-minute wind power production (BPA, 2013), are particularly admirable. On the other hand, the U.S. Federal Energy Regulatory Commission recently declined to obligate transmission system operators to share data from variable energy resources with other entities (FERC, 2012), a setback to data availability.

17.2.1.1 Historical wind plant data

Historical data from existing wind plants give an accurate picture of the statistical properties of real plant production, including the effects of wind variability and the effects of curtailment. High-sample-rate (1-minute or faster) data make it possible to more accurately model the effects of wind on balancing reserves. Of course, historical wind production data are limited to sites that have existing wind plants. Due to the spatially and temporally specific nature of the wind, results from a few plants are not easily generalizable, except in the important area of understanding the general character of the ratio of fast to slow variations in wind (Apt, 2007). In addition, wind plants in many locations are frequently curtailed (required to operate at less than full capacity) for reliability reasons, making the data from these hours nonrepresentative.

17.2.1.2 Empirical wind speed data

An alternative to gathering wind plant power production data is to gather wind speed data and translate speeds into power. Anemometer or LIDAR wind speed measurements are gathered at most current and planned wind plant locations but are costly to obtain for other locations at wind turbine hub-height elevations. In the United States, large quantities of data gathered 10 meters above ground level are publically available (see, e.g., ncdc.noaa.gov), but accurately estimating wind speeds at hub height from 10 m data is potentially unreliable because of its reliance on assumptions about atmospheric stability and surface roughness.

Even if wind speed data are available, they need to be converted into wind power data. While it is straightforward to convolve time-series wind speed data with a manufacturer's power curve, the results may not accurately reflect the production from actual wind plants (see Figure 2.2). An anemometer measures wind speeds at a single point in space, whereas a plant produces power based on many speed vectors across its area. Also, wind turbines create shadowing that is not easily modeled. Given these challenges, there is need for more research to develop speed-to-power translation methods based on the statistical characteristics of wind plant power production data. The EWITS and WWSIS studies (EnerNex Corporation, 2011; GE Energy, 2010) are notable for providing detail and validation on the methods used for speed-to-power translation (Pennock, 2012; Potter et al., 2008).

17.2.1.3 Data from numerical weather prediction models

The final alternative is to generate data for potential wind plant locations using NWP models, calibrated with weather data (wind speeds, temperatures, etc.) from a historical time period, using a process known as data assimilation (e.g., Stauffer & Seaman, 1994). Because of the ability to generate data for wind plants at any on- or offshore location, almost all of the studies used NWP models in some form. A majority of the U.S. studies (Charles River Associates, 2010; EnerNex Corporation, 2011; EnerNex et al., 2010; GE Energy, 2005a, 2008; NYISO, 2010; U.S. DOE, 2008) used data from the commercial firms AWS Truepower (formerly AWS TrueWind) and 3Tier. Mesoscale models have the advantage of producing wind speed data for locations without the need for costly anemometer installations. However, NWP data must be used cautiously. A number of studies have shown that mesoscale models underestimate the high-frequency variability in wind speed data (e.g., Skamarock, 2004). In a few cases, this difference has been noted in detailed literature about the generation of wind datasets (EWEA, 2009; Potter et al., 2008), but it is relatively uncommon for wind integration studies to address the underestimation of short-term wind variability in any depth.

To illustrate this reduction in high-frequency variance, we calculated the average power spectral density (PSD) for the wind speed data released with the EWITS and WWSIS studies. To do so, we computed the PSD using a fast Fourier transform (FFT) of the 10-minute data for 100 randomly selected EWITS and WWSIS sites and averaged the spectral power at each frequency over all 100 sites. Since the PSD of wind speed data generally follow the Kolmogorov spectrum ($PSD \mu f^{5/3}$ for frequencies in the inertial subrange (Pope, 2000) in our case for $f > (1/24$ hrs)), we show the data before and after dividing the PSD by $f^{-5/3}$ (see Figure 17.1). The results show that at frequencies greater than 5×10^{-4} Hz (periods of 30 minutes or shorter), data from the model have almost an order-of-magnitude less spectral power in wind speeds than in observed wind data. While this reduced variability is particularly notable for wind speed data, the effect on wind plant power variability is somewhat less than that implied by Figure 17.1, since (as previously mentioned) wind plant production is the average of many wind speed vectors convolved with a particular turbine's power curve. Katzenstein and colleagues (2010) show that increasing the number of turbines or wind plants does reduce the high-frequency variability. In their study, the spectrum from 20 wind plants was proportional to $f^{-2.56}$.

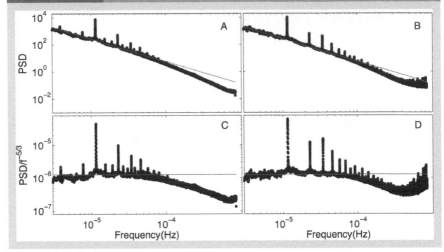

FIGURE 17.1 *Average power spectral density of 10-minute wind speed data generated for the WWSIS (GE Energy, 2010; left) and EWITS (EnerNex Corporation, 2011; right) studies. Panels A and B show the data with lines showing the Kolmogorov spectrum ($f^{-5/3}$). Panels C and D show the same data after dividing by ($f^{-5/3}$). Note that the EWITS graphs come from the updated dataset posted in June 2012. The original dataset (Brower, 2009) showed less PSD at higher frequencies.*

To the extent that reduced wind speed variability affects wind power variability, this reduced PSD could have an effect on conclusions regarding grid reliability and integration costs. The costs of variability come from a combination of increased requirements for regulation, load following, and unit commitment. Since unit commitment occurs over longer time scales (1-to 3-hour time intervals), this reduced variability is unlikely to substantially affect unit commitment cost estimates. However, estimates of load following and regulation requirements (balancing reserves) are likely to be at least somewhat affected by this phenomenon. In order to correct for this issue in the WWSIS dataset, Potter and colleagues (2008) used empirical data from several power plants to develop stochastic power curves with added variability; to our knowledge, this is the only reviewed study that did so. It is possible that the method used in Potter and colleagues (2008) was an overcorrection, since Milligan and colleagues (2012) reported that data resulting from this method include more high-frequency variability than they found in historical plant data.

Another challenge in the production of representative mesoscale data is stitching together data from many shorter model runs in a way that does not produce discontinuity along the seams. Problems with seams were observed in the public WWSIS (Lew, 2009) and EWITS datasets and, in the latter case, ultimately corrected (Pennock, 2012).

Finally, NWP models are computationally intensive and limited in spatial and temporal resolution. The NWP models used in the earliest wind integration studies had a spatial resolution of 8 to 10 km and a 60-minute sampling rate (see, e.g., GE Energy, 2005a, 2008). More recent NWP models produce output with 2-km resolution and a 10-minute sampling frequency (see, e.g., EnerNex Corporation, 2011; GE Energy, 2010). Wind data sampled at a rate of only a few (or one) samples per hour are useful for some applications, such as unit commitment modeling and capacity value estimation, but are insufficient for estimating the effect of higher-frequency variability on balancing reserves.

For this reason, several studies supplemented NWP model data with minute-by-minute or even second-by-second historical plant output data to create hybrid, minute-by-minute datasets (EnerNex Corporation, 2011; GE Energy, 2005a, 2008, 2010). The hybridization process involves extracting detrended variability from historical plant data and adding the resulting time series to the NWP outputs. While this approach was used in several of the studies to estimate balancing reserves requirements, none of the studies (perhaps due to the nonpublic nature of power plant data) provide detailed validation of the statistical properties of the hybrid high-resolution data. Validation of hybrid data is crucial since historical wind data tend to be limited in geographic scope, coming from a single wind plant (GE Energy, 2005a) or small number of wind facilities in a limited geographical region (EnerNex Corporation, 2011; GE Energy, 2008). An alternative to combining the model data with plant data, proposed in Rose and Apt (2012), is to synthesize data using frequency-domain statistical methods.

17.2.1.4 Combining wind and load data

Because wind and load are both closely connected to weather more generally, most of the studies reviewed collected load data (P_d) from the same time period as the wind data (P_w) and subtracted them to create a net load (P_n) profile for each time point k, as shown in Eq. (17.1).

$$P_n[k] = P_d[k] - P_w[k], \forall k \qquad (17.1)$$

The net load profile provides a reasonably accurate profile of the variability of the wind/load combination, with the caveat that it does not capture the effect of wind turbine pitch control systems, which are increasingly being employed to reduce variability (Xie et al., 2011). In order to examine future scenarios, historical load data are typically scaled based on the expected load in the study year.

17.2.2 Statistical analysis of wind and net load data

Once produced, wind and net load time-series data are a source of useful information when appropriate statistical methods are used to analyze them. Understanding the variability of wind and net load as well as wind-power forecast data on different time scales provides some insight into the reliability effects of large-scale wind integration.

The most common statistical method, used in almost all of the reviewed studies, is to measure statistical properties of changes in wind or net load over different time intervals (typically 10-minute or 1-hour intervals). Step changes are typically calculated by assembling a discrete time series of power (wind or net load) data, $P[k]$ (with a total of K observations), and then computing differences, $\Delta P[k]$, according to Eq. (17.2).

$$\Delta P[k] = P[k] - P[k-1], \forall k \in \{2 \ldots K\} \tag{17.2}$$

Many of the studies computed the standard deviation of $\Delta P[k]$ and draw conclusions based on this measure (e.g., EnerNex Corporation, 2011; GE Energy, 2005a, 2010; NYISO, 2010). However, the standard deviation does not necessarily convey information about the frequency of low-probability, dramatic changes in wind or net load, which are the primary reliability risks – the main reason that reserves are required.

Using the standard deviation as a measure of variability is a valid assumption if step changes,$\Delta P[k]$, or forecast errors are distributed according to a Gaussian probability density function. However, as noted in Hodge and Milligan (2011) and Holttinen and colleagues (2008) and illustrated here and in Chapter 9, wind data do not follow Gaussian distributions. To illustrate the difference between the Gaussian distribution and wind data, we compared the statistical properties of wind power data to Gaussian probability density functions (PDFs). To do so, we measured the 10-minute changes $\Delta P[k]$ in wind power production using Eq. (17.2) for three datasets. The empirical probability density function for $\Delta P[k]$ was computed by counting the number of intervals k for which

$$x - 0.05 \leq \frac{\Delta P[k]}{P_{cap}} \leq x + 0.05 \tag{17.3}$$

for all x in the set $\{-0.95, -0.94, \ldots, 0.94, 0.95\}$, where P_{cap} is the net generation capacity for the plant. The resulting counts were scaled to give a total probability mass of 1, which resulted in an empirical probability density function for each dataset. Comparable Gaussian probability density functions were determined by computing the mean and standard deviation of the time series $\Delta P[k]$.

This calculation was performed for three distinct datasets. The first was the output from all 10 GW of wind generation operating in the Bonneville Power Authority (BPA) area in 2010; the second was power output from a 300 MW wind plant in the central United States over 1 year; and the third was power output from a 120 MW wind plant also in the central United States over 2 months. The BPA dataset

FIGURE 17.2	*Comparison of empirical and fitted Gaussian probability density functions of wind power step changes (A–C) and forecast errors (D). Panel A shows 5-minute step changes from the aggregated wind production in BPA (BPA, 2013). Panels B and C show 10-minute step-change data from a 300 (B) and a 120 (C) MW wind plant. Panel D shows the distribution of wind power forecast errors for the BPA data.*

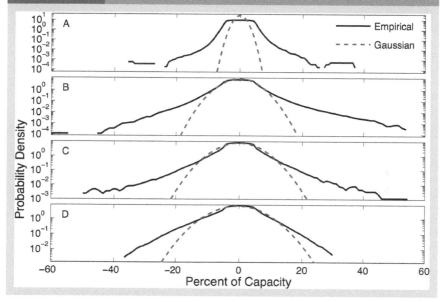

reflects the combined power production from many wind plants and thus includes substantial spatial averaging. In all three cases, the tails of the empirical distributions show far greater probability density than one would expect from the fitted Gaussian distributions (see Figure 17.2). For the BPA dataset, we repeated the Gaussian/empirical comparison for the difference between the 24-hour-ahead wind power production forecast and actual wind power production. Again, the data show vastly greater weight in the tails than one would expect from Gaussian statistics.

In order to understand how probability densities translate into actual probabilities and step-change occurrence frequencies, Figure 17.3 shows the complementary cumulative distribution function (CCDF) of $|\Delta P[k]|$ for the 5-minute step changes in the BPA dataset and the equivalent Gaussian cumulative distribution. The CCDF for the absolute value of a zero-mean random variable x, such as $\Delta P[k]$, if it is Gaussian distributed, is:

$$\Pr(|x| \geq X) = 2(1 - \Pr(x < X)), \text{ for } X \geq 0 \tag{17.4}$$

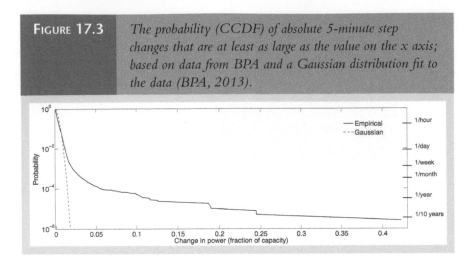

FIGURE 17.3 The probability (CCDF) of absolute 5-minute step changes that are at least as large as the value on the x axis; based on data from BPA and a Gaussian distribution fit to the data (BPA, 2013).

where $\Pr(x < X)$ is the standard cumulative distribution function (CDF) for x. The results in Figure 17.3 illustrate the dramatic extent to which the Gaussian function underestimates the frequency of extreme events.

To further illustrate the difference between Gaussian and observed wind data, we generated a 30-day time series using a mean-reverting Gaussian random-walk stochastic process, with the same 5-minute step-change standard deviation. The random-walk parameters were chosen to produce roughly the same capacity factor as that of the BPA data, and the output was not allowed to go below zero or above the maximum from the BPA data. As shown in Figure 17.4(A), the empirical wind data show substantially faster ramp rates over longer time periods than do the synthesized data. For comparison purposes, we also show two randomly selected samples of wind power data from the EWITS and WWSIS studies in Figure 17.4(B). While the studies have somewhat reduced high-frequency variability, the differences between the EWITS/WWSIS data and the empirical plant data are not obvious.

While load deviations are not as extreme as wind deviations, the heavy-tailed nature of wind data will cause net-load data to be heavy tailed for high-wind-penetration scenarios. A few of the more recent studies (Charles River Associates, 2010; GE Energy, 2008, 2010) noted the heavy-tailed nature of these step changes, but most did not.

In spite of the divergence between the Gaussian statistics and wind data, many of the studies in this review establish reserves rules based on the standard deviation (σ) of wind or net load variability. Because the reliability effects of wind production depend critically on the severity and frequency of relatively low-probability events, these statistical differences are very likely to have an effect on conclusions in these studies.

The Gaussian assumption reflects, in part, existing regulatory requirements. NERC requirements for reserves margin, for example, explicitly state that errors within 2σ of expected load variation must be covered without the use of contingency reserves (GE Energy, 2010), despite the fact that load step changes also have

FIGURE 17.4	*Panel A compares the power production of wind plants in BPA to synthesized data generated from a Gaussian random process, with the same standard deviation in step changes. Panel B shows two randomly selected wind power time series from the WWSIS and EWITS mesoscale data.*

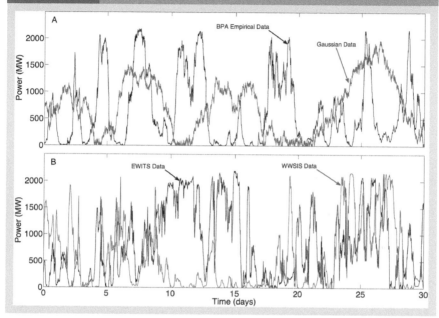

heavy-tailed distributions. Holttinen and colleagues (2008) also studied the use of the standard deviation to estimate regulation requirements, arguing that scheduling balancing reserves to cover 2σ to 6σ of net load variability would provide sufficient reliability; however, they do not substantiate this claim with reliability modeling.

The SPP study (Charles River Associates, 2010) is commendable for estimating reserve requirements without an implicit Gaussian assumption. Rather than using standard deviations of wind or net load for their estimates, they compute the 95th and 5th percentile step change magnitudes to establish up- and down-regulation requirements. A similar approach is suggested in the ERCOT study (GE Energy, 2008). The data presented in this section suggest that future studies should explicitly quantify the magnitude of low-probability ramp events for which reserves are needed rather than basing the estimations on standard deviations.

Another statistical technique found in several studies (EnerNex Corporation, 2006, 2011; EnerNex et al., 2010) and suggested in Holttinen and colleagues (2008) is to develop a combined standard deviation of wind and load by adding the variance (σ^2) of the wind and load step-change data and using the combined variance to represent that of the net load. This technique is valid only if there are no correlations between wind and load and if the statistics of each are Gaussian, neither of which is accurate.

A much better approach is to compute net load data using equation (17.1) and then compute the sizes of the extreme changes (i.e., the 1st and 99th, or 0.1th and 99.9th, percentile values for $\Delta P[k]$). While such extreme values seem to be low probability, a 99.9th percentile 5-minute step change will occur once every 3.5 days – a relatively frequent event. Effective planning for low-probability events could reduce the risk of reliability problems due to wind integration and may allow wind power to be integrated with less curtailment, an increasingly common problem in many U.S. regions (e.g., Dillan, 2013; Gu et al., 2011).

17.2.3 Power system modeling methods

One of the most important functions of wind integration studies is to determine the capacity of the existing transmission and generation infrastructure to support a proposed quantity or configuration of wind power production while keeping grid reliability at or above a specified level. Examining step change frequencies in net load can provide some insight into grid reliability, but a detailed understanding of the costs and effects of wind integration, as well as the relative benefits of technologies like grid-scale storage, requires at least some power system modeling.

While there are many ways in which wind generation might affect grid reliability, we can roughly divide these effects into two broad categories. The first is electrical effects on the transmission system that might result in equipment damage, trigger instability, or cascading blackouts. Examples of electrical analyses include voltage stability analysis (Vittal, O'Malley, & Keane, 2010), fault ride-through analysis (Seman, Niiranen, & Arkkio, 2006), and contingency analysis. Since detailed engineering studies of these issues are often performed on a site-by-site basis for individual wind plants, only a few of the reviewed studies (e.g., Charles River Associates, 2010) include detailed electrical systems analysis. The second type of analysis is of imbalances between supply and demand and the resources needed to maintain this balance.

17.2.3.1 An overview of power system models

Power system operators (including independent system operators, balancing authorities, regional transmission authorities, and vertically integrated utilities) must monitor the balance between supply and demand on a second-by-second basis because of the physics of the system and dearth of electricity storage. When this balance is not maintained, the result is that generators in the system speed up or slow down according to the differential equation known as the swing equation:

$$P_m = P_e + D\omega + M\frac{d\omega}{dt} \tag{17.5}$$

where ω is the rotational speed of the generator, P_m is the mechanical power input to a generator, P_e is the electrical power output of a generator, D is a damping constant that includes friction in the rotating machinery, and M is an inertia constant (Sauer & Pai, 1997). When P_m is not equal to P_e, machine speeds (ω) deviate from nominal, causing the frequency of AC voltages to deviate from nominal (50 or 60 Hz). The challenge of power system operations is to maintain the balance between P_m and P_e while keeping network flows and voltages within acceptable limits. The challenge of power system modeling is to accurately capture the many methods used by system operators to solve this problem.

The ways in which variable energy resources (VER) affect grid reliability and the methods used by system operators to maintain reliability differ along different time scales. Ideally, one would use the same grid model, accurately capturing both generator dynamics and power flows, to estimate the effects across all time scales. However, the most sophisticated power system models (dynamic/transient generator models coupled to nonlinear transmission network models) are generally too complicated to support analysis covering longer time scales. Even if detailed, fine-grained generator simulations could be run with sufficient computational speed, the results could still be misleading due to the large number of parameters, many of which are not accurately known *a priori*. Therefore, different types of models are needed to understand the behavior of power grids along different time scales. The types of models, the times scales on which they typically operate, and the degree to which they address transmission and/or generation effects is summarized in Figure 17.5. These model types are then discussed separately in Sections 17.2.3.2 through 17.2.3.7.

FIGURE 17.5	*An illustration of the time resolution and scope of different types of power system models used in wind (and solar) integration studies.*

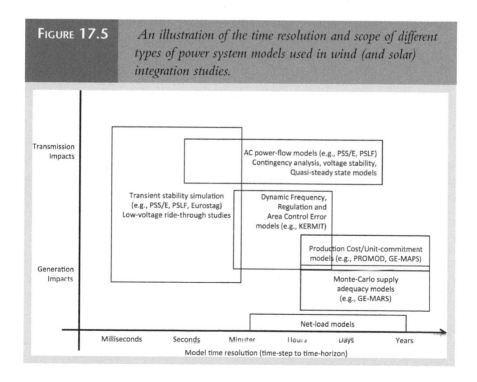

17.2.3.2 The simple net load model

The simplest model used in many integration studies is, as previously discussed, the net load model in equation (17.1). A net load profile gives a rough estimate of how quickly generator power production (P_m) will need to ramp in order to maintain the balance between P_m and P_e.

The net load model has some significant limitations, particularly for short-time-scale analysis. The first challenge is to obtain accurate wind and load data, with variability at the right time scales. A second challenge comes from the fact that the balance between supply and demand does not need to balance perfectly at every instant within a given area. The inertia and damping in rotating machines, M and D in equation (17.5), in addition to automatic generator controls, have the effect of averaging frequency deviations across time scales. If one were to estimate that reserves or load following resources needed to be procured to follow the fastest observed change in net load, one might overestimate the costs of wind integration. As previously discussed, most of the studies do not look at the worst-case changes but rather use the standard deviation of the step changes in wind, load, and/or net load to estimate the amount of regulating reserves that would be required to balance supply and demand. This may indicate that additional balancing resources, beyond those projected in these studies, will eventually be required to mitigate large changes in wind production. Finally, net load models do not represent the transmission system and thus cannot account for the ability of an operator to quickly exchange power with neighboring operators. When one area has a momentary surplus, power is exported to neighboring areas and vice versa. In the United States, NERC regulates the accuracy with which utilities maintain their internal supply–demand balance through their balancing control performance standards, known as CPS1 and CPS2 (NERC, 2011b).

17.2.3.3 Quasi-steady-state network models

An improvement on the net load model that allows one to understand the effect of wind on the transmission system is to explicitly model power flows between locations over a sequence of time intervals. Assuming that one can estimate the load, generation, and wind power at each node for a sequence of time periods, this type of quasi-steady-state (QSS) simulation can be used to model regional effects of wind plants on power flows. QSS models were used in several of the studies, including GE Energy (2005a, 2010).

The most accurate QSS models estimate both load and generation for time intervals and use these values as inputs to an AC power flow model. DC models can be a reasonable approximation if voltages or power losses are not important to the outcome. In U.S. DOE (2008), power flows were modeled assuming that power flows on individual transmission lines are fully controllable. This is not an accurate representation of real power networks, except for the rare case of flow-controlled transmission lines (e.g., FACTS devices). Since few such lines exist, AC power flow QSS models are substantially more accurate.

However, even the most accurate QSS models have limitations. Estimating which power plants are likely to be operating at what levels for particular time intervals requires production cost modeling (see Section 17.2.3.4), and capturing second-by-second variations in frequency requires a dynamic model (see Section 17.2.3.6).

17.2.3.4 Production cost simulation models

Estimating the hourly (or subhourly) state of power plants with various fuel types and costs for plausible wind penetration scenarios is both important and difficult. Most of the recent wind integration studies use production cost simulation (PCS) software for this purpose. Most of the studies used proprietary PCS tools such as GE MAPS (Charles River Associates, 2010; GE Energy, 2005a, 2008, 2010), Ventyx PROMOD (EnerNex Corporation, 2006, 2011), or Global Energy's PROSYM (EWEA, 2009). The most sophisticated of these model generator ramp rates, startup/shutdown costs, and transmission limits (typically using the DC power flow model). One study (EWEA, 2009) used a research PCS tool, WILMAR (Weber, Meibom, Barth, & Brand, 2009), which was specifically designed for wind integration studies. Each of these tools uses a unit-commitment generator cost model. The most sophisticated models (e.g., PROMOD) solve for power flows and account for transmission constraints.

The evolution of production cost simulations is evident in these studies. The NREL 2030 study (U.S. DOE, 2008) used an internally developed PCS model, with detailed cost data for a large area (the whole of the eastern United States) but (as noted previously) a transportation model of electrical flows. The newer studies include substantial detail regarding generator and transmission constraints. Since most of the newer studies used PCS models for 1 to 3 years of data, with 1- to 3-hour time increments, the results provide reasonable estimates of the effect of wind power on dispatch and unit commitment costs. Because the time scales for unit commitment calculations are somewhat longer, the reduced higher-frequency variance found in mesoscale data (Sec. 3.2) should have little effect on unit commitment calculations.

One area in which PCS technology is rapidly evolving (for both PCS and the industrial unit commitment systems that PCS simulates) is the ability to deal with the stochastic nature of renewable generation. The science of stochastic unit commitment is rapidly maturing (see, e.g., Takriti, Birge, & Long, 1996; Wu et al., 2007), and there is increasing evidence that effective use of this technology can reduce wind integration costs (Madaeni & Sioshansi, 2013). As stochastic methods are effectively adopted in the PCS software used in integration studies, future studies are likely to provide more detailed insight into the benefits, costs, and challenges of wind integration.

17.2.3.5 Supply-adequacy reliability modeling

One of the critical responsibilities of a system operator is to ensure that there is sufficient generation capacity to supply load during future periods of high demand for many years into the future. The most common standard for supply adequacy is to ensure that there are sufficient supply resources to reasonably expect that shortages will occur not more frequently than 1 day in 10 years, or not more than 2.4 hours of shortage per year (NERC, 2011a).

The most common method for determining the adequacy of a given supply portfolio is to use Monte Carlo simulation for time intervals (typically 1 hour) to determine the annual probability of a supply shortage (Billinton & Jonnavithula, 1997; Ghajar & Billinton, 1988). Many of the reviewed integration studies used commercial Monte Carlo models (most commonly GE-MARS; see EnerNex Corporation, 2006, 2011; GE Energy, 2005a, 2010) to estimate the effect of wind on supply adequacy. One of the weaknesses in GE-MARS is the limited ability to account for transmission constraints within the model. As with WindDS (U.S. DOE, 2008), GE-MARS uses a transportation model of the transmission network (see EnerNex Corporation, 2011, p. 90). Technology for composite transmission and generation adequacy is relatively mature in the research literature (Billinton & Li, 1994) but was not, to our knowledge, employed in any of the integration studies reviewed here.

One of the most important reasons for supply adequacy analysis in wind integration studies is to establish rules for setting the capacity credit due to wind plants, since many system operators provide wind plants with financial payments based on their contributions to system adequacy. Since this is a subject of ongoing research (Hasche et al., 2011; Keane et al., 2011), methods for establishing capacity credits varied from one study to another. One particular challenge for setting capacity payments, discussed in Chapters 11 and 12, is that, because of the potential for substantial variations in wind speeds from one year to the next (Katzenstein et al., 2010), at least 4 to 5 years of data are needed to establish accurate estimates of ELCC (Hasche et al., 2011). Also, the siting of renewables and transmission can have a dramatic effect on the capacity value of a wind plant. The EWITS study (EnerNex Corporation, 2011), for example, found that expanded transmission increased the capacity contribution of wind by approximately 50% (from 16% to 24% of nameplate capacity) by enabling capacity that is not required for resource adequacy in one area to contribute to resource adequacy in another area. This conclusion differed from that of the WWSIS study (GE Energy, 2010), which argued that resources outside of a given capacity market should not be considered when determining resource adequacy.

17.2.3.6 Power system dynamics

The power system modeling methods in Sections 17.2.3.1 through 17.2.3.5 make the assumption, at least implicitly, that there is no imbalance between supply and demand and that voltages in the network are nominal. In reality, momentary imbalances between supply and demand cause small fluctuations in node voltages and small deviations in frequency, which are governed by equation (17.5). Both of these phenomena can, in extreme cases, trigger instabilities in the system, potentially resulting in large blackouts. It is therefore difficult to make confident conclusions about phenomena that occur along short time scales (seconds to minutes) using the steady-state models described. Since unit commitment and system adequacy are long-time-scale calculations, dynamic models are largely unnecessary for these issues. However, power system stability calculations and estimating the effect of wind on balancing (particularly regulating) reserve requirements depend critically on short-time-scale phenomena. While detailed nonlinear, transient power system models and 1-second data are probably not always required in wind integration studies, there is need for model validation in order to quantify the uncertainty associated with simpler models.

The California Energy Commission (CEC) study (KEMA, 2010) presented an illustrative alternative to the net load and QSS modeling approaches to estimating balancing reserve requirements. In KEMA (2010), researchers developed a composite dynamic model of generators in each of four areas, which together form the U.S. Western Interconnection. Each area model captured the empirically estimated behavior of generators in that area using a version of equation (17.5). A simplified transmission model was used to connect the four areas. Because KERMIT was substantially more tractable than a fully dynamic model with detailed representation of every generator, it was possible to simulate many different scenarios and estimate the regulating and load-following reserve requirements for several high-penetration renewables scenarios. Because of the explicit way that system dynamics were captured in KERMIT, the CEC study could differentiate between the behavior of fast-ramping storage resources and slower-ramping fossil fuel plants. As a result, the study was able to draw important quantitative conclusions about the economic value of fast-ramping storage. To our knowledge, this is the only wind integration study to use dynamic modeling to estimate reserve requirements. While some of the modeling assumptions in KERMIT limit the policy conclusions that can be drawn from the model outputs, using dynamic models to understand the effects of wind is an important area for research (e.g., Mackin et al., 2013; Wang & McCalley, 2013), which should be incorporated into future integration studies.

17.3 Summary of studies

In Sections 17.3.1 through 17.3.11, we summarize the 11 studies (ordered chronologically) reviewed in this chapter. For ease of comparison among the studies, each summary follows a standard template. The first paragraph states when and by whom the study was conducted, the wind penetration levels assessed (in terms of the percentage of total energy that is generated by wind annually), and the major research questions that the study addresses. The second paragraph focuses on the study scenarios and the key assumptions about the scenarios, such as changes to the transmission system, changes to the non–wind-generating portfolio, and changes in market operations and balancing authority size. The third paragraph describes the development of wind and net load profiles used in the study, while the fourth paragraph describes the analytical methods employed by the study. The fifth paragraph presents the study's key findings, with particular emphasis on changes in reserve requirements. In several cases we add a sixth paragraph discussing notable methodological innovations or challenges associated with each study.

Following the description of the individual studies, Table 17.1 summarizes the source of the wind data used in each study, the sample rate for this data, the statistical methodology for characterizing the distribution of changes in wind/net load, as well as the power-production and power-flow models used to represent the electric systems for each study.

17.3.1 New York (NYSERDA) 2005

In 2005, GE conducted a study for the New York Independent System Operator (NYISO) and New York State Research and Development Authority (NYSERDA) that modeled 3,300 MW of installed wind capacity, approximately 6% wind penetration on an energy basis, for a 2008 study year (GE Energy, 2005a). The study focused on assessing net load variability, determining operational and reliability effects over a range of time scales, evaluating the effect of wind forecast accuracy on the value of wind generation, and quantifying the effective capacity of wind generation.

The 3,300 MW of wind capacity modeled were sited primarily in upstate New York, with a smaller amount of offshore wind capacity. As such, this study covered a relatively small geographic area compared to several of the other studies. Non–wind-generating facilities and transmission infrastructure were derived from NYISO's 2008 system model and were not optimized for wind. Concurrent wind penetration in neighboring regions was not modeled, potentially affecting the analysis of the interchange with these systems. The NYISO operates a day-ahead unit commitment market based on forecasted load. Economic dispatch commands are issued at 5-minute intervals to balance generation with load, so much of the subhourly balancing activity is managed through this economic redispatch process. Regulation adjustments, made by the AGC system at 6-second intervals, are used to balance variability at the sub–5-minute time scale.

As with several other studies in this review, the study used AWS TruePower meteorological models to generate wind data. These models simulated hourly wind speed data with 8-km grid resolution using historical weather records for 2001 through 2003. Later studies used AWS data with higher spatial and temporal resolution. The researchers applied the hourly wind speeds to the power curve for a generic 1.5 MW turbine and then smoothed the resulting power output with a moving-average filter. The filter slightly reduces the variability in power outputs and was intended to reflect the smoother power output of a multiturbine wind plant relative to a single wind turbine (GE Energy, 2005b). AWS also provided day-ahead and hour-ahead wind forecasts using a Markov chain to generate data with forecast error. In addition, GE created 108 3-hour blocks of wind data with 1-minute output data and six 10-minute blocks of 1-second wind output data. These higher-resolution datasets were created by extracting 1-minute/1-second deviations from the hourly trend in plant output at a 105 MW wind project in northwestern Iowa. Scaling these deviations to match the size of the modeled study sites and applying these extracted deviations to the hourly AWS output produced

1-minute/1-second plant output datasets (GE Energy, 2005b). Load data from 2001 through 2003 were scaled upward to match projected 2008 load levels.

The authors used a variety of analytical methods in the study. First, they computed standard deviations of net load variability at hourly, 5-minute, and 6-second intervals. As with many of the other studies, GE applied Gaussian statistics to characterize this variability and to assess reserve requirements. Specifically, regulation was set equal to three standard deviations of 6-second net load variability. Second, GE used a suite of simulation programs to evaluate the effect of wind integration on system operating costs, transmission congestion, and system reliability on an hourly time scale. GE's Multi-Area Production Simulation (MAPS) program was employed to evaluate the change in system operating costs, and its Multi-Area Reliability Simulation (MARS) program was used to assess system reliability effects. Finally, several stability analyses were performed using a QSS model based on GE's Positive Sequence Load Flow (PSLF) and dynamic simulation using GE's Positive Sequence Dynamic Simulation (PSDS) software.

The study concluded that existing NYISO procedures and generators would be sufficient to accommodate the increased variability caused by the 3,300 MW of new wind capacity at all time scales, including the incremental uncertainties cause by errors in wind forecasts. According to the study, maintaining existing levels of control performance would require an increase of 36 MW of regulating capacity, which would constitute an increase in regulating capacity of 0.1% of peak load. As a result of new wind, the standard deviation of net load step changes increased by approximately 6% at the hourly level and 3% at the 5-minute level relative to load alone. These changes are well within the existing dispatch capability of the system. Contingency and spinning reserve requirements were not affected by the introduction of wind power. The study also found that power factor correction at the wind generation sites, enabling the plants to operate in the +/− 0.95 power factor range, had significant reliability benefits. Though the study did not focus on economic effects, hourly production cost simulations suggested that new wind generation will result in fuel cost savings in the range of $40 to $50 per MWh of wind production and a decrease in zonal spot prices by as much as 10%. The exact value of these savings depends in part on the ability to incorporate wind into the day-ahead market, with not more than 10% day-ahead forecast error. Given the observed anticorrelation between wind and load, the study estimated the ELCC of inland wind generation to be approximately 10% of their nameplate capacity.

Overall, this study separated the analysis among different time scales in a reasonable manner and provided specific conclusions about the different types of reserves needed for each time scale. As with other studies that created wind data with fine temporal resolution by hybridizing outputs from NWP models with empirical data from existing wind plants, questions remain about the representativeness of the empirical data, in the case of the output of a single wind plant in Iowa, and how accurately scaling and applying the variability from these data to the NWP outputs reproduces actual wind output behavior. As is the case in many of these studies, the use of Gaussian methods is problematic for modeling low-probability, high-effect events and could result in an underestimation of regulation requirements, particularly given the small size of the 6-second dataset.

17.3.2 Minnesota 2006

In 2006, the Minnesota Public Utilities Commission contracted with EnerNex to conduct a wind integration study looking at 15%, 20%, and 25% wind penetration on an energy basis in Minnesota and the eastern Dakotas for a 2020 study year (EnerNex Corporation, 2006). The study estimated the costs associated with integrating wind power into the existing generating mix as well as the ability of wind to contribute to supply adequacy in Minnesota.

New wind sites for the 15% and 20% penetration scenarios were distributed throughout the study area, while the additional wind necessary to reach 25% penetration was concentrated in a relatively small area with high wind potential. No additional wind capacity was modeled in the larger Midwest Independent System Operator (MISO)[1] footprint, although the entire MISO market was included in the study model runs. As with the majority of wind integration studies, the transmission and non–wind-generating infrastructure modeled in the study were those that existed in the ISO's existing model rather than a system that was optimized for wind. In addition, the entire study area was modeled as a single balancing authority, because MISO was in the process of consolidating a number of balancing authorities. The MISO market structure includes a day-ahead market and a real-time market that clears in 5-minute intervals. Variability over time scales shorter than 5 minutes is managed by AGC regulation.

The wind data used in this study were produced by WindLogics using the MM5 mesoscale NWP model with a 5-minute sampling interval (WindLogics, 2006). The model conducted three separate year-long simulations initialized with National Center for Environmental Prediction (NCEP) data from 2003, 2004, and 2005. The model was run using "telescoping" spatial resolution to maintain computational efficiency with a base 12-km resolution and a finer 4-km resolution at potential wind installation sites. Wind speeds were translated to power, using a "power plant" (as opposed to single-turbine) model, based on empirical data from a 30 MW installation. Wind forecast data were generated using the NCEP NAM and Rapid Update Cycle models used by the National Weather Service. Load data were gathered from Minnesota utility archives for 2003 through 2005 and scaled to reflect expected load growth for the year 2020.

The study used basic statistical analysis of net load data to estimate reserve requirements. It estimated regulation requirements based on a hybrid dataset, which combined the mesoscale model data with higher-resolution data (the details of which are not specified) from NREL. Analysis of the hybrid data showed the output fluctuations of wind to be less than 2% of wind nameplate capacity over an unspecified

---

1. MISO changed its name to the Midcontinent Independent Systems Operator, Inc. on April 26, 2013.

"regulation time frame." The authors used Ventyx PROMOD, a production cost simulation model that optimizes plant outputs based on hourly load, transmission, available generating units, and required reserve margins, to assess the operational effects of wind integration. Reliability analysis in this study was conducted using GE MARS and a program called Marelli by New Energy Associates to ensure that system adequacy remains above a loss-of-load probability (LOLP) target of 2.4 hours per year. Wind power was represented in MARS as a load modifier. The ELCC of wind was estimated by comparing runs with and without this load modifier.

One of the main conclusions in this report was that increasing spatial diversity dramatically reduced the number of periods without significant wind generation and thus reduced requirements for reserve generation capacity, given the assumptions in the study. Nonetheless, the study found that wind integration required additional balancing and regulating reserves. Overall balancing reserves increased by approximately 2% of peak load in the 25% wind case, including an increase in regulation equal to 0.1% of peak load. The annual cost imposed by wind variability and uncertainty depended on the level of wind penetration and on the specific wind profile for the study year. The study found that, relative to dispatchable generation, wind imposed additional costs ranging from $2.11 per MWh (15% wind, 2003 baseline weather data) to $4.41 per MWh (25% wind penetration, 2005 baseline weather data). The authors found that combining balancing authorities significantly decreased certain ancillary services requirements, including balancing reserves (both regulation and load following), by approximately 50%. The study suggested that wind development would have modest effects on changes in net load along regulation (5- to 10-minute) time scales and larger effects on hourly step changes. The study also suggested that ELCC varies substantially (between 5% and 20% of nameplate capacity) from year to year.

This study had many positive attributes. It obtained 5-minute wind data from a respected weather simulation model, included a relatively accurate transmission system model (through PROMOD) in its simulations, and the unit commitment method was appropriate for hourly analysis. However, there are a number of shortcomings to the subhourly analysis, which depended heavily on Gaussian statistics to calculate balancing reserve requirements. The reliability models were hourly models, whereas the study modeled 5-minute data, leaving a short-term analysis gap. The assumption that wind fluctuations on the regulating time scale were less than 2% of nameplate capacity and uncorrelated to load fluctuations is poorly supported in the report. Finally, as the authors note, the results obtained in this study depend heavily on Minnesota's participation in the larger MISO market, for which increased wind was not modeled; therefore the effective wind penetration levels could be considered to be lower than the stated 15%, 20%, and 25% penetration levels.

17.3.3 Texas (ERCOT) 2008

In 2008, GE produced a wind integration analysis for the Electric Reliability Council of Texas (ERCOT; GE Energy, 2008). The study examined five scenarios with between 0 and 15,000 MW of wind capacity, representing 0 to 17% wind penetration on an energy basis. The study assessed the level of ancillary services required, as well as the cost of procuring those services, for each wind scenario.

The five wind scenarios in the study consisted of a 0 MW, a 5,000 MW, two 10,000 MW, and a 15,000 MW scenario. The two 10,000 MW cases differed in the spatial distribution of selected wind sites. The study assumed that the non–wind-generation portfolio remained constant (i.e., generators are added to or retired from the model). Since Texas policy is to develop the transmission system to support new renewable generating capacity (GE Energy, 2008), existing transmission constraints were not considered in this analysis. Additionally, ERCOT is not synchronously interconnected with other systems, so wind penetration outside the study area was not a consideration. At the time the study was performed, the ERCOT ancillary service market consisted of regulation, responsive reserve, nonspinning reserve, and replacement services. Both ERCOT regulation and responsive reserves provide services that fall into the balancing reserve category described in the introduction. As with many of the studies, the ERCOT study illustrated the lack of consistency in reserve nomenclature and categorization; in the study, ERCOT's "responsive reserves" also act as contingency reserves. The study assumed 5-minute economic dispatch at the nodal level, a procedure ERCOT adopted in 2009.

Based on 2005 and 2006 meteorological data, AWS TruePower developed 2 years of hourly wind data for this study using a mesoscale meteorological model with 10-km spatial resolution. These data were converted into hourly power output using performance curves for typical wind turbines after adjusting for wake interference. Minute-by-minute variability in power output was extracted from 1-minute-resolution, historical data from wind plants in Texas and applied to the simulated hourly dataset to create a hybrid 2-year, minute-by-minute wind production dataset. Because they appeared much more frequently than would be seen if changes in wind output followed a normal distribution, changes in the 1-minute historical power data greater than 5% of nameplate capacity were assumed to reflect curtailment events or other nonwind phenomena and were excluded from this process. AWS also provided next-day and 4-hour-ahead wind forecast for the study period. Analysis of load was based on actual minute-by-minute load data for 2005 and 2006.

The study used statistical analysis to characterize net load, the correlation between wind and load, and regulation deployments. Gaussian statistics were used to characterize variability, but regulation requirements were specified based on the larger of the 98.8th percentile of regulating events in the same month in the preceding year or of the preceding month, after removing fast ramps from the data as specified earlier. Regulating events were defined as the difference between net load and the economic dispatch for each 5-minute increment. Hourly production cost simulation was conducted with GE MAPS. Outputs from the MAPS simulations were used to determine the generating capacity available to provide regulation in each hour. The study did not include any transmission modeling or identify transmission constraints in the system.

The authors concluded that load and wind generation forecast errors were essentially independent and that severe errors in both forecasts were unlikely to occur in the same hour. Moreover, while net load forecast accuracy decreased as wind penetration increased, the largest wind forecast errors tended to underestimate wind generation. This increased operating costs by increasing reserve requirements but did not decrease system security. Because of the increased variability in net load, regulation and reserve requirements also increased with increased wind penetration. At the 1-minute time frame, net load variability was found to increase linearly with wind penetration. The average increase in regulation deployment with 15,000 MW of wind was 18 MW, but this increased to 54 MW of up-regulation and 48 MW of down-regulation at the 98.8th percentile, which represent an increase in regulation of approximately 0.08% of peak load, which is quite small. Sufficient capacity was available to provide up-regulation in all hours, but changes in commitment and dispatch procedure may be needed to meet all down-regulation needs. Overall, the study concluded that 15,000 MW of wind could be added to ERCOT without dramatic changes to operating procedures. It is worth noting that, as of 2013, ERCOT has more than 10,400 MW of installed wind capacity (ERCOT, 2013). The authors noted that this conclusion assumed that the current mix of thermal generation will remain constant and that if wind penetration caused plant retirements, the ancillary services market could be negatively affected.

This study was one of only two, along with Charles River Associates (2010), that set regulation requirements based on a threshold rather than establishing the regulation criteria based on the standard deviation of net load. Given the non–Gaussian nature of wind output, this approach is likely to produce more accurate results relative to using the standard deviation approach. The creation of the hybrid wind dataset, however, systematically eliminated the most extreme wind ramping events from the historical plant data. While the authors correctly stated that many of these events are due to curtailment or other non–wind-speed factors, the data in Section 17.2 show that this outlier-removal process may exclude operationally important wind variability from the final dataset. Conversely, since the geographic diversity of the selected wind sites is somewhat limited, variability may be lower than was modeled in the study.

17.3.4 United States 20% Wind (NREL) 2008

Sponsored by U.S. DOE, NREL's *20% Wind Energy by 2030* (U.S. DOE, 2008) was a very broad report on the feasibility of 20% wind penetration at the national level. Much of the study was a high-level discussion of wind power technology, manufacturing, and environmental effects. The quantitative portions of the study focused on identifying the wind resources and transmission capacity additions needed to supply 20% of electric energy from wind and the costs and benefits associated with this scenario rather than detailed analysis of the operational effects of wind power. Because of its broad goals, it is not an "integration" study in the same sense as other studies in this review. However, given that this report is widely cited in the literature, and to illustrate the evolution in wind modeling approaches, it is included in this review.

As stated, the study considered a single scenario with 20% penetration on an energy basis, which required the construction of 293 GW of wind capacity. The study used a large optimization model to choose an optimal subset of potential sites, resulting in 50 GW of new offshore wind capacity, primarily along the eastern seaboard, and 243 GW of new land-based wind capacity. The 20% scenario was compared to a base case that assumed no expansion of wind or other renewables beyond the installed capacity in 2006.

The wind data for this study came from state-by-state, seasonal, and diurnal capacity factor estimates from AWS True Power and National Commission on Energy Policy/National Center for Atmospheric Research (NCEP/NCAR) Reanalysis data. These capacity factors were applied to state wind-speed maps from a variety of sources to estimate available wind power at different wind-speed classes.

The study used an engineering-economic optimization approach to reach its conclusions. The study included a broad analysis of the costs associated with building new transmission systems to meet increased demand for energy. Transmission system modeling was conducted with NREL's Wind Deployment System (WinDS), which models U.S. capacity expansion using a large linear optimization model. WinDS assumed that transmission flows could be fully directed, based on a transportation-style model of transmission connections among U.S. states. WinDS did not model existing flows in the power network but rather assumed that 10% of the existing interstate transmission capacity could be used to move new wind power.

The study concluded that the 20% wind scenario would require the construction of an extensive 765 kV transmission backbone across the country. However, it was not transparent how the modeling results lead to this conclusion. The study proposed only two lines linking the western Mid-Atlantic states to the east coast, and no transmission was added to link to the southeastern states. Given the transmission bottlenecks in the eastern United States, this imbalanced transmission footprint was somewhat surprising. It is possible that this could be a result of the approximate transmission system model used for this study. In addition, the study found that the additional cost of including 20% wind penetration would only be 2% over the base-case cost, though they conceded that this assumed rather optimistic cost and performance assumptions for both wind and conventional units.

This study was notable for being the first to catalog, in a systematic way, the available wind power at various wind-speed-class levels at a national level. Relative to more recent integration studies, however, this study had a number of weaknesses. The wind data, derived from a variety of sources using different techniques, lacked the consistency and spatial and temporal resolution of the data used in other studies. Additionally, using a transportation model of the power grid, as included in WindDS, made it difficult to determine the extent to which the transmission construction results are accurate.

17.3.5 California (CEC) 2010

In 2010, the California Energy Commission (CEC) released a report produced by KEMA, Inc. (now DNV KEMA) that examined scenarios in which renewable energy resources (wind and solar) contributed 20% of total energy for 2012 and 33% of total energy in 2020 (KEMA, 2010). The study's primary focus was to determine the optimal use of grid-connected storage to provide ancillary services and meet NERC standards when renewable energy resources provide a significant portion of the energy used within the California ISO.

In order to consider the effect of seasonality, the 20% and 33% penetration scenarios were modeled for one day in each of February, April, July, and October in both 2012 and 2020. The model treated each of California, the Southwest, the Northwest, and the mountain region (Colorado and Wyoming) as tightly connected groups of generators. The model assumed no additional renewable generation outside of California. Planned generating unit retirements and planned unit repowering were also included in order to provide a relatively accurate picture of the scenario years.

The California ISO provided historical demand, photovoltaic, and concentrating solar generation, wind generation, and conventional generation, as well as frequency and interchange data at 4-second resolution, for the 4 base days in the study: Wednesday, July 9, 2008; Monday, October 20, 2008; Monday, February 9, 2009; and Sunday, April 12, 2009. The 4-second wind data came from two hubs within California and were scaled up to match the studied penetration scenarios. Approximately one third of the future wind capacity was assumed to be in the BPA control area in the neighboring Northwest market, and 50% of this wind power was assumed to be levelized (with no variability) prior to import into the California ISO. Wind forecast data were synthesized based on historical forecast errors for the base days. For the purpose of model calibration, the California ISO also provided 4-second data on a large number of parameters, including system frequency, area control error (ACE), interchange schedules, and total system generation, for all areas modeled in the analysis. Forecast errors were assumed to be similar to those for existing wind power plants and unchanging over time.

The central modeling tool used in this study was a dynamic model developed by KEMA, known as KERMIT, a calibrated dynamic simulation, which was designed to provide second-by-second frequency and interchange data for 24-hour periods (see Section 17.2.3.6). Within each area, generators were modeled using dynamic models that accounted for generator ramp rates, rotational speeds, frequency control systems, and the control actions of balancing authorities to regulate flows in and out of each area (regulation). Using KERMIT, the study estimated the need for regulation services and the role of storage in supporting these regulation requirements. This was the only study in this review that estimated regulating reserves using dynamic modeling.

The study found that sustained but opposite ramping of load and wind in the mornings and evenings had the largest effects on system performance. Overall, the authors concluded that California would need to add balancing reserves equal to approximately 0.7% of peak load for the 20% case and 2.1% of peak load in the 33% case. Notably, the study concluded that to reach 20% or 33% renewable penetration, California would need to invest in substantial amounts of high-ramp-rate power resources. The report argued that storage could provide balancing services with lower greenhouse gas emissions than would result from the use of thermal plants for balancing. Because of faster ramp rates and the ability to both generate and consume power, the study found that a 30- to 50-MW storage device could be as effective, in terms of regulating frequency to within limits, as a 100-MW combustion turbine used for regulation. The study did not conduct an economic or benefit-cost analysis for these options, however.

As the first integration study to use detailed dynamic modeling of regulating reserves, the dynamic modeling approach used in this study has substantial methodological value. The generalizability of the study's conclusions about balancing reserves was fairly limited, however, due to the very limited quantity of wind data used.

17.3.6 Nebraska 2010

The 2010 Nebraska Statewide Wind Integration Study examined varying penetration levels of wind in the area covered by the Nebraska Power Association (NPA) for the year 2018. NPA conducted the study on behalf of NREL, with help from the EnerNex Corporation and Ventyx, in order to quantify the effect of wind integration costs and determine the merit of alternative transmission and penetration scenarios, particularly with respect to the necessary amount of balancing reserves. The study examined four scenarios, with wind penetration levels ranging from 10% of total energy sales to 40% of total energy sales, and was designed to quantify both reserves and wind integration costs in the NPA and greater SPP region, of which the NPA is a part.

The study included a 10% wind scenario, two 20% scenarios, and one 40% scenario. To obtain the desired penetration levels, the study added additional sites to existing Nebraska wind plants based on both the wind resources available and geographic diversity. The first two scenarios (10% and 20%) included only existing and currently planned transmission, although a few transmission constraints were removed without specifying what infrastructure improvements were used to remove these constraints. In the other two scenarios (20% and 40%), an extra-high-voltage (765 kV) expansion overlay was added. The structure of the overlay was derived from a previously developed proposal for the SPP territory. For each of the four scenarios, penetration levels in the rest of the SPP area were assumed to match the Nebraska levels (i.e., 10%, 20%, or 40%), in order to keep pricing within the region comparatively consistent. Wind penetration outside SPP was set to approximately 6% for all four scenarios, which the authors noted would increase the amount of wind energy exported from SPP.

The data used for the study were taken from NREL's 10-minute resolution mesoscale data estimated for 2004 through 2006 for wind plants east of the Rocky Mountains, the same data used for the EWITS (EnerNex Corporation, 2011) study. Regulating reserve requirements were estimated from a hybrid higher-resolution wind and load dataset. The study did not provide details of the methods used to develop this dataset. The study used hourly load forecast data for Nebraska from 2004 through 2006, as well as actual hourly load data for all of SPP for the same time period.

The study used the Ventyx PROMOD IV regional production cost simulation model, which included the transmission network, for cost and dispatch analysis. The four scenarios were each run three times, once for each year of data. Day-ahead forecast data were used to produce unit commitment schedules. Additionally, with regard to the cost analysis, the study attempted to model day-ahead, real-time, and ancillary services markets (which were not present in the SPP at the time of the study) to represent future market conditions. Within the study, three balancing areas of NPA were fully represented, while the remaining Nebraska utilities were included in the model in aggregate.

The study found that the 10% scenario required additional balancing reserves of about 1% of peak load and the 20% scenarios required additional balancing reserves of just below 2% of peak load. The 40% scenario required additional reserves of close to 4% of peak load. In the 40% penetration scenario, the inclusion of realistic forecast errors resulted in an 18% increase in combined-cycle plant usage relative to a perfect information case. The study determined that no significant wind curtailment was necessary for any of the scenarios, as all scheduled wind generation was accommodated by redispatching other generation sources and exporting excess wind. However, the authors noted that additional transmission would be necessary regardless of curtailment due to a number of constraint violations within the transmission model but that these constraints were largely eliminated by the transmission overlay. Based on the study's modeling efforts, wind integration costs, including production costs due to wind forecast error and balancing reserves, ranged from $1.32 to $1.75 per MWh (in 2009 dollars) for the four scenarios. The authors noted that this estimate could be low if wind penetration outside of SPP exceed the modeled 6% penetration level.

17.3.7 New York (NYISO) 2010

Building on the earlier New York wind integration study performed by GE (GE Energy, 2005a), NYISO conducted its own study looking at the effect of 3,500 to 8,000 MW of in-state wind capacity, approximately 5% to 12% penetration on an energy basis, on operational costs, market operations, and transmission requirements. When the study was conducted, the NYISO Interconnection included 1,275 MW of wind capacity.

The study conducted analysis for 3 study years, 2011, 2013, and 2018, with progressively higher wind penetration levels. All existing wind sites were included in the analysis, and new sites were selected from projects in the NYISO interconnection queue. Wind generation in neighboring regions was not modeled, which may have influenced the accuracy of power import/export results in this study. NYISO uses, and the study simulated, 5-minute dispatch and AGC regulation to provide subhourly balancing.

NYISO and AWS True Power developed wind plant data for 1-minute, 10-minute, and hourly intervals based on the wind profiles generated for EWITS (EnerNex Corporation, 2011). Due to the fact that NYISO updates its dispatch schedules every 5 minutes and that short-term wind forecast data were not included in their dataset, there was a need to simulate a 5-minute-ahead forecast. This short-term simulation was added to the NWP-derived simulated wind by using the assumption that the wind production in the next 5 minutes would be equal to the current wind production. Combining this with load data produced both a net load time series as well as a 5-minute-ahead net load forecast. The changes in net load, which were used to determine regulation, were calculated by subtracting the changes in wind over 10-minute increments from changes in load over 5-minute increments. NYISO developed load profiles internally based on 2005 and 2006 historical load data.

As in most of the studies, the authors used Gaussian statistics to characterize wind variability. The standard deviation of 5-minute net load step changes was multiplied by three to produce monthly regulation requirements. Production cost simulation was conducted using ABB's GridView, which modeled both security-constrained unit commitment and security-constrained economic dispatch based on transmission network data. To analyze network constraints, transmission path limits were modeled using the PSS/MUST simulation program. PSS/MUST calculated the transmission path capacities, which accounted for uncertainties in interactions between sources and sinks of power and variability in the dispatch pattern. Once the grid constraints were identified, the group developed several transmission investment plans and performed a benefit-cost analysis for each plan.

The NYISO study concluded that net load variability increases linearly, on all time scales, with wind penetration. It estimated that for each 1,000 MW of wind power added from between 4,250 MW and 8,000 MW, the regulation requirement would increase by 9% due to higher-magnitude ramping events. This is a relatively modest increase in regulation requirements, equal to 0.3% of peak load in the 12% penetration case. The study also suggested that the existing NYISO dispatch processes could handle the higher-magnitude ramping events that wind generation would bring. To achieve this, some current fossil fuel generation would need to be committed for use in regulation. This study concluded that the 8 GW of wind power would offset 1.6 to 2 GW of fossil-fuel generation. The authors estimated that 9% of potential wind generation would need to be curtailed due to transmission limitations. To reduce curtailment, the study suggested upgrades to the existing high-voltage system rather than the construction of new EHV lines, as other studies have suggested.

The strength of the study was in its analysis of transmission constraints, the upgrades needed to accommodate increased wind penetration, and how wind affects current fossil fuel generation. Its primary weakness came, as with many of the studies, from it use of standard deviations to estimate regulation requirements.

17.3.8 Southwest Power Pool (SPP) 2010

In 2010, Charles River Associates (CRA) conducted a fairly detailed study of wind integration in the SPP region (Charles River Associates, 2010). This study looked at 10%, 20%, and 40% regional wind deployment, by energy, although the analysis was less detailed at the 40% level. The study conducted power flow analysis in order to identify the transmission upgrades required to minimize wind curtailment. The study also approximated the operational and market effects of wind integration via PCS, which analyzed congestion patterns, unit commitment and dispatch, and the effect of forecast errors.

Wind plants for each scenario were selected from the SPP generation interconnection queue. Because plants in the queue clustered in regions of high wind resource potential, the scenarios included less geographic diversity than some other studies. The baseline scenario included all of the commercial wind generators operating as of February 2009, about 4% wind penetration on an energy basis. For each of the 10%, 20%, and 40% scenarios, comparable wind penetration was assumed to occur in the neighboring regions with high wind potential, which improved the credibility of interregional modeling. The SPP region was modeled as a single balancing authority rather than the 13 balancing authorities that operated within the region at the time the study was conducted, reflecting operation changes anticipated by the end of 2013. This allowed the study to simulate a cooptimized energy and ancillary service market. The simulated market included day-ahead unit commitment scheduling and real-time dispatch signals sent to generators at 5-minute intervals. In addition to the existing transmission infrastructure, several transmission upgrades were considered. These upgrades were selected purely on the basis of improved transmission system operations; no economic analysis was conducted surrounding these upgrades.

As with several of the studies, this study drew on 10-minute simulated wind power data from AWS TruePower. Data from wind plants were scaled to reflect plant-size differences between the sites selected from the current generation interconnection queue and those in the original dataset. SPP provided historical hourly load profiles for the years 2004 through 2006, corresponding to the base years used to generate the wind profiles. Hourly load data for neighboring regions were drawn from FERC filings for 2002.

As with other studies, this study estimated reserve requirements based on the statistical analysis of wind and net load step-change data. However, the study reported more detailed statistics than did most other studies, including 5th and 95th percentile step changes and maximum hourly increase/decrease in net load as well as the more commonly reported mean and standard deviation.

To assess the effects of wind integration on the SPP transmission system, the authors conducted AC contingency analyses for the peak load hour in the summer and winter, the minimum net load in the spring, and the peak wind, peak load hour in the fall for the base case and 10% and 20% penetration scenarios. When the contingency analysis found violations of SPP criteria that could not be avoided by shifting dispatch among generators, additional transmission was added to the system. The resulting power flow cases were used to analyze voltage and transient stability as well as available transfer capability. Production cost simulation was performed using GE MAPS, which was used to identify potentially costly transmission constraints, the effect of wind on ancillary service requirements, and the effects of wind forecasting error.

The study's analysis of wind and net load data revealed several interesting points. Because the SPP service area is comparatively flat, wind power production from different plants had a higher correlation than was reported for other regions. Possibly related to this, hourly changes in wind output exhibited a heavier-tailed distribution relative to that reported in other studies. Net load variability was also shown to vary nonlinearly with wind capacity at both the hourly and 10-minute time frame. Rather than using standard deviations to estimate regulating reserve requirements, the study proposed a new heuristic method for calculating the quantity of regulation required to meet NERC control performance standards, based on the assumption that wind and nonwind contributions to regulation are minimally correlated. The proposed heuristic for estimating the amount of up-regulation needed, R_{up}, is as follows:

$$R_{up} = ((0.01L_{peak} + L_{10})^2 + a(\Delta W_{95})^2)^{1/2} - L_{10} \qquad (17.6)$$

where L_{peak} was the peak load, L_{10} was a constant used in NERC reliability criteria (CPS2), a was a constant to adjust the relative contributions of wind to regulation requirements, and ΔW_{95} is the 95th percentile step change (increase) in wind. Using this metric, the study estimated that additional regulation equal to approximately 0.05%, 0.2%, and 0.6% of peak load would be required for the three scenarios, respectively. While this approach seems preferable to using standard deviations, additional research is needed to determine if this approach would meet NERC requirements for ACE and if ignoring larger step changes (beyond the 5th and 95th percentiles) would have adverse effects on system reliability. CRA found that there was a need for additional transmission upgrades to transport wind power from the western region to load centers in the eastern portion of SPP. The study noted that selecting sites based on geographic diversity could reduce the need for new transmission. Finally, although wind facilities do not fail as a single unit, the study recommended that the loss of connection to large wind plants be considered when scheduling contingency reserves.

This study stood out as one of the most detailed and thorough in this review. It was one of only a few that included detailed AC contingency analysis and used percentiles rather than standard deviations to set regulation requirements. While the 5% to 95% 10-minute forecast error thresholds match NERC's CPS2 standard, these thresholds still excluded more extreme events that may be operationally important. In addition, the estimation of regulation requirements based on 10-minute rather than 1-minute wind data indicates that these results provide an estimate of what is needed to satisfy CPS2 but may understate the regulating reserves necessary to meet CPS1.

17.3.9 Western United States (WWSIS) 2010

The Western Wind and Solar Integration Study (WWSIS; GE Energy, 2010) was prepared by GE for NREL. The study was designed to assess the operational effects of up to 35% (30% wind, 5% solar) renewable energy penetration for the 2017 study year. Key research objectives included quantifying the potential benefits of geographically dispersed wind sites, balancing area cooperation and improved forecasting/integration of forecasting into the unit commitment process, and developing reserve requirement guidelines that adequately account for wind variability.

Geographically, WWSIS focused on the WestConnect service area, which consists of northern California, Nevada, Arizona, New Mexico, Colorado, and parts of Wyoming. The study examined 12 scenarios with different wind penetrations and siting criteria. These scenarios were divided into four cases, one with 10% wind penetration throughout the Western Interconnection, one with 20% wind penetration in the WestConnect but 10% in the rest of the Western Interconnection, one with 20% wind throughout the Western Interconnection, and a case with 30% wind in WestConnect but 20% wind in the rest of the Western Interconnection. For each of these penetration levels, the study examined three different wind-siting heuristics. The first, the "In Area" scenario, required that each state in the WestConnect achieve the targeted wind penetration level using the best wind resources within that state. This scenario did not include any additional transmission. The second, the "Mega Project" scenario, achieved the wind penetration targets using the best wind resources in the WestConnect without any consideration of state lines. In this scenario, the majority of wind generation was sited in Wyoming and new transmission was added to transport power from this area. Last, the "Local Priority" scenario used the best wind sites within the WestConnect but included a 10% capital cost advantage to resources within each state, resulting in a scenario that was a combination of the In Area and Mega Project scenarios. Wind in the Western Interconnection but outside of WestConnect was always modeled using the In Area criteria. Transmission and non–wind-generation infrastructure were modeled based on 2017 projection and not optimized for wind integration. The authors noted that increased flexible generation and transmission would likely be available in a 35% renewable penetration case, thus making their results conservative. On the market side, WECC was modeled with 5 balancing authorities rather than the 37 that are currently operating in WECC. In addition, all generators were assumed to be available for least-cost economic dispatch and not encumbered by any power purchase agreements. Generation and interstate exchange within WECC were scheduled on an hourly basis, and economic dispatch was modeled at 5-minute intervals.

The 3TEIR Group produced the base wind speed data for WWSIS using the Weather Research and Forecasting (WRF) NWP, based on measurements from 2004 through 2006. The WRF model provided 10-minute wind speed data with a 2-km spatial resolution, along with day-ahead hourly wind forecasts. To avoid unrealistically accurate forecasts, forecast wind profiles were created using weather data from the Global Forecast System rather than WRF. Unfortunately, the aggregate annual energy from the forecast wind profiles exceeded that of the WRF wind profiles, resulting in forecast errors that were biased upward. To mitigate this problem to some extent, all forecasts were reduced by 10%. Due to the size of the area being modeled, the WRF model was run in 3-day blocks and the data subsequently merged. Though some smoothing was performed at the seams of the 3-day blocks, these days still exhibited more variability than days that did not have seams (GE Energy, 2010). For this reason, these days were excluded from the analyses of hourly and 10-minute variability (GE Energy, 2010). Because mesoscale meteorological models, like WRF, tend to understate the wind speed variability at short time scales, the modeled wind speed data were converted into power data using a statistical model (SCORE-lite) intended to replicate empirically observed turbine ramping characteristics (Potter et al., 2008). Once the 10-minute power data were generated, analysts developed hybrid 1-minute data based on empirical data from an unspecified number of existing wind plants. Ventyx provided hourly load data for the western region, which were interpolated using a cubic spline to create 1-minute and 10-minute load data. It is important to note that this interpolation smooths out intrahour variability, which could affect the conclusions drawn from these data.

The study used three main analytical approaches. First, statistical analysis of step changes was used to characterize the frequency and magnitude of wind and net load ramping events over a variety of time scales and determine regulation and balancing reserves requirements. Second, the GE MAPS PCS model and GE MARS reliability model were used to simulate production and reliability on an hourly basis. Finally, a MATLAB-based quasi-steady-state (QSS) model with 1-minute time steps was used to validate their regulating reserve estimates for selected periods with large changes in wind output.

Along with EWITS (EnerNex Corporation, 2011), WWSIS (GE Energy, 2010) is one of the most comprehensive integration studies. The study concluded that the 35% renewable case is feasible but that increased net load variability would require additional balancing resources. The study suggests that balancing reserves should equal three times the standard deviation of 10-minute net load or, more heuristically, 3% of load plus 5% of short-term forecast wind. Of these reserves, one standard deviation of the 10-minute net load variability should be available for regulation. In practice, these requirements equated to additional regulation of less than 0.5% of peak load and additional balancing reserves equal to approximately 1% of peak load for the 30% case. Because existing thermal units would be dispatched less frequently rather than decommitted to accommodate renewables, the study suggested that up-reserves could be provided from existing generation. In case of a severe overforecast, spinning reserves may be required to provide regulation resulting in shortfalls in the contingency reserves, which could be managed through a variety of means, including increasing spinning reserves, storage, or demand-side management.

The study also suggested several operational changes in order to facilitate wind integration. In particular, it found that greater cooperation would be needed between balancing areas, that subhourly scheduling should be incorporated into system operation, and that existing transmission infrastructure must be utilized at a higher rate. Subhourly scheduling could reduce ramping by load following generators by approximately 50%. The study also suggested incorporating day-ahead wind forecasts into unit commitment procedures, increasing the flexibility of dispatchable generation and requiring wind plants to provide down reserves. In the

35% renewable case, the results suggested a 40% decrease in system operating costs due to reduced fuel consumption and emissions, neglecting the increased operational cost due to increased net load variability. Finally, the study found that wind resources in the study area had a capacity value in the range of 10% to 15% of nameplate capacity using a capacity valuation method that is a few percent more conservative than ELCC.

The WWSIS study (as with EWITS; EnerNex Corporation, 2011) was commendable because the 10-minute wind datasets upon which they were based are publically available. The study was also quite clear and thorough in documenting the assumptions behind the analyses. Though issues with the data persist, such as the seam effect discussed earlier, considerable effort, such as the application of SCORE-lite in WWSIS, went into replicating observed wind outputs over large geographic areas. Unfortunately, the higher-resolution 1-minute data are not publically available, and the methods used to validate the hybrid data were less well documented. WWSIS included analyses across a wide range of time scales, for the most part using appropriate modeling tools. While the study used standard deviations to estimate reserve requirements, the use of QSS simulation provides some validation of its approach.

17.3.10 Eastern United States (EWITS) 2011

The Eastern Wind Integration and Transmission Study (EWITS) covered the U.S. Eastern Interconnection (EnerNex Corporation, 2011) and was in many ways a companion to the Western U.S. study (GE Energy, 2010). Led by EnerNex Corporation, the study modeled the Eastern Interconnect in 2024 as though it were managed by seven large balancing authorities with a market structure that used day-ahead bidding. The analysis focused on developing the transmission capacity necessary to support the modeled wind penetration scenarios, assessing the operational effects of wind generation and determining the capacity value that wind generation could provide.

The study examined four wind penetration scenarios for the 2024 study year. The first three scenarios all considered 20% wind energy penetration with differing wind plant siting criteria. The first scenario focused on onshore generation at high-capacity-factor sites. This resulted in substantial wind development in the Great Plains. The second scenario shifted more generation eastward and had limited offshore wind development. The third scenario focused on siting wind plants near load centers. The third scenario had the lowest level of wind development in the Great Plains and the highest offshore wind development. Finally, the fourth scenario considered 30% penetration with over 300 GW of on- and offshore development.

This study, as with EnerNex Corporation and colleagues (2010) and Charles River Associates (2010), used wind data developed by AWS TruePower using the propriety Mesoscale Atmospheric Simulation System (MASS) model (Brower, 2009) based on historical weather data from 2004 through 2006. The MASS model data had a 10-minute temporal and 2-km spatial resolution. The authors noted that the model results tend to underestimate wind speed at most sites and that at several sites, nighttime wind speeds were overestimated while daytime speeds were underestimated

(Brower, 2009). For each site, the raw output from the NWP model was filtered to adjust for model bias, the proportion of the modeled plants that fell within each grid cell, wake effects, and other adjustments. Next, the adjusted wind speeds were applied to power curves for IEC class one, two, and three turbines. Each of these power curves was a composite of two to three commercial turbines. Finally, the results were filtered to reduce variability, reflecting the effect of spatial diversity in turbines within large wind plants (Brower, 2009). Wind power forecasts were produced for 1-day-ahead, 6-hour-ahead, and 4-hour-ahead time frames (EnerNex, 2011). In addition, 1-minute-interval power output data were synthesized for selected periods based on historical plant data from 17 locations in ERCOT. To create this hybridized dataset, minute-to-minute variability was extracted from the ERCOT data by calculating 10-minute trends using a bicubic fitting process and applying the residuals to the simulated power outputs (Brower, 2009). In addition, the study used 2004 through 2006 load data from the PowerBase database and 2006 power-flow case data from FERC (EnerNex Corporation, 2011).

Transmission requirements for each scenario were modeled using Ventyx PRO-MOD IV, a deterministic PCS. From the PCS model, hourly transmission flows were calculated, using a DC power flow model. While the DC model neglected voltage control and reactive power in the network, it was a substantial improvement over the transportation model used in U.S. DOE (2008), and the report explicitly identified the limitations in the DC model. From the resulting transmission flow data, the study suggested that significant transmission investment would be required to support the wind-power deployment in the proposed scenarios. To estimate the costs for this transmission investment, the study chose several potential transmission overlays, including high-voltage DC lines and ultra-high-voltage AC lines. As with most of the studies reviewed, EWITS uses the standard deviation of step-change data to estimate regulation requirements. Specifically, they estimate regulation requirements for hour $h(R_h)$ as follows:

$$R(h)=3\sqrt{(((0.01L(h))/3)^2+\sigma_{ST}(h)^2)} \tag{17.6}$$

where $L(h)$ is the load in hour h and $\sigma_{ST}(h)$ is the expected standard deviation of the wind step changes during hour h. This is internally consistent only if wind and load are uncorrelated and only if the standard deviation fully characterizes the variability, as is the case with a Gaussian distribution, neither of which is accurate.

The study concluded that 20 to 30% wind penetration was feasible, given a substantial expansion of the existing transmission infrastructure and aggregation of many small balancing areas into a few large ones. The study estimated overall integration costs for the 20% scenarios at between $3.10 and $5.13 per MWh of wind production depending on where wind plants were sited. The integration costs were lowest in Scenario 3, where wind capacity was spread fairly evenly across the study area, and highest in Scenario 1, where wind capacity was more geographically concentrated in the Great Plains. Integration costs in the 30% penetration scenario equaled $4.54 per MWh. The study found wind curtailment of 2% to

10%, given substantial transmission expansion. The authors also presented a range of capacity credit values for wind that vary substantially depending on the wind siting and transmission scenarios modeled. Assuming current transmission infrastructure, the study found ELCC values between 16% and 30% of nameplate capacity. With expanded transmission, these values increased to 24% to 33% of nameplate capacity. In all cases, the scenario with wind sited close to load centers produced the highest ELCC estimates. It is worth noting that these values were calculated across multiple-capacity markets, a technique that resulted in comparatively high ELCC values and was criticized by the WWSIS team (GE Energy, 2010).

In many ways, this study was among the most technically sound of those reviewed here. It analyzed a broad set of issues related to wind integration quite well. The authors made a commendable effort to explicitly consider many of the issues that were not included in other studies, most notably transmission constraints. However, as in other studies, the use of standard deviations to estimate reserve requirements without substantial validation was problematic.

17.3.11 United States 80% Renewables (NREL) 2012

NREL's Renewable Energy Futures report studied the technical barriers to increasing renewable energy penetration in the United States to 80% by 2050 (NREL, 2012). As with U.S. DOE (2008), this study was not intended as a complete integration study but rather focused on the ability of renewable resources to match a high proportion, up to 80%, of total load.

The study modeled several different penetration levels but focused on reaching an 80% renewable target that included wind and other renewable resources. Different 80% scenarios were developed based on the potential evolution of renewable energy technology, levels of transmission investment, types of storage and demand-side balancing resources, and different supply portfolios. In these scenarios, wind power was projected to provide from 32% to 43% of overall electricity demand. Sensitivity analysis was performed on both fossil fuel costs and demand growth assumptions. The majority of the scenarios expanded the transmission infrastructure and access to existing transmission capacity to support renewable energy deployment, and the models included the retirement of thermal generation without allowing for the construction of new units. No assumptions were made about future policy measures, such as tax credits, that could potentially affect renewable deployment.

Wind speed data were assembled from several sources, including mesoscale data from AWS TruePower and anemometer data from Pacific Northwest National Laboratory and other regional entities. Capacity factors for potential wind plants were estimated based on the wind resource class and projected improvements in wind turbine designs – primarily from larger rotors and advanced tower designs. The simulated wind data had an hourly resolution.

This study used two different models for cost estimates: the Regional Energy Deployment System (ReEDS) and ABB GridView. ReEDS is a capacity- and transmission-expansion model developed by NREL that used linear programming to simultaneously compute optimal dispatch and transmission-expansion plans. Like WinDS from U.S. DOE (2008), ReEDS used a transportation model to model the effect of new transmission. The model also captured some policy issues such as emissions, siting constraints, and reserve requirements. For wind and solar power, the study estimated that balancing reserves were needed to cover two standard deviations of the expected forecast error. ReEDS did not differentiate between different ancillary services, and therefore the study noted that it may underestimate the need for short-term storage. GridView was used to supplement ReEDS because it had a more robust simulation of real-time grid operation. The generation and transmission capacity output from ReEDS was run through GridView to make sure the developed scenario was feasible. GridView also incorporated unit commitment and dispatch and had a DC transmission model.

The study concluded that achieving 80% renewable energy penetration was technically feasible and consistent with renewable resource availability. These conclusions were predicated on significant investments in renewable capacity and transmission infrastructure and increased flexibility in the electric system through a combination of storage, demand-side response, and flexible dispatch and ramping of conventional generating units. Because of increased distances from generation sites to load centers, transmission and distribution losses were projected to range from 8.4% to 9.5%, up from 6.4%, without the addition of new renewables. In the 80% scenarios, 8.1% of wind, solar, and hydropower generation were curtailed. In the higher renewable penetration scenarios, thermal plants were assumed to change roles in the system; natural gas plants were used entirely as peaking plants, and coal plants had increased diurnal and seasonal ramping. However, no market systems were suggested to ensure these plants stayed profitable. Depending on the trajectory of renewable technologies, the study suggests that the retail price of electricity will increase between 21% and 45% in the 80% renewable scenario.

This study was unique among the others in this chapter because of the extremely high penetration levels considered. Given the challenge, it does a reasonable job of describing what changes would likely be necessary to reach such a high penetration level. However, the study's transportation model of the transmission network raises questions about the reliability of the transmission investment plans.

17.3.12 Additional integration studies

While the 11 studies reviewed are representative of most large-scale integration studies, this list is by no means comprehensive. Here we briefly mention a few additional, notable recent studies.

Two studies looked at wind integration in the U.S. state of Idaho. In a 2007 study (EnerNex Corporation, 2007), EnerNex looked at the effect of up to 1,200 MW of new wind capacity in the Idaho system and estimated the operating cost of wind integration to be approximately $10/MWh. A follow-up study in 2013 (Idaho Power, 2013) found similar results for the average integration costs but suggested that the incremental cost of additional wind energy at the higher levels of wind penetration ranged from $15 to $50/MWh.

Using a similar method to the EWITs study (EnerNex Corporation, 2011), NREL published a wind integration and transmission study for the Hawaiian island of Oahu (Corbus et al., 2010). This study was unique in that it looked at the feasibility of a very high renewables penetration scenario (40% of energy from renewables) in a location without the ability to share resources across regions. The study identified the need for storage, substantial operational changes (such as ramp-rate limits and wind curtailment), and interisland DC transmission links to support the proposed level of wind integration.

17.3.13 Related academic studies

In addition to the industry studies reviewed, wind integration has also received considerable attention in the peer-reviewed academic literature. While a complete review of all related literature is beyond the scope of this chapter, it is useful to note a few particularly relevant research articles.

Several articles have used simplified power system models to study the broader effect of large-scale wind on power system economics and operations. DeCarolis and Keith (2006) used a greenfield model (no existing transmission capacity) to evaluate optimal low-carbon generation scenarios and found that the variability of wind increased combined fuel and capital costs by about 10%. They also found that there is no threshold beyond which integration costs sharply increase. Denny and O'Malley (2007) used a detailed power system model, including PCS, to model total system benefits, in terms of fuel savings and emissions benefits, of wind integration. They found that there was a threshold (about 25% penetration by energy) above which the marginal benefits of wind capacity begin to decrease. On the other hand, a more recent study (Budischak et al., 2013) used a capacity-expansion model of the PJM territory, without transmission constraints, to look at the optimal combinations of generation and storage required to achieve 30, 90, and 99.9% renewable scenarios. They concluded that it was feasible to achieve very high penetration levels and, interestingly, that it was more cost effective to overbuild and overgenerate and curtail renewable generation than it was to build large amounts of storage.

A number of peer-reviewed research articles suggest and evaluate various methods for wind integration studies. Holttinen and colleagues (2008) suggest the use of standard deviations for establishing the reserve requirements for systems with high levels of wind power, a method that was used in many of the studies reviewed. More recently, Papavasiliou and colleagues (2011) studied a stochastic optimization approach to estimating reserve requirements and found that this approach substantially outperforms heuristic methods. A number of recent articles suggest and apply methods for quantifying costs associated with wind power variability. Mauch and colleagues (2013a) estimated reserves requirements based on day-ahead load and wind forecast uncertainty. They found that the forecast uncertainty is greatest on days when the wind is forecast to be blowing strongly (see also Mauch et al., 2013b); thus, there are some days when significant reserves are required and other days when much smaller reserves are needed. Because of the additional uncertainty associated with estimating the reserves that are needed to cover 95% of the day-ahead forecast errors, Mauch and colleagues (2013a) find reserve requirements substantially higher than those reported in many of the studies reviewed here. They suggest that a dynamic method should be used to schedule reserves based on the day-ahead forecast values, noting that ERCOT's requirement that operational reserves be procured to cover 95% of the day-ahead forecast errors may not be the most cost-effective method to handle day-ahead uncertainty. Fertig and colleagues (2012) show that interconnecting wind plants among regions provides a reduction in variability very dependent on the time scale considered. They demonstrate that there is only modest benefit in smoothing the large variability at time scales of approximately 12 hours; the smoothing from interconnection is largely in the approximately 1-hour variability. Lueken and colleagues (2012) used regulation prices and data from wind, solar PV, and solar thermal plants to estimate the cost of variability from each technology. They conclude that, given their modeling approach, which schedules up- and down-regulation to match the variability of wind/solar power, variability can add $15 to 40/ton CO_2 to the cost of abating emissions using these technologies. Katzenstein and Apt (2012) use a similar approach and conclude that wind variability adds $3 to $10/MWh in variability costs and that, as a result, the marginal benefits from wind plants diminish quickly after the addition of only a few additional plants.

A paper by Soder and Holttinen (2008) highlighted different modeling approaches and assumptions that could be used in wind integration studies and made several recommendations about which of these approaches were most desirable. This article suggested, for example, that wind data used in integration studies should reflect the smoothing effect of aggregating large numbers of individual turbines and that dispatch models should include the ramp constraints for thermal generators. Many of the recommendations in Soder and Holttinen (2008) have been widely incorporated into recent integration studies. The application of other recommendations, such as the full-scale dynamic analysis of the power system, remains relatively rare.

Finally, a paper on best practices for wind integration studies by Holttinen and colleagues (2013) provided recommendations for wind integration studies based on the authors' experience with such studies (including several of those reviewed here). Holttinen and colleagues outlined five steps for integration studies – data collection, system configuration and reserve estimation, capacity estimation, system flexibility, and transmission simulation.

17.4 Comparison of reserve estimation results

While all studies found that increased wind generation increases the variability in net load, several pointed out that existing power system technology and practices

TABLE 17.1 Summary of data and methods for reviewed studies.

Study	Wind Speed/Power Data Sources	Sample Interval (minutes)	Statistical Methods for Characterizing Net Load Variability and Reserve Requirements	Power System Models
NYSERDA 2005	AWS 8-km met. model Historical plant data from IA	60 1/60	Gaussian methods for reserve calculations	GE MAPS GE MARS GE PSLF and PSDS
MN 2006	MM5 4-km met. model	5	Gaussian methods for reserve requirements	PROMOD GE MARS
ERCOT 2008	AWS 10-km met. model Historical plant data from TX	60 1	Regulation requirements set to the 98.8th percentile of regulation events.	GE MAPS
NREL 2008	AWS, state mapping programs	60	N/A	WinDS
CEC 2010	96 hours of plant data from CA	1/60	Small sample – not characterized	KERMIT
NE 2010	Same dataset as EWITS 2011		Gaussian methods for reserve requirements	PROMOD IV
NYISO 2010	Same dataset as EWITS 2011		Gaussian methods for reserve requirements	ABB GridView PSS/MUST
SPP 2010	Same dataset as EWITS 2011		Reserve requirements based on the 5th and 95th percentile deviations between wind forecast and output.	GE MAPS
WWSIS 2010	3TEIR Group 2-km met. model and probabilistic power output model Historical wind plant data	10 1	Gaussian methods for reserve requirements	GE MAPS GE MARS QSS
EWITS 2011	AWS 2-km met. model Historical plant data from TX	10 1	Net load variability characterized by standard deviations.	PROMOD IV GE MARS
NREL 2012	AWS, state mapping programs	60		ABB GridView ReEDS

were designed to manage load variability and that managing the additional variability from wind was not fundamentally different from managing load variability (GE Energy, 2008). In all of the studies, high levels of wind penetration were found to be technically feasible but frequently to require some modifications in system characteristics (e.g., generating mix, transmission capacity, and balancing authority extent) or operating practices (plant commitment scheduling frequency, wind participation in down-regulation, etc.). In addition, geographic diversity in wind resources and improved wind forecast accuracy were found to offer substantial benefits. Wind integration was consistently found to require an increase in the reserves required to maintain operational reliability as well as to have an associated economic cost.

Given the substantial focus of almost all of the reviewed studies on estimating operating reserves, this section compares the quantitative reserves estimates from the studies, as well as some key contributors to those findings.

17.4.1 Reserve requirements

As noted in Section 17.1 and in EnerNex Corporation (2011), Holttinen and colleagues (2011), and NERC (2011a), the terminology used to define different reserve types has evolved over time and varies from country to country and region to region. NERC defines two reserve types, which are commonly estimated in wind integration studies: regulation reserves that must be responsive to AGC and contingency reserves that are available to cover the unexpected loss of a generating unit (NERC, 2011b). In addition to regulating and contingency reserves, the concept of load-following reserves was also used frequently in these studies. Many of the studies refer to load-following resources (sometimes termed reserves), which are not defined by NERC guidelines and which adjust for changes in net load of periods of several minutes to hours in response to subhourly economic dispatch commands (EnerNex Corporation, 2011). Several studies suggested the formalization of a load-following or "variability" reserve category (for example, Charles River Associates, 2010; GE Energy, 2010). For this review, we use the term "balancing reserves" to refer to all balancing activities inclusive of load following and regulation.

Under NERC standards, regulating reserves are responsible for maintaining system balance in the period between economic redispatch 95% of the time. These standards do not define a specific amount of regulating reserves that must be procured, but this requirement is often heuristically implemented as 1% of peak load. Those studies that estimated regulation (Charles River Associates, 2010; EnerNex Corporation, 2006; GE Energy, 2005a, 2008, 2010; NYISO, 2010) did so based on statistical characterization of wind power, net load, or wind forecast errors. Figure 17.6 shows the annual average of the hourly increase in regulating capacity estimated by these studies. These estimates were broadly consistent across studies and relatively modest. Even at 40% wind penetration, the increased regulating reserves were on the order of 0.8% of peak load (Charles River Associates, 2010), less than twice the baseline regulation level. The NYSERDA study (GE Energy, 2005a), which examined 6% wind penetration, estimated the need for additional regulation as three times the standard deviation of 6-second net load variability. In EnerNex Corporation (2006), regulating reserves were estimated based on the assumption that wind output fluctuation would be less than 2% of wind nameplate capacity. These findings are consistent with the observation made in Chapters 2 and 8 that the variability of wind and solar power at short periods is several orders of magnitude less than the low-frequency variability.

Current procedure in ERCOT is to procure regulation based on the 98.8th percentile of historical regulation deployments for same hour of the day in the prior month and the same month in the prior year (GE Energy, 2008). This same method was found to be adequate for the wind penetration cases studied, although it was suggested that incorporating wind forecasts into the procurement process could improve the accuracy of regulation procurement (GE Energy, 2008). The SPP study approximated the amount of regulation based on the 5th and 95th percentile step changes in 10-minute wind power output using the equation described in Section 17.3.9 (Charles River Associates, 2010). WWSIS estimated regulation at one standard deviation of 10-minute net load variability based on the observation that 10-minute variability is approximately twice 5-minute variability; thus, one standard deviation of 10-minute variability was equal to two standard deviations of 5-minute variability and therefore assumed to cover 95% of 5-minute variability (GE Energy, 2010). QSS modeling in WWSIS generally supported this approach. The NYISO study set a regulation requirement equal to three standard deviations of 5-minute net load variability (NYISO, 2010). EWITS (EnerNex Corporation, 2011) described a method for calculating regulation reserves similar to that used in EnerNex Corporation (2006) and described in Section 17.3.11 but did not provide an numerical estimate of regulation requirements.

With the exception of KEMA (2010), which used a scenario-based dynamic modeling approach, the studies reviewed here estimated the additional balancing reserves required to manage wind variability and uncertainty statistically (EnerNex Corporation, 2006, 2011; EnerNex et al., 2010; GE Energy, 2005a, 2010). Note that KEMA (2010) referred to its estimate as an estimate of increased regulation but defines this as "a proxy for the net amount of capacity capable of fast ramping to follow system changes via regulation and balancing energy" and thus is more closely aligned with our definition of balancing reserves. The average increase in balancing reserves estimated by each study is shown in Figure 17.6. As with regulation, the estimates are relatively linear and relatively consistent across the studies. The estimated total balancing reserves required were, of course, higher than the reserves required for regulation alone. The EWITS study (EnerNex Corporation, 2011) produced the highest estimated reserves. The Minnesota study (EnerNex Corporation, 2006) estimated the additional load-following and operating reserves components of balancing reserves at two times the standard deviation of 5-minute net load variability and at two times the standard deviation of the next-hour forecast error, respectively. Exploring forecast error, both this study and EnerNex Corporation (2011) concluded that wind output variability, and therefore forecast error, was greatest in the mid-range of the aggregate production curve. Thus, in keeping with the recommendation in Holttinen and colleagues

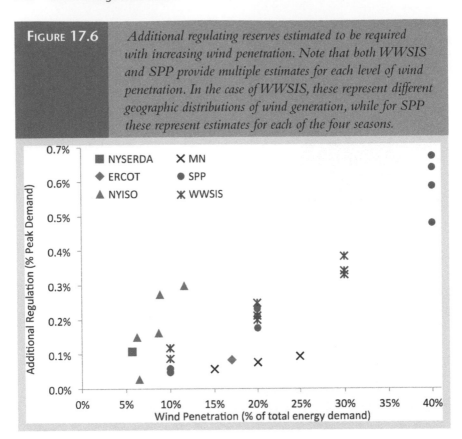

FIGURE 17.6   *Additional regulating reserves estimated to be required with increasing wind penetration. Note that both WWSIS and SPP provide multiple estimates for each level of wind penetration. In the case of WWSIS, these represent different geographic distributions of wind generation, while for SPP these represent estimates for each of the four seasons.*

(2013), these studies produced reserve requirements that varied with meteorological conditions. WWSIS estimated total balancing reserves at three times the standard deviation of 10-minute net load variability.

The WWSIS study also suggested two heuristics for grid operators to use to procure these reserves (GE Energy, 2010). The first was that reserves should equal 3% of load plus 5% of *forecasted* wind power. The second of these was a regionally specific variation of this rule with different load and wind components to maximize the fit with the three-standard-deviation rule (GE Energy, 2010). The EWITS study estimated total balancing reserves as the sum of its regulation estimate and one standard deviation of next-hour wind forecast error, as determined by a regionally and weather-specific function (EnerNex Corporation, 2011). The Nebraska study (EnerNex et al., 2010) used the same methodology as EnerNex Corporation (2011). Both Charles River Associates (2010) and GE Energy (2008) suggested that load following would increase with wind penetration but did not estimate the required reserves explicitly.

Figures 17.6 and 17.7 both show a surprising level of consistency in the reserves estimates among the various studies. This may reflect the fact that the heuristic methods used for reserves estimation were remarkably similar among the studies

FIGURE 17.7 — *Total additional reserves estimated to be required with increasing wind penetration. Note that both WWSIS and EWITS provide multiple estimates for each wind penetration level. These estimates represent different geographic distributions of wind generation.*

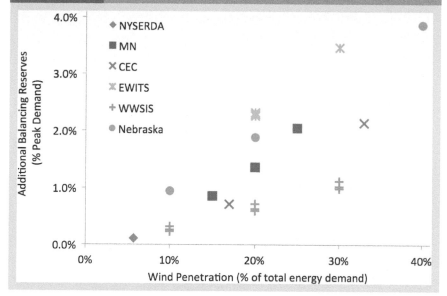

and that the data used for these studies come from a small number of mesoscale modeling groups. The CEC study alone (KEMA, 2010) used a substantially different methodology but came to similar conclusions regarding the absolute quantity of reserves, though with additional analysis regarding the need for fast-ramping storage.

Figure 17.8 compares the balancing reserves required for study-level results with the balancing reserves required at the state level in the WWSIS assessment (GE Energy, 2010). These results are for the Mega Project scenario, in which large wind sites were concentrated at the sites with the greatest wind potential, resulting in dramatically higher levels of wind penetration in New Mexico and Wyoming than in the other West Connect states. For the 10%, 20%, and 30% penetration scenarios, wind generation in Wyoming, for example, reached approximately 45%, 125%, and 195% of the total load in the state. These results pointed out the necessity of aggregating wind over large balancing areas.

For contingency reserves, NERC requires available reserves equal to the most severe single contingency (NERC, 2012). In WECC, this is implemented as the larger of the most severe contingency or the sum of 5% of hydro generation and 7% of thermal generation, at least half of which must be spinning reserves (GE Energy, 2010). WWSIS reported that this requirement was generally put into practice as

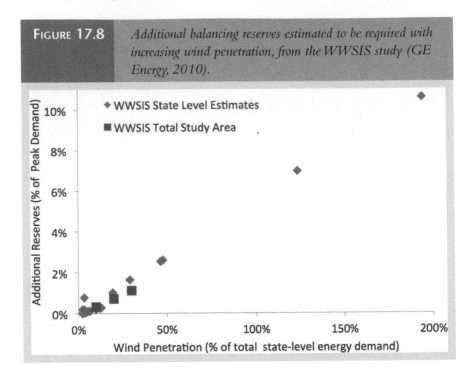

FIGURE **17.8** *Additional balancing reserves estimated to be required with increasing wind penetration, from the WWSIS study (GE Energy, 2010).*

spinning contingency reserves equal to the larger of 3% of load or 50% of the worst contingency (GE Energy, 2010). Generally, wind plants are not large enough to constitute the worst contingency in a system. In addition, they do not fail as a single unit; therefore they do not affect the contingency reserves requirements (EnerNex Corporation, 2006, 2011; GE Energy, 2005a; NYISO, 2010). Several studies did suggest that the wind could affect contingency requirements at the highest penetration levels (Charles River Associates, 2010; EnerNex Corporation, 2011; GE Energy, 2010).

17.4.2 Effects of geographic diversity

Geographically dispersed wind sites were also commonly cited as a factor that reduces wind variability (EnerNex Corporation, 2006, 2011; EWEA, 2009; GE Energy, 2010). The studies did relatively little, however, to determine the benefits and costs of diversity. Differing wind siting scenarios in EWITS were intended to capture the benefits of geographic diversity, and the authors noted that the integration costs tend to be lower when wind is spread more evenly across the study area, but the extent to which this is attributable to geographic diversity was not quantified. The WWSIS study (GE Energy, 2010) found that for 30% wind penetration, the standard deviation of 1-hour net load variability in Wyoming was 87% higher than for load alone; for the WestConnect service area, this increase was only 4% of the variability of load alone. They attributed this to the effects of "temporal averaging, geographic diversity and wide area aggregation" but did not separate out the benefits provide by geographic diversity from those of aggregation. Moreover, WWSIS also concluded that there were not significant differences among the siting scenarios, and the location of wind resources was not critical so long as there was adequate transmission infrastructure. One study notes that the effect of geographic diversity was likely to be lower in areas with relative flat, uniform terrain (Charles River Associates, 2010). There is also a growing academic literature on the effects of wind diversity and aggregation discussed in Chapters 3 and 12.

17.4.3 Wind forecast accuracy

Improved wind forecasting techniques also mitigate the difficulty of incorporating wind into the power system (see Chapter 9). Several studies provided estimates of the benefits of perfect wind forecast relative to state-of-the-art wind forecast and found significant operational improvements with improved forecasting (EnerNex Corporation, 2011; GE Energy, 2005a, 2010). EWITS determined that perfect, day-ahead wind forecast would reduce the cost of wind generation by between $2.26 and $2.84/MWh. WWSIS concluded that perfect wind forecast would decrease WECC operating cost by $1 to 2/MWh of wind energy relative to current state-of-the-art forecasts and prevent the need for any wind curtailment (GE Energy, 2010). The NYSERDA study found perfect wind forecasts to be worth approximately $1.50 per MWh of wind generation (GE Energy, 2005a). Underforecasting wind does not pose significant challenges to system stability, as it can be managed by curtailing wind if necessary, but it does reduce the economic competitiveness of wind production (GE Energy, 2008).

17.4.4 Suggested system changes and operating practices

The effect of wind integration depends heavily on the system in which it is being integrated. These studies made numerous recommendations that could reduce the cost of wind integration. First, several of the studies concluded that larger balancing authority areas were better suited to managing wind variability than smaller balancing authorities (Charles River Associates, 2010; EnerNex Corporation, 2006, 2011; GE Energy, 2008, 2010). Large balancing areas reduced variability in net load and provided a large pool of generating units from which to manage deviations from forecasted net load. Balancing area consolidation is discussed in Chapter 5.

Second, wind benefits from systems with greater flexibility in the generating portfolio, as more startups and shorter cycles are required to accommodate net load variability caused by wind (Charles River Associates, 2010; KEMA, 2010).

Third, robust transmission is also important, as this allows for the aggregation of wind and regulating units over large geographic areas (Charles River Associates, 2010; EnerNex Corporation, 2011; EnerNex et al., 2010; EWEA, 2009; GE Energy, 2010; NYISO, 2010). Without significant transmission enhancements, wind curtailment could be significant (EnerNex Corporation, 2011).

Finally, a number of market mechanisms have the potential to reduce the cost of wind integration, including increasing the frequency of scheduling in intraday and subhourly markets (Charles River Associates, 2010; EnerNex Corporation, 2011; GE

Energy, 2010), incorporating wind forecasts into the day-ahead markets (GE Energy, 2005a, 2008), and having wind participate in regulatory markets by providing regulation down through wind curtailment (Charles River Associates, 2010; GE Energy, 2008, 2010).

Several studies noted that wind generation reduced the amount of thermal generation that was used, freeing up additional generating units to respond to net load variability. This process reduced the profitability of these thermal units, however, and could lead to plant retirements that would reduce the plants available for regulation and load following, as noted in EnerNex Corporation (2011), GE Energy (2008), and NREL (2012). This issue requires further study.

17.5 Discussion and conclusions

We have reviewed 11 large North American wind integration studies. This review highlights a number of areas in which wind integration studies have evolved to provide valuable insight, as well as a few areas in which improvements in methods and additional research are needed to facilitate greater insight from future studies. We now summarize our observations and conclusions from this review.

17.5.1 Wind data sources

The quality of the wind data used for wind integration studies has improved substantially over time. Almost all of the wind integration studies used data from mesoscale numerical weather prediction models. Early studies used models with a 10-km resolution, whereas the most recent models used 2-km spatial and 10-minute temporal resolution. As shown in Section 17.2, mesoscale models can produce wind speed data that underestimate variability at small time scales. However, the more recent of the two datasets reviewed (the updated EWITS [EnerNex Corporation, 2011] data) showed less evidence of this reduction than the data from the WWSIS model, although the methods used in the updated EWITS data are not documented. The decision to release the data for these two studies publically is commendable, making it possible to perform a variety of validation analyses, which should motivate further improvements in the quality of these data.

Accurately translating wind speed data to statistically representative wind power output data continues to be a challenge. One of the studies (GE Energy, 2010) used a statistical technique (SCORE/SCORE-lite; Potter et al., 2008) to add additional variability to 10-minute resolution power output data to compensate for the reduced variability in the mesoscale data, but there is some evidence that this process may produce data with too much rather than too little variability (Milligan et al., 2012). Several of the studies combined high-resolution data from power plants with 10-minute simulated data in order to look at faster-time-scale phenomena, such as regulating reserves. However, it is not possible to independently verify that the data have the correct statistical properties, since most of the methods and all of the data for this hybridization process are not public.

Finally, there is a need for more research to develop methods that reliably combine high-resolution measured plant data with mesoscale model data to produce hybrid data with statistical properties similar to those of actual wind plants. Some progress has been made in this area. For example, Rose and Apt (2012) introduce a method for synthesizing high-resolution data from low-resolution data, based on the spectral properties of wind power. A significant barrier to this research is the lack of publicly available wind power production data with which to validate candidate methodologies. While it is feasible to obtain proprietary data from some generators, it is difficult to publish reproducible methods using proprietary data. Making some historical, high-resolution power plant production datasets pubic would be tremendously valuable to the electricity industry.

17.5.2 Statistical modeling and balancing reserves

Almost all of the studies in this review used net load step-change statistics to estimate the need for additional balancing (regulation and load-following) reserves. Most of the studies implicitly or explicitly assumed that load and wind are uncorrelated and that the data fit Gaussian statistical models, neither of which is accurate. Methods such as the one proposed in Charles River Associates (2010) that use the magnitude of low-probability ramping events rather than standard deviations are likely to produce balancing resource estimates that more accurately predict what will be needed to maintain system reliability. An even more useful improvement would be to build on the methods in KEMA (2010), which used a dynamic power system model to simulate the effect of different amounts and types of balancing resources.

High-penetration wind scenarios, as they move from concept to deployment, are very likely to motivate substantial changes in the ways that ancillary resources are scheduled, purchased, and dispatched. Operational changes, such as the large-scale use of pumped hydro in high-renewable penetration countries such as Portugal, Germany, and Ireland, illustrate the types of changes that are likely to be needed. Many of the studies in this review explicitly assumed that the structure of the energy and ancillary services markets will be largely unchanged, with the exception of more frequent dispatch intervals and balancing area aggregation. In future studies, it may be valuable to think more broadly about and model more explicitly the ways in which different types of balancing services can be purchased from different types of power plants. For example, storage and demand response could provide highly responsive balancing services but will be deployed only if electricity markets reward power plants for their responsiveness. Modeling studies of fast-ramping resources, such as in KEMA (2010), pushed the state of the art, making it increasingly feasible to quantify the benefits of responsive resources.

A growing number of research papers estimate balancing reserve requirements from historical data. For example, Mauch and colleagues (2013a) estimated total balancing reserve requirements in Texas and the U.S. Midwest and estimated that 0.07 to 0.30 MW of day-ahead reserves are needed per MW of wind. These reserve requirements are based on day-ahead load and wind forecast uncertainty. Forecast uncertainty is greatest on days when the wind is forecast to be blowing strongly (Mauch et al., 2013b); thus, there are some days when significant reserves are required and other days when much smaller reserves are needed. Because of the additional uncertainty associated with estimating the reserves that are needed to cover 95% of the day-ahead forecast errors, the reserves estimates in Mauch and colleagues (2013a) are substantially higher than those reported in many of the studies reviewed here and in Figures 17.7. Mauch and colleagues (2013a) suggest that a dynamic method should be used to schedule reserves based on the day-ahead forecast values. Additional research is needed into methods that can dynamically predict the reserve requirements for high-wind scenarios. For example, one could imagine scheduling reserves at a level of half the uncertainty in the day-ahead forecast and rescheduling after an intraday unit commitment when uncertainties are lower.

Another area for which improvements in methods are needed is in the modeling of wind forecast errors. Some early work (e.g., Doherty & O'Malley, 2005) assumed that wind and load are uncorrelated Gaussian random variables, which, as shown in Section 17.2, is not supported by the data. More recent research in this area (e.g., Hodge et al., 2012; Mauch et al., 2013b) points out the heavy-tailed nature of the distributions. Appropriate use of these results should allow for increasing statistical accuracy and thus more insightful results in future integration studies. This topic is also discussed in Chapter 9.

17.5.3 Production cost simulation (PCS)

The technology for power system PCS has improved substantially across the studies in this review. While some early studies either did not include PCS (e.g., GE Energy, 2008) or did not include accurate transmission system models (e.g., WindDS in U.S. DOE, 2008), the most recent studies used detailed PCS models that modeled unit commitment, generator ramp rate constraints, and transmission limits. Given that these costs can be important contributors to wind integration costs (Mount et al., 2012), including these details is important. Since most PCS is performed on hourly data, the fine time-scale data challenges described in Sections 17.5.1 and 17.5.2 are unlikely to have a significant effect on PCS results.

Going forward, there is a need to improve the ways in which uncertainty and wind forecasts are handled in PCS. The vast majority of the studies used deterministic PCS models, whereas uncertainty, particularly in wind and solar forecast data, become increasingly important to unit commitment decisions as renewable penetration increases. Also, a number of the studies assumed that the entire fleet of existing power plants would continue to be available for unit commitment, even under 20 to 40% wind scenarios. Clearly, at least some economically uncompetitive power plants would be retired in high renewables cases. This retirement process should be simulated in order to understand which plants or types of plants need to be incentivized to remain operational in order to manage the costs of transition to high-renewables-penetration scenarios.

17.5.4 Power system reliability modeling

Power system reliability modeling is important to wind integration studies, particularly given the potential for long periods of low wind and the need to accurately compensate wind plants for their contribution to system adequacy. Regarding the computation of capacity credits for wind, most of the studies used some variant of the ELCC method from Keane and colleagues (2011), but with only 2 or 3 years of data. As suggested in Hasche and colleagues (2011), accurate estimates of ELCC require 5 or more years of data because substantial interannual variations in wind resources are possible. Going forward, capacity credit calculations in wind integration studies should be based on at least 5 years of wind data.

A majority of the studies reviewed used GE MARS for reliability modeling. While this tool is useful for generation adequacy assessment, reliability is a function of both generation and transmission. The transmission network models used in GE MARS do not capture the physics of actual power flows, which can be very important to reliability. The methodology for composite generation and transmission reliability modeling is relatively mature in the research literature, with specialized methods available specifically for wind integration studies (e.g., Billinton, Yi, & Karki, 2009). We suggest that future wind integration studies incorporate composite system adequacy assessment methods with at least DC transmission system models in order to provide more useful insight into the reliability effects of large-scale wind integration. This will be particularly important in studies that estimate the effect of power-plant retirement and transmission system overlays, since both of these can have important reliability effects.

17.5.5 Analysis of transmission system investments

Several of the studies suggested the need for substantial transmission expansion in order to facilitate high-renewables-penetration scenarios. While some transmission expansion is certainly warranted, large-scale expansion of the bulk transmission network is costly and will face substantial siting barriers. Thus there is, in our opinion, a need for creative thinking about how to most effectively use existing transmission resources, with perhaps a minimum amount of network expansion, to facilitate high-renewables scenarios. As an example of the type of analysis that is needed, Hoppock and Patiño-Echeverri (2010) compared the levelized cost of energy from distant, high-capacity-factor wind sites to energy from near, lower-capacity-factor sites. They found that transmission investment costs could make the lower-quality sites less expensive. The existing technology available for studying optimal transmission expansion is a significant barrier to further progress in this area. Only in the most recent study (NREL, 2012) did analysts attempt to optimize their transmission expansion plans, and in this case a transportation model of the grid was used rather than a power-flow model. Research on optimization methods for transmission switching (Barrows & Blumsack, 2012; Fisher et al., 2008) and transmission expansion (Alguacil, Motto, & Conejo, 2003) suggest approaches to this problem, but substantial research is needed before these algorithms can be applied to large-scale problems.

As methods for optimal system expansion planning improve, future large-scale studies should examine creative combinations of offshore wind, strategically located onshore wind, solar (which is often more easily located near load centers and is increasingly cost competitive), storage, controllable AC or DC transmission lines, and demand response resources. There may be synergies among these technologies that enable higher-penetration scenarios while minimizing curtailment, ancillary service costs, and reliability effects.

17.5.6 Looking forward

The future state of the electricity industry always differs from the scenarios analyzed in large-scale integration studies. Given this fact, studies should focus more on quantifying the relative effect of particular changes in operating policy or technologies than on seeking to precisely quantify the economic or reliability effect of a particular penetration scenario. For example, a conclusion that using fast-ramping storage will reduce ancillary service costs by 10% is likely to be more useful than one that says that the ancillary service costs will be $1.52/MWh for scenario X. This is particularly the case given the fact that new technology and new policies are likely to cause substantial changes in the way that power systems operate in the 10- to 20-year time horizons that are typical in integration studies.

References

Alguacil, N., Motto, A.L., & Conejo, A.J. (2003). Transmission expansion planning: A mixed-integer LP approach. *IEEE Transactions on Power Systems, 18*(3), 1070–1077.

Apt, J. (2007). The spectrum of power from wind turbines. *Journal of Power Sources, 169*(2), 369–374.

Barrows, C., & Blumsack, S. (2012). Transmission switching in the RTS-96 test system. *IEEE Transactions on Power Systems, 27*(2), 1134–1135.

Billinton, R., & Jonnavithula, A. (1997). Composite system adequacy assessment using sequential Monte Carlo simulation with variance reduction techniques. *IEE Proceedings – Generation Transmission and Distribution, 144*(1), 1–6.

Billinton, R., & Li, W. (1994). *Reliability assessment of electrical power systems using Monte Carlo methods*. New York: Plenum Press.

Billinton, R., Yi, G., & Karki, R. (2009). Composite system adequacy assessment incorporating large-scale wind energy conversion systems considering wind speed correlation. *IEEE Transactions on Power Systems, 24*(3), 1375–1382.

BPA. (2013). *Wind generation and total load in the BPA balancing authority.* http://transmission.bpa.gov/business/operations/wind

Brower, M. (2009). *Development of eastern regional wind resource and wind plant output datasets.* National Renewable Energy Laboratory report, NREL/SR-550-4676. www.nrel.gov/docs/fy10osti/46764.pdf

Budischak, C., Sewell, D., Thomson, H., Mach, L., Veron, D.E., & Kempton, W. (2013). Cost-minimized combinations of wind power, solar power and electrochemical storage, powering the grid up to 99.9% of the time. *Journal of Power Sources, 225*, 60–74.

Charles River Associates. (2010, January 4). *SPP WITF wind integration study*. CRA Project No. D14422. Report for Southwest Power Pool, Little Rock, AR. www.uwig.org/CRA_SPP_WITF_Wind_Integration_Study_Final_Report.pdf

Corbus, D., Schuerger, M., Roose, L., Strickler, J., Surles, T., Manz, D., Burlingame, D., & Woodford, D. (2010). *Oahu wind integration and transmission study*. National Renewable Energy Laboratory report, NREL/TP-5500-48632.

DeCarolis, J.F., & Keith, D.W. (2006). The economics of large-scale wind power in a carbon constrained world. *Energy Policy*, *34*(4), 395–410.

Denny, E., & O'Malley, M. (2007). Quantifying the total net benefits of grid integrated wind. *IEEE Transactions on Power Systems*, *22*(2), 605–615.

Dillan, J. (2013). Grid constraints mean less power output from wind projects. *Vermont Public Radio*, 30 January, 2013.

Doherty, R., & O'Malley, M. (2005). A new approach to quantify reserve demand in systems with significant installed wind capacity. *IEEE Transactions on Power Systems*, *20*(2), 587–595.

EnerNex Corporation. (2006). *Final report – 2006 Minnesota wind integration study: volume one*. www.uwig.org/windrpt_vol%201.pdf

EnerNex Corporation. (2007). *Operational impacts of integrating wind generation into Idaho Power's existing resource portfolio*. www.idahopower.com/pdfs/AboutUs/PlanningForFuture/wind/Petition_ReviseAvoidedCostRates1.pdf?id=238&.pdf

EnerNex Corporation. (2011). *Eastern wind integration and transmission study*. National Renewable Energy Laboratory report, NREL/SR-5500-47078. www.nrel.gov/docs/fy11osti/47078.pdf

EnerNex, Ventyx, and Nebraska Power Association. (2010). *Nebraska statewide wind integration study*. National Renewable Energy Laboratory report, NREL/ SR-550-47519.

ERCOT. (2013). *Winter wind propels ERCOT to a new wind record of 9,481 MW*. www.ercot.com/news/press_releases/show/26398

EWEA. (2009). *Integrating wind: Developing Europe's power market for the large-scale integration of wind power*. European Wind Energy Association/Trade Wind, EIE/06/022/SI2.442659.

FERC. (2012, June 22). Order no. 764: *Integration of variable energy resources*. Federal Energy Regulatory Commission. Docket No. RM10-11-000. www.ferc.gov/whats-new/comm-meet/2012/062112/E-3.pdf

Fertig, E., Apt, J., Jaramillo, P., & Katzenstein, W. (2012). The effect of long-distance interconnection on wind power variability. *Environmental Research Letters*, *7*(3), 034017.

Fisher, E.B., O'Neill, R.P., & Ferris, M.C. (2008). Optimal transmission switching. *IEEE Transactions on Power Systems*, *23*(3), 1346–1355.

GE Energy. (2005a). *The effects of integrating wind power on transmission system planning, reliability and operations, report on phase 2: System performance evaluation*. Albany: New York State Energy Research and Development Authority.

GE Energy. (2005b). *NYSERDA phase 2: Appendices*. New York State Energy Research and Development Authority. www.nyserda.ny.gov/Publications/Research-and-Development-Technical-Reports/Wind-Reports.aspx

GE Energy. (2008). *Analysis of wind generation impact on ERCOT ancillary services requirements*. http://variablegen.org/wp-content/uploads/2013/01/AttchB-ERCOT_A-S_Study_Final_Report.pdf

GE Energy. (2010). *Western wind and solar integration study*. National Renewable Energy Laboratory report, NREL/SR-550-47434.

Ghajar, R., & Billinton, R. (1988). A Monte Carlo simulation model for the adequacy evaluation of generating systems. *Reliability Engineering and System Safety*, *20*(3), 173–186.

Gu, Y., Xie, L., Rollow, B., & Hesselbaek, B. (2011). *Congestion-induced wind curtailment: Sensitivity analysis and case studies*. North American Power Symposium (NAPS), Boston, MA.

Hasche, B., Keane, A., & O'Malley, M. (2011). Capacity value of wind power, calculation, and data requirements: The Irish power system case. *IEEE Transactions on Power Systems, 26*(1), 420–430.

Hodge, B.-M., Ela, E., & Milligan, M. (2012). Characterizing and modeling wind power forecast errors from operational systems for use in wind integration planning studies. *Wind Engineering, 36*(5), 509–524.

Hodge, B., & Milligan, M. (2011). *Wind power forecasting error distributions over multiple timescales.* Presented at the Power and Energy Society General Meeting. http://dx.doi.org/10.1109/PES.2011.6039388

Holttinen, H., Flynn, D., Abildgaard, H., Lennart, S., Keane, A., Dillon, J., & O'Malley, M. (2013). Steps for a complete wind integration study. *46th Hawaii International Conference on System Sciences,* 2261–2270.

Holttinen, H., Meibom, P., Orths, A., Lange, B., O'Malley, M., Tande, J.O., Estanqueiro, A., Gomez, E., Söder, L., Strbac, G., Smith, J.C., & van Hulle, F. (2011). Impacts of large amounts of wind power on design and operation of power systems, results of IEA collaboration. *Wind Energy, 14*(2), 179–192.

Holttinen, H., Milligan, M., Kirby, B., Acker, T., Neimane, V., & Molinski, T. (2008). Using standard deviation as a measure of increased operational reserve requirement for wind power. *Wind Engineering, 32*(4), 355–378.

Hoppock, D.C., & Patiño-Echeverri, D. (2010). Cost of wind energy: Comparing distant wind resources to local resources in the Midwestern United States. *Environmental Science and Technology, 44*(22), 8758–8765.

Idaho Power. (2013). *Wind integration study report.* www.idahopower.com/pdfs/AboutUs/PlanningForFuture/irp/2013/windIntegrationStudy.pdf

Katzenstein, W., & Apt, J. (2012). The cost of wind power variability. *Energy Policy, 51,* 233–243.

Katzenstein, W., Fertig, E., & Apt, J. (2010). The variability of interconnected wind plants. *Energy Policy, 38*(8), 4400–4410.

Keane, A., Milligan, M., Dent, C.J., Hasche, B., D'Annunzio, C., Dragoon, K., Holttinen, H., Samaan, N., Soder, L., & O'Malley, M. (2011). Capacity value of wind power. *IEEE Transactions on Power Systems, 26*(2), 564–572.

KEMA. (2010). *Research evaluation of wind generation, solar generation, and storage impact on the California Grid.* Public Interest Energy Research Program. California Energy Commission, CEC-500-2010-010.

Lew, D. (2009). *Western data set irregularity.* National Renewable Energy Laboratory. www.nrel.gov/electricity/transmission/pdfs/western_dataset_irregularity.pdf

Lueken, C., Cohen, G., & Apt, J. (2012). Costs of solar and wind power variability for reducing CO_2 emissions. *Environmental Science and Technology, 46*(17), 9761–9767.

Mackin, P., Daschmans, R., Williams, B., Haney, B., Hunt, R., Ellis, J., & Eto, J. (2013). Dynamic simulation study of the frequency response of the western interconnection with increased wind generation. *46th Hawaii International Conference on System Sciences,* 2222–2229.

Madaeni, S.H., & Sioshansi, R. (2013). The impacts of stochastic programming and demand response on wind integration. *Energy Systems, 4*(2), 109–124.

Mauch, B., Apt, J., Carvalho, P.M.S., & Jaramillo, P. (2013a). What day-ahead reserves are needed in electric grids with high levels of wind power? *Environmental Research Letters, 8*(3). DOI: 10.1088/1748-9326/8/3/034013

Mauch, B., Apt, J., Carvalho, P.M.S., & Small, M.J. (2013b). An effective method for modeling wind power forecast uncertainty. *Energy Systems, 4*(4), 393–417.

Milligan, M., Ela, E., Lew, D., Corbus, D., Yih-huei, W., & Hodge, B. (2012). Assessment of simulated wind data requirements for wind integration studies. *IEEE Transactions on Sustainable Energy, 3*(4), 620–626.

Mount, T.D., Maneevitjit, S., Lamadrid, A.J., Zimmerman, R.D., & Thomas, R.J. (2012). The hidden system costs of wind generation in a deregulated electricity market. *The Energy Journal, 33*(1), 161–186.

NERC. (2011a). *Planning resource adequacy analysis, assessment and documentation, standard BAL-502-RFC-02.* www.nerc.com/files/BAL-502-RFC-02.pdf

NERC. (2011b). *Balancing and frequency control: A technical document prepared by the NERC resources subcommittee.* North American Electric Reliability Corporation. www.nerc.com/docs/oc/rs/NERC%20Balancing%20and%20Frequency%20Control%20040520111.pdf

NERC. (2012). *Standard BAL-002-1a – disturbance control performance.* North American Electric Reliability Corporation. www.nerc.com/files/BAL-002-1a.pdf

NREL. (2012). *Renewable electricity futures study.* National Renewable Energy Laboratory. www.nrel.gov/analysis/re_futures/

NYISO. (2010). *Growing wind: Final report of the NYISO wind generation study.* www.uwig.org/GROWING_WIND_-_Final_Report_of_the_NYISO_2010_Wind_Generation_Study.pdf

Papavasiliou, A., Oren, S.S., & O'Neill, R.P. (2011). Reserve requirements for wind power integration: A scenario-based stochastic programming framework. *IEEE Transactions on Power Systems, 26*(4), 2197–2206.

Pennock, K. (2012). *Updated eastern interconnect wind power output and forecasts for ERGIS.* National Renewable Energy Laboratory technical report, NREL/SR-5500-56616.

Pope, S.B. (2000). *Turbulent flows.* Cambridge, UK: Cambridge University Press.

Potter, C.W., Lew, D., McCaa, J., Cheng, S., Eichelberger, S., & Grimit, E. (2008). Creating the dataset for the Western Wind and Solar Integration Study (USA). *Wind Engineering, 32*(4), 325–338.

Rose, S., & Apt, J. (2012). Generating wind time series as a hybrid of measured and simulated data. *Wind Energy, 15*(5), 699–715.

Sauer, P.W., & Pai, M. (1997). *Power system dynamics and stability.* Upper Saddle River, NJ: Prentice Hall.

Seman, S., Niiranen, J., & Arkkio, A. (2006). Ride-through analysis of doubly fed induction wind-power generator under unsymmetrical network disturbance. *IEEE Transactions on Power Systems, 21*(4), 1782–1789.

Skamarock, W.C. (2004). Evaluating mesoscale nwp models using kinetic energy spectra. *Monthly Weather Review, 132*(12), 3019–3032.

Soder, L., & Holttinen, H. (2008). On methodology for modelling wind power impact on power systems. *International Journal of Global Energy Issues, 29*(1), 181–198.

Stauffer, D.R., & Seaman, N.L. (1994). Multiscale four-dimensional data assimilation. *Journal of Applied Meteorology, 33*(3), 416–434.

Takriti, S., Birge, J.R., & Long, E. (1996). A stochastic model for the unit commitment problem. *IEEE Transactions on Power Systems, 11*(3), 1497–1508.

U.S. DOE. (2008). *20% wind energy by 2030: Increasing wind energy's contribution to U.S. electricity supply.* www.nrel.gov/docs/fy08osti/41869.pdf

Vittal, E., O'Malley, M., & Keane, A. (2010). A steady-state voltage stability analysis of power systems with high penetrations of wind. *IEEE Transactions on Power Systems, 25*(1), 433–442.

Wang, C., & McCalley, J.D. (2013). Impact of wind power on control performance standards. *International Journal of Electrical Power & Energy Systems, 47*, 225–234.

Weber, C., Meibom, P., Barth, R., & Brand, H. (2009). Wilmar: A stochastic programming tool to analyze the large-scale integration of wind energy. *Optimization in the Energy Industry*, Springer, 437–458.

WindLogics. (2006). *Final report – Minnesota wind integration study: Volume II*. St. Paul, MN: WindLogics, Inc.

Wu, L., Shahidehpour, M., & Li, T. (2007). Stochastic security-constrained unit commitment. *IEEE Transactions on Power Systems, 22*(2), 800–811.

Xie, L., Carvalho, P.M.S., Ferreira, L.A.F.M., Liu, J., Krogh, B.H., Popli, N., & Ilic, M.D. (2011). Wind integration in power systems: Operational challenges and possible solutions. *Proceedings of the IEEE, 99*(1), 214–232.

INDEX

Printed in Canada